Gas Lasers – Recent Developments and Future Prospects

NATO ASI Series

Advanced Science Institutes Series

A Series presenting the results of activities sponsored by the NATO Science Committee, which aims at the dissemination of advanced scientific and technological knowledge, with a view to strengthening links between scientific communities.

The Series is published by an international board of publishers in conjunction with the NATO Scientific Affairs Division

A Life Sciences	Plenum Publishing Corporation
B Physics	London and New York
C Mathematical and Physical Sciences	Kluwer Academic Publishers
D Behavioural and Social Sciences	Dordrecht, Boston and London
E Applied Sciences	
F Computer and Systems Sciences	Springer-Verlag
G Ecological Sciences	Berlin, Heidelberg, New York, London,
H Cell Biology	Paris and Tokyo
I Global Environmental Change	

PARTNERSHIP SUB-SERIES

1. **Disarmament Technologies**	Kluwer Academic Publishers
2. **Environment**	Springer-Verlag / Kluwer Academic Publishers
3. **High Technology**	Kluwer Academic Publishers
4. **Science and Technology Policy**	Kluwer Academic Publishers
5. **Computer Networking**	Kluwer Academic Publishers

The Partnership Sub-Series incorporates activities undertaken in collaboration with NATO's Cooperation Partners, the countries of the CIS and Central and Eastern Europe, in Priority Areas of concern to those countries.

NATO-PCO-DATA BASE

The electronic index to the NATO ASI Series provides full bibliographical references (with keywords and/or abstracts) to more than 50000 contributions from international scientists published in all sections of the NATO ASI Series.
Access to the NATO-PCO-DATA BASE is possible in two ways:

– via online FILE 128 (NATO-PCO-DATA BASE) hosted by ESRIN,
Via Galileo Galilei, I-00044 Frascati, Italy.

– via CD-ROM "NATO-PCO-DATA BASE" with user-friendly retrieval software in English, French and German (© WTV GmbH and DATAWARE Technologies Inc. 1989).

The CD-ROM can be ordered through any member of the Board of Publishers or through NATO-PCO, Overijse, Belgium.

Series 3: High Technology – Vol. 10

Gas Lasers –
Recent Developments
and Future Prospects

edited by

W. J. Witteman

Department of Applied Physics,
University of Twente,
Enschede, The Netherlands

and

V. N. Ochkin

Low Temperature Plasma Optics Department,
P.N. Lebedev Physical Institute,
Moscow, Russia

Kluwer Academic Publishers

Dordrecht / Boston / London

Published in cooperation with NATO Scientific Affairs Division

Proceedings of the NATO Advanced Research Workshop on
Gas Lasers – Recent Developments and Future Prospects
Moscow, Russia
July 2–5, 1995

A C.I.P. Catalogue record for this book is available from the Library of Congress

ISBN-13: 978-94-010-6588-7   e-ISBN-13: 978-94-009-0235-0
DOI: 10.1007/978-94-009-0235-0

Published by Kluwer Academic Publishers,
P.O. Box 17, 3300 AA Dordrecht, The Netherlands.

Kluwer Academic Publishers incorporates the publishing programmes of
D. Reidel, Martinus Nijhoff, Dr W. Junk and MTP Press.

Sold and distributed in the U.S.A. and Canada
by Kluwer Academic Publishers,
101 Philip Drive, Norwell, MA 02061, U.S.A.

In all other countries, sold and distributed
by Kluwer Academic Publishers Group,
P.O. Box 322, 3300 AH Dordrecht, The Netherlands.

Softcover reprint of the hardcover 1st edition 1996

TABLE OF CONTENTS

Preface

This volume contains the proceedings of the NATO Advanced Research Workshop 950443 on "Gas lasers-recent development and future prospects". The workshop was held in Moscow, July 2-5, 1995. During the workshop 22 oral presentations and 23 posters have been presented.

Among the continuously expanding research on new laser systems in the extending spectrum range gas lasers are unique in many ways: the availability of high (average) power in all parts of the spectrum from the far infrared to the vacuum ultraviolet, the homogeneity of the active medium with the potential of high beam quality even at high power and their relatively low costs. In the gas laser development one can distinguish the research towards new or improved laboratory devices and the efforts that are devoted to the development of characteristics like reliability, low costs and versatility that make the laser more suitable for industrial purposes. The industrial applications with dedicated devices are not only a natural extension of the laser development itself but moreover they have nowadays a strong stimulating effect on this development.

The workshop offered the participants many opportunities to discuss fundamental and technological problems of different types of lasers connected with beam proporties, excitation technology, new pumping schemes, pulsed power, construction materials and new codes for the description of laser operation. The interest was especially directed towards high power systems operating in the ultraviolet and vacuum ultraviolet, the radio·frequency discharge physics for waveguide structures and the achievement in molecular CO and CO_2 systems. Other striking results of gas laser developments are a table-top kilowatt CO_2 slab laser, high continuous output power in the order of watts in the wavelength region of 1 to 3 microns with radio frequency discharges in atomic gases at relatively high gas densities and the efficient optical pumping of excimer lasers with ferrite flashes.

The high pulse power as well as high average power in the visible and ultraviolet region obtained with excimers are by far superior to what can be obtained with other systems. Although the early results were mainly achieved with complicated e-beam pumped machines nowadays most excimers operate with relative simple fast gas discharges. In view of the wide spread applications the reliability and technical performance were successfully undertaken in international collaboration. Average powers of 1 kW in the ultraviolet are now available. The drive to go to shorter wavelength and high power is also very promising for media with electronic transitions of diatomic molecules like the F_2 laser at 157 nm.

A diversity of gas laser applications is discussed: leading industrial applications with molecular systems; high resolution lithography with excimer lasers; efficient X-ray production with high power excimer lasers; fast switching techniques of infrared laser beams.

We wish to express our appreciation to the NATO Scientific Affairs Division, whose financial support made the workshop possible. We also gratefully acknowledge the contribution of the Russian Ministry of Science and Technology Policy, Russian Foundation for Fundamental Research, Russian Academy of Engineering of Science and P.N. Lebedev Physical Institute.

W.J. Witteman
V.N. Ochkin

PROSPECTS FOR GAS LASERS

W.J. WITTEMAN
University of Twente
Department of Applied Physics
P.O. Box 217
7500 AE Enschede
The Netherlands

1. Abstract

Gas lasers are in many aspects superior to other types of lasers. They are capable of high power generation and the developed systems cover the spectral range from the far infrared to the VUV. The steady progress of gas lasers is accompanied by new discharge technologies. Especially high pressure discharges with excimers and molecular gases have shown considerable improvements over the last decade. Recently very good results are obtained with efficient high density RF discharges. It will be discussed that short wavelength systems prefer pulsed excitation and that for continuous operation at short wavelengths the gas lasers have the best prospects. The paper deals with the prospects of UV and VUV lasers, RF excitation of continuous and the feasibility to reach the multi-terawatt regime with gas lasers.

2. Introduction

Reflections on the prospects of gas lasers bring our thoughts involuntarely to the comparison with solid state devices which are compact and have made considerable progress during the last decade, especially with respect to new crystals and the all solid state diode pumped systems. Those systems operate in the infrared region. In spite of these remarkable progresses the gas lasers are still in several aspects superior or competitive for many application purposes. From a general point of view one might remark that gas lasers cover the spectral region from the far infrared to the VUV and that practically all gas lasers are attractive for building fairly economical high-aperture, large volume and high energy systems. Further the optical quality of the beams in terms of line width and phase front is by far superior because of the homogeneity and small non-linear effects of the active media. Also the world-wide annual returns of gas lasers are still the highest. However, there is more.

The developments in the last decade have confirmed that the new class of excimer lasers have unique properties to reach high powers in the UV and VUV and long term operation with average output powers above 1 kW [1,2,3]. This new class of

1

lasers has large bandwidths and several systems have been successfully used to amplify femtosecond pulses at power levels unknown before [4,5,6].

Progress of gas lasers is strongly connected with new developments in the field of gas discharges. In fact new areas of discharge physics e.g. in multi-atmospheric systems, excimer gases and molecular gases were stimulated in the past by the drive for new or improved laser systems. The most striking recent result is the efficient homogeneous laser excitation by RF excitation at considerably higher atomic or molecular gas densities.

In the following we shall restrict the subject to the prospects of UV and VUV lasers, the potential of RF discharge technology for enhanced power generation and the feasibility of multi-terawatt gas lasers.

3. Excimer lasers

For obtaining high pulse power as well as high average power in the visible and ultraviolet region the excimer gas lasers are by far superior to other systems. In the early years, e-beam pumping of rare gas halide mixtures producing the so-called excimers was the opening activity to reveal the prospects of a new class of gas lasers. Most excimers are nowadays successfully operated by a relatively simple fast gas discharge. This achievement has strongly stimulated the development of self-sustained high pressure discharges that are characterized by inherent instabilities caused by the formation of negative ions and streamers. At present an experienced technique for XeCl with 308 nm radiation applies a X-ray preionization directly followed by a well-balanced prepulse avalanche discharge and a critically damped mainpulse. In view of wide spread applications the reliability and technical performance for high repetition rate was also successfully undertaken in international collaboration like Eureka (Europe) and Ammtra (Japan). Average powers of 1 kW in the UV are now available.

This success challenges to reach similar performances for excimers with shorter and shorter wavelength. The first follower is most likely ArF at 193 nm that so far lases with a relatively short pulse duration, low average output power and with an unmatched discharge. The same remark can be made with respect to KrF (248 nm) and XeF (B$\rightarrow$X, 351 nm). These developments will be stimulated by unique applications like X-ray source development, lithography and processing of compound materials.

The drive to go to shorter wavelength is also very promising for media where the inversion is obtained on electronic transitions of diatomic molecules like the molecular F_2 laser that lases at 157 nm. For this laser the results obtained with e-beam pumping are by far superior to those obtained with discharge pumping. The present e-beam pumped devices operating up to 8 bar deliver energies up to 10 J/liter and intensities of several MW/cm^2. Also this system has the potential to be developed for a high average output power at high repetition rate. Much will depend on the progress with gas discharge technology to reach a stable glow in an impedance matched circuit as already reached for XeCl.

Apart from the need for new discharge technology the major problem for these short wavelength (VUV) lasers is the high pumping power requirement.

It is straightforward to show that the required power density P_{th} to reach the threshold gain g_{th} per unit length is given by the relation:

$$P_{th} \geq \frac{8\pi g_{th} h v^3 \Delta v}{c^2} \tag{1}$$

where v is the radiation frequency and Δv the line width of the transition. The excimer molecules have due to the bound-free-transition a relatively large line width. For Doppler broadening Δv is already proportional to v so that the required power is at least proportional to the fourth power of v. This strong dependence on v indicates the practical difficulties to reach threshold and that only pulsed excitation is feasible.

In this respect it is interesting to note that the minimum required total power is independent on the length L of the active medium if the total required minimum gain $g_{th}L$ is related to constant system losses. Further, the cross section of the active medium may be small, so that the required total power decreases with this cross section. This means that the prospects to obtain continuous VUV lasers are the best for narrow channels (waveguides) of media that can stand elevated temperatures. For that reason it can be argued that continuous solid state lasers at these short wavelengths are very unlikely and that gas lasers have the best prospects. Also frequency doubling with non-linear crystals are very inefficient for continuous processes and do not offer an option for substantial continuous power generation.

4. Novel RF excited systems

In the context of the last mentioned remark it is challenging to develop new discharge technologies with efficient continuous energy deposition in lasing VUV transitions. A possible way is the further development of high pressure RF discharges that are classified as α and γ type and that have shown so far stability in several gases at densities above 100 torr and input powers over 100 W/cm^3 [7,8,9].

During the last ten years these discharges have shown to be very fruitful to improve considerably the waveguide and slab CO and CO_2 lasers. Compact table top CO_2 systems deliver now output powers in the range of 1 kW [10]. This technology with dense discharge power in molecular gases is accompanied by relatively high molecular dissociation. A further improvement was then obtained by catalytic regeneration of the dissociation products. By means of heating of one of the gold plated electrodes an increase of output power of 40% was obtained [8]. It is expected that a new generation of compact and efficient high power CO_2 and CO lasers will be used for a broad range of applications.

Another striking development in gas laser research with perspectives for efficient high power generation is the so called recombination laser were ions and electrons recombine volumetrically by three body collisions at high pressure. In this system the recombined species decay to the lasing upper level and the lower level decays to the metastable level. The metastable atoms are directly ionized by a glow discharge. Quantum efficiencies of 19% are available in Ar-Xe-mixtures and lasing efficiencies as high as 9 percent have been observed [11]. These promising results were obtained with pulsed e-beam sustained discharges at gas densities up to 9 bar. The basic requirement to come to continuous lasers with high output powers is the discharge stability at gas densities where three body collisions are effective. Fortunately, this requirement can be met by applying the above mentioned technology for RF discharges in waveguides. It has been shown that the atomic Xe laser with about 300 W γ-type RF excitation at 125 MHz yields a cw output of 1.5 W for a slab configuration of 2.25x8 mm^2 and an active length of 37 cm [9]. This development will continue in the directions of both more power and detailed understanding of the physics involved. It is expected that this technology with γ-type discharges in waveguide structures will be studied for other gases too, including excimers.

5. Terawatt class gas laser systems

High power laser development for strong-field physics in various applications, e.g. multiphoton processes, pumping of XUV lasers and higher harmonic generation, has been an active area of quantum electronics since the advent of lasers. Early progress in the development of short pulses was based on mode-locking techniques. The more bandwith available the shorter the pulse. Most gas lasers developed in the past have, even with molecular an rotational structures, limited bandwidths so that these lasers have been overshadowed in recent years by the development of picosecond and femtosecond solid state lasers having bandwiths of 10^{11} to 10^{14} Hz. These systems operate in the infrared. The limitations to the available power from the short pulse amplifiers are not determined by the stored energy but by the effects of optical non-linearities causing self-focusing and detrimental effects such as the breakdown threshold of optical crystal and mirror. Further, the challenge is not only to generate laser beams with extreme high power, the highest reported values are at terawatt-level, but to obtain these beams with low phase front distortions i.e. at diffraction limit. This is important for obtaining high focused intensity but also because the threshold breakdown value of the material increases with the uniformity of the beam. The maximum irradiance I of an aberration-free optical system at the focus is for a Gaussian beam given by $I=4P/\lambda^2F^2$ where F is the relative aperture of the focusing element and P the beam power.

Well-known advances in solid state lasers such as Nd:glass, Ti:sapphire and Cr:LiSrAlF$_6$ to circumvent the optical non-linearities in the development of ultrahigh brightness laser beams in the infrared have been made by using the chirp-pulse amplification technique [12]. Although subpicosecond pulses with peak powers in excess of 1 TW have been reported and in some cases near-diffraction limited divergency is claimed with brightness in the order of 10^{20} W cm^{-2} sr^{-1} the inherent limitation for the power density and diffraction is still determined by the optical non-linearities of the solid state amplifiers. Moreover, due to the availability of rod diameters, in the order of mm's, the total power is limited. Also the damage threshold and the availability of gratings with a sufficient large size and efficiency for recompression the pulse after amplification imposes an upper limit for high brightness amplification using this chirp-pulse amplification technique.

5.1. Terawatt pulses with excimer laser systems

In the context of these limitations excimer laser amplifiers operating in the visible and ultraviolet region have proved to be very successful to generate output powers at the terawatt level [4,5,6]. Because of the λ^{-2} dependence, ultrahigh brightness can be achieved with excimer systems at significantly lower power. Further, the optical non-linearities of the gaseous gain medium are several orders of magnitude smaller than those of solid state lasers, so that the gaseous gain medium permits the direct amplification of ultra short pulses to high power densities. The gain medium is, in principle, homogeneous and scalable to large diameters so that high powers can be achieved without large phase and amplitude fluctuations. However, experiments have shown that the saturation depends on the pulse duration because of finite rotational-vibrational relaxation time of the active medium [5].

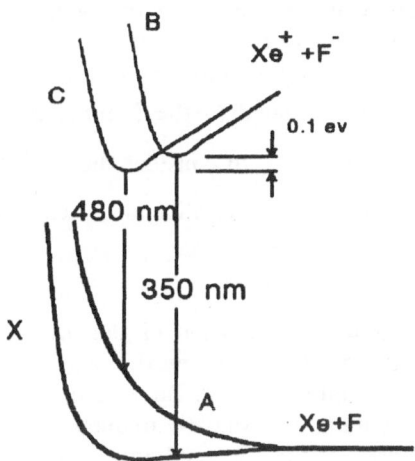

Figure 1. Schematic XeF energy level diagram.

An interesting system in this respect , is the bleu-green XeF (C→A) laser because of its bound-free transition with a highly repulsive lower state, which results in a broad line width (80 nm), high saturation energy (50 mJ/cm^2) and a small cross section for stimulated emission (10^{-17} cm^2). See fig. 1. These properties allows one to obtain high output power with a large system and large aperture. The broad gain band allows to amplify ultra short pulses down to 10 femtosecond Fourier transform limited duration. The XeF (C→A) excimer operating successfully with e-beam pumping [5,13] is therefore an attractive candidate for direct amplification of femtosecond pulses as already have been shown [5].
Instead of e-beam pumping one can also apply photodissociation of XeF$_2$ to produce large densities of XeF molecules in the excited C-state. With this

technique it is possible to generate multi-joule laser pulses of several microseconds [14,15]. By means of ferrite surface plasma's several performances can be obtained with radiation production in a spectrum region below 220 nm that coincides with the absorption spectrum of XeF_2. In this photolysis of XeF_2 the components of XeF^* (B) and F are obtained. By means of atomic or molecular collisions there is an efficient transfer of the excitation from the B to the C state which is 700 cm^{-1} lower. This C state will be the upper laser state in the bound free C→A transition.

5.2. Terawatt pulses with CO_2 laser systems

CO_2 Laser systems have also the potential of generating picosecond pulses, since the gain spectrum, although periodically modulated by the rotational structure, has a width of about 10^{12} Hz. Increased pressure and an isotopic gas mixture smoothes the discrete gain spectrum.

The limitations and prospects of picosecond laser pulse amplification in CO_2 can be analyzed as follows. When a pulse propagates in a amplifying medium two characteristic time constants should be considered. First of all the time τ_l that corresponds to a frequency interval between two adjacent rotational-vibrational lines, which is for example about 16 picosecond for the P (20) and P (22) of the 10.4 μm band. The other one is the collisionally induced dephasing time T_2 by collision broadening which is related to the Lorentzian half width Δv by $T_2 = (\pi \Delta v)^{-1}$. It is now instructive to consider the amplification for three cases with different initial pulse duration τ_0 relative to τ_ℓ an T_2 [16].

a) $\tau_0 < \tau_\ell < T_2$

In this case the frequency spectrum of the original pulse obtained from slicing a small part from an oscillating line may cover several neighbouring discrete rotational-vibrational lines. Because of the periodic modulation of the amplifying medium, the smooth frequency spectrum of the incident pulse is after amplification changed into a spectrum of discrete peaks corresponding to neighbouring rotational transitions. Its Fourier transform in the time domain is then a pulse train consisting of a main pulse and a few satellites separated at a distance τ_ℓ. If the original pulse is obtained from slicing a single pulse out of a multiline mode-locked laser containing, for instance, 8 rotational lines [17] then the satellite pulses are much weaker and even absent. At higher pressure the broadening effect smoothes this discrete gain spectrum of the amplified pulse. As an example the amplification of a 5 picosecond single line CO_2 laser pulse (see fig. 2a) through a 10 bar amplifier results in a short train of 5 picosecond pulses, shown in fig. 2b. A further smoothening of the gain spectrum of the amplifier is obtained by using an isotopic gas mixture. If one of the oxygen atoms is replaced by an isotope the symmetry of the CO_2 molecule is lost

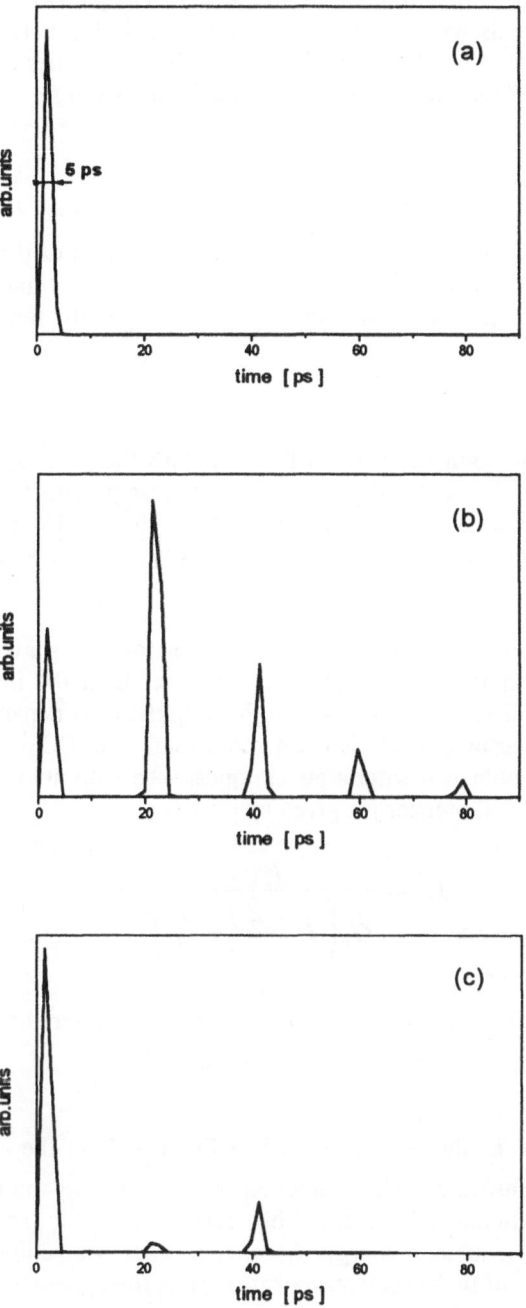

Figure 2. Amplification of picosecond CO_2 laser pulses.

and thus odd rotational transitions are present. The gain spectrum is then twice as dense as with a regular CO_2 mixture. Using a 10 bar mixture of $^{12}C\ ^{16}O_2 : ^{12}C\ ^{16}O\ ^{18}O : ^{12}C\ ^{18}O_2 = 1 : 2: 1$, computer modelling shows a considerable less short-pulse distortion as is shown in fig. 2c

b) $\tau_\ell < \tau_0 < T_2$

Since the laser pulse τ_0 is longer than τ_ℓ only one rotational line is involved in the amplification. The frequency spectrum of the pulse is larger than the gain spectrum of the transition line so that there is spectrum narrowing of the amplified pulse and consequently the pulse duration increases.

c) $\tau_0 > T_2$

In this case the gain spectrum will not disturb the pulse spectrum and the pulse duration is unaltered. Since T_2 is inversely proportional to the gas density, it becomes about 10 picosecond at 10 bar. Computer modelling shows that pulses longer than 20 picosecond are amplified without distortion.

For high power short pulse amplification pulse distortion is avoided when the frequency spectrum of the incident pulse is narrower than the line width of the particular transition. Then the saturation of the amplification depends on the pulse duration and the molecular relaxation time constants. The CO_2-laser has for the short pulses in principle two saturation energies. The saturation energy of a P-transition with rotational number j is given by [18]:

$$E_s' = \frac{h\nu}{\sigma_s\left(1 + \frac{2j-1}{2j+1}\right)} \tag{2}$$

This saturation energy is relevant for pulse durations shorter than the rotational relation time τ_r but longer than τ_ℓ $\left(\tau_\ell < \tau_0 < \tau_r\right)$.

If τ_0 is longer than τ_r the amplification benefits also from the energy stored in other rotational transitions. The depletion of the lasing transition is then replenished by rotational relaxation. This means that the saturation energy increases with the available energy from the other transitions. This larger saturation energy E_s'' of the considered P-transition is then given by [18]:

$$E_s'' = \frac{h\nu}{\sigma_s\left[P(j-1) + \frac{2j-1}{2j+1}P(j)\right]} \tag{3}$$

where

$$P(j) = \left(\frac{2hcB}{kT}\right)(2j+1) \, exp\left[-Bj(j+1)\frac{hc}{kT}\right] \qquad (4)$$

where B is the rotational constant of the CO_2-molecule. Similar results can be derived for R-transitions. For short pulse high power generation high gas pressure is essential because the τ_r is inversely proportional to the pressure which makes rotational relaxation more effective and also because the saturation energy $E_s{''}$ is proportional to the pressure. Full profit of the available energy stored in the rotational transitions can also be obtained with picosecond pulses where $\tau_0 < \tau_\ell$ so that the pulse spectrum covers several rotational lines.

Output powers in the order of several terawatt peak power can be obtained in this way with a large aperture high pressure amplifier. For instance taking a 10 bar X-ray or e-beam preionized CO_2 laser discharge the energy loading may be as high as 100 J/bar liter. The stored laser energy can be as high as 10 J/bar liter. Amplifying a pulse of 20 picoseconds there is already some spectrum overlap with neighbouring rotational transitions and in addition substantial rotational relaxation so that a great deal of the available energy can be extracted. The maximum saturation energy at this pressure is about 1 J/cm^2. However, the limitations to the available power from the amplifier in the picosecond regime are not set by the available stored energy but by detrimental effects as optical component damage and gas breakdown, which is roughly about 1 J/cm^2. Taking an aperture of 10 cm diameter an output energy of 100 J in 20 picosecond or multi-terawatt is feasible.

6. References

1. Witteman, W.J., Ekelmans, G.B., Trentelman, M. and van Goor, F.A. (1990) Discharge technology for excimer lasers of high average power, Spie Vol. 1387, Eighth International Symposium on Gas Flow and Chemical Lasers, 37-45 (1990).
2. Godard, B., Murer, P., Stehle, M., Bonnet, J., Pigache, D. (1993) First 1 kW XeCl laser, CLEO '93, Baltimore, Md, May 2-7, paper CThI1.
3. Witteman, W.J., van Goor, F.A., Timmermans, J.C.M., Couperus, J., van Spijker, J. (1993) Development of a 1 kW excimer Eurolaser, CLEO '93, Baltimore, Md, May 2-7, paper CThI3.
4. Taylor, A.J., Gosnell, T.R. and Roberts, J.P. (1990) Ultrashort-pulse energy extraction measurements in XeCl amplifiers, *Opt. Lett.* **15**, 118-120.
5. Hofmann, Th., Sharp, T.E., Dane, C.B., Wisoff, P.J., Wilson, W.L., Tittel, F.K. and Szabo, G. (1992) Characterization of an ultrahigh peak power XeF (C→A) excimer laser system, *IEEE J. Quantum Electr.* QE-**28**, 1366-1375.
6. Watanabe, S., Endoh, A., Watanabe, M., Sarukura, N. and Hata, K. (1989) Multitera watt excimer laser system, *J. Opt. Soc. Am.* B**6**, 1870-1876.
7. He, D. and Hall, D.R. (1983) A 30 W radio frequency excited waveguide CO_2 laser, *Appl. Phys. Lett.* **43**, 726-728.

8. Heeman-Ilieva, M.B., Udalov, Yu.B., Hoen, K. and Witteman, W.J. (1994) Enhanced gain and output power of a sealed-off RF-excited CO_2 waveguide laser with gold plated electrodes, *Appl. Phys. Lett.*, **64**, 673-675.

9. Tskhai, S.N., Udalov, Yu.B., Peters, P.J.M., Witteman, W.J. and Ochkin, V.N. (1995) Continuous wave near-infrared atomic Xe laser excited by a radio frequency discharge in a slab geometry, *Appl. Phys. Lett.*, **66**, 801-803.

10. Colley, A.D., Baker, H.J. and Hall, D.R. (1992) Planar waveguide, 1 kW CW, carbon dioxide laser excited by a single transverse rf discharge, *Appl. Phys. Lett.*, **61**, 136-138.

11. Botma, H., Peters, P.J.M. and Witteman, W.J. (1993) Saturation studies of the e-beam sustained discharge atomic xenon laser, *IEEE J. Quant. Electr. QE-***29**, 2519-2524.

12. Maine, P., Strickland, D., Bado, P., Pessot, M. and Mouron, G. (1988) Generation of ultrahigh peak power pulses by chirped pulse amplification, *IEEE J. Quantum Electr., QE-***24**, 398-403.

13. Peters, P.J.M., Bastiaens, H.M.J., Witteman, W.J., Sauerbrey, R., Dane, C.B. and Tittel, F.K. (1990) Efficient XeF (C→A) laser excited by a coaxial electron beam at intermediate pumping rate, *IEEE J. Quantum Electr. QE-***26**, 1569-1573.

14. Eckstrom, D.J. and Walker, H.C. (1982) Multi-joule performance of the photolytically pumped XeF (C→A) laser, *IEEE J. Quantum Electr., QE-***18**, 176-181.

15. Zuev, V.S., Kashnikov, G.N., Kirilenko, V.V., Mamaev, S.B., Sorokin, V.A. and Sukhorukov, V.F. (1989) Photodissociation XeF laser emitting visible and ultraviolet radiation when pumped with radiation from a sectioned surface discharge", *Sov. J. Quantum Electr.*, **19**, 748-750.

16. Pogorelsky, I.V., Fischer, J., Kusche, K., Babzien, M., Kurnit, N.A., Bigio, I.J., Harrison, R.F. and Shimada, T. (1995) Subnanosecond multi-gigawatt CO_2 laser, *IEEE J. Quantum Electr., QE-***31**, 556-566.

17. Rooth, R.A., van Goor, F.A. and Witteman, W.J. (1983) An independently adjustable multiline AM mode-locked TEA CO_2 laser, *IEEE J. Quantum Electr., QE-***19**, 1610-1612.

18. Witteman, W.J. (1987) *The CO_2 laser*, Springer Series in Optical Science, volume 53, Springer-verlag, Heidelberg.

PHYSICS OF HIGH-POWER CO LASERS
A RECENT PROGRESS

A. P. NAPARTOVICH
Troitsk Institute for Innovation and Thermonuclear Research
142092 Troitsk, Moscow region, Russia

1. ABSTRACT

A short overview is given of recent studies on physical processes in high power CO lasers excited by an e-beam sustained discharge in cryogenic $CO:N_2$ mixtures at moderate and high gas density. Processes determining the high efficiency of the CO laser pumped by a gas discharge are discussed and the problem of the diverse influences of the spatial inhomogeneities of plasma parameters on the output beam quality is evaluated with the help of mathematical modeling. Results of 2-D self-consistent calculations of kinetic master equations for vibrational populations along with a system of paraxial wave equations for wave fields on molecular lasing transitions are presented. A strong influence of the nonuniform gas refraction on the laser beam quality was found. This influence plays a decisive role at a gas density higher than 1/3 of the normal one. A 2-D diffraction model allows one to evaluate the accuracy and applicability of the widely used simplified model based on the condition of the laser gain to be equal to threshold. The spread of the e-beam entering into the gas chamber is the reason the plasma parameters and in particular the gas density of the discharge to be nonuniform in the direction from the cathode to the anode. To describe such effects an approximate 2-D model was developed in which the large set of kinetic equations (about 100 equations) was replaced by several analytical relationships allowing to calculate the key quantities in the vibrational kinetics. Application of this simplified model to simulations of experimental devices demonstrated that the spatial nonuniformities in both directions (along the gas flow and the discharge current) cause a strong increase in the laser beam divergence. As a possible way to overcome these problems the phase conjugation effect (4-wave mixing) in the laser medium is considered.

11

W. J. Witteman and V. N. Ochkin (eds.), Gas Lasers - Recent Developments and Future Prospects, 11–22.
© 1996 *Kluwer Academic Publishers.*

2. INTRODUCTION

Comparing the state of development of high power gas lasers for industrial applications the CO_2 laser is a well established tool while the CO laser is still remaining a laboratory device. At low temperatures, $T \leq 100$ K, the laser efficiency for the CO laser is about 2 times greater than that of CO_2 laser. However, this prominent advantage only could be realized by the development of a simple and effective cooler. At present this is a difficult problem. At temperatures more easily attainable technically, the efficiency for both lasers is almost the same. To answer the question about the need to develop CO lasers in the multi-kilowatt power range one has to take into account the advantages of the better focussability and transmission of the CO laser beam in optical fibers, the reduced plasma shielding and different absorption depth. For some applications the primary interest is the wave-length range which covers the spectral bands for many chemical compounds. Here, the application of the CO laser for oxygen isotopes enrichment [1] should be mentioned. At present, several national and international projects are in progress aimed at the development of a compact, efficient, high-power CO laser for industrial applications [2, 3, 4]. Because of the comparatively more complicated physical processes in CO lasers the approaches to the design and the development of devices with the power in kilowatt range are still not established. Self-sustained dc and RF discharges and e-beam sustained discharges as well were employed for pumping CO lasers. There were lab-scale CO lasers with subsonic and supersonic gas flow (see the overview of early publications in [5]). This paper presents results of recent studies on the physical processes in high power CO lasers with particular emphasis on a fast-flow CO laser pumped by an e-beam sustained discharge in cryogenic gas mixtures at moderate and high gas density.

3. KINETIC MODEL

The formation mechanism of the partial inversion between the rotation-vibrational levels in the CO molecule is very specific. The key process is so called anharmonic pumping resulting from the vibration-vibration (V-V) exchange between molecules in low and high vibrational levels. Because of the anharmonicity of molecular vibrations the vibrational quantum number diminishes with the level number. Then at a low gas temperature the more probable is the process shown in Fig. 1, where the molecule with a lower level comes down and with the higher level comes up. The necessary conditions for this process to prevail against the reverse one are: $Q_{VV} \gg Q_{VT}$ and $T_v > T$, where Q_{VV} and Q_{VT} are the rates of the V-V exchange and V-T relaxation respectively, T is the gas temperature and T_v is the effective vibrational temperature for the lower levels. The resulting evolution of the vibrational distribution function (VDF) after the discharge was switched on is shown schematically in Fig. 2. (An analytical theory giving explicit expressions for the VDF shape as a function of time was formulated in [6]). As one can see from Fig. 2, the effective vibrational temperature describing the relative population in the neighboring levels is very high in the range v

≥ 10. For this particular case (generally, for v greater than the so called Treanor number [7]) in this range inversion may be realized between rotational sublevels in adjacent vibrational bands. It is clear that the inversion takes place simultaneously on

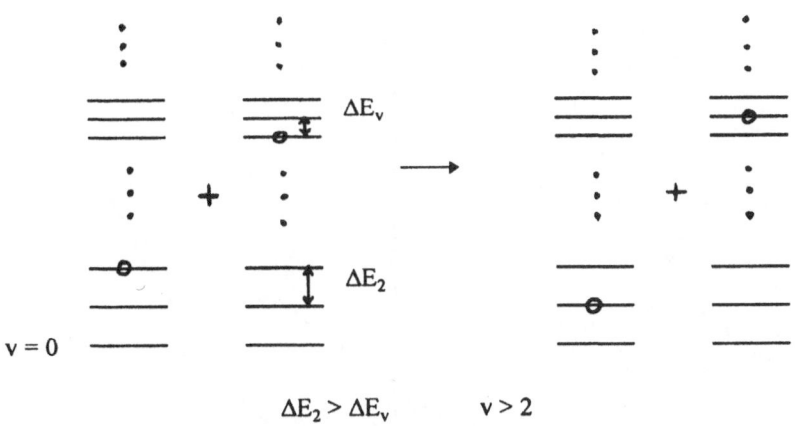

$\Delta E_2 > \Delta E_v \qquad v > 2$

Figure 1. Anharmonic pumping process

many transitions. This is the underlying reason for the multifrequency lasing of the CO laser.

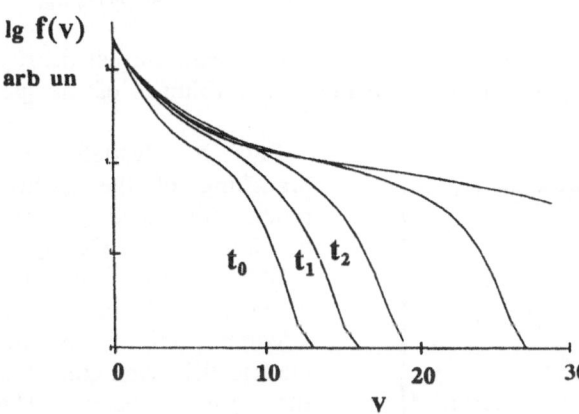

Figure 2. Evolution of the VDF in a course of excitation. $t_0 < t_1 < t_2$

The laser gas is excited by discharge electrons. To realize a sufficiently high discharge power density it is necessary to produce a certain ionization. Experimental studies of the mechanisms of the plasma decay for conditions of the CO laser demonstrated [8,9] the utmost importance for the laser mixture being very pure. In particular, the addition of CO_2 to nitrogen at a gas temperature of 80 K and a gas density of $N = 10^{19}$ cm^{-3} in an amount ≥ 10 ppm causes the electron attachment processes to be faster than the electron-ion recombination (see Fig. 3). It was assumed in the ref. 8 that electrons attach to the large molecular clusters from CO_2 molecules which are formed in a cryogenic gas. Fig. 3 demonstrates that at a CO_2 concentration of 100 ppm the plasma decay rate could increase 10 times. Authors of [8] have found also that water vapor accelerates the

14

electron-ion recombination very strongly. Fig. 4 illustrates this effect. In experiments [8], the discharge current at a water vapor content of 100 ppm became almost 10 times less than in pure nitrogen. Taking care about a very good gas cleanup, an e-beam current density on the order of 100 μA/cm^2 provides the plasma conductivity sufficient for an effective operation of a multi-kilowatt laser at a gas density of about one half of

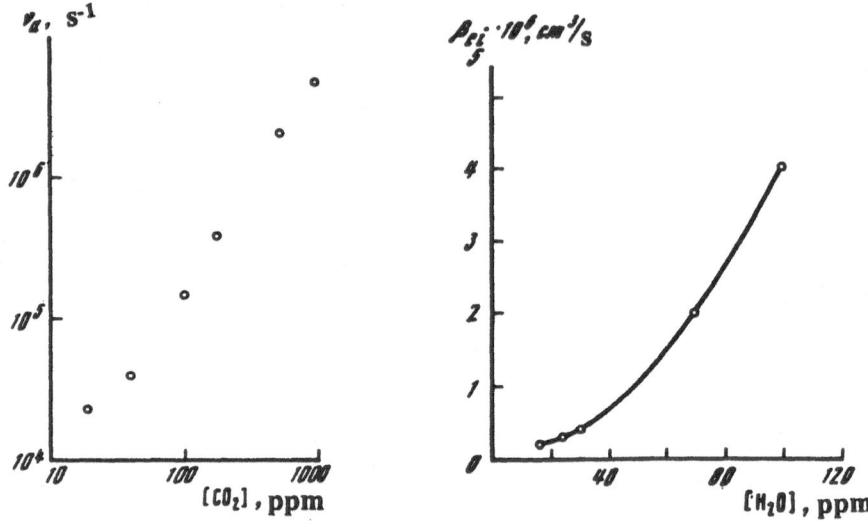

Figure 3. Attachment frequency as a function of CO_2 concentration in N_2

Figure 4. Electron-ion recombination coefficient as a function of water vapor content in N_2

the normal one, [10]. It should be emphasized that for this e-beam current density there is no problem with stability of the foil separating the volumes of the gas discharge and electron gun.

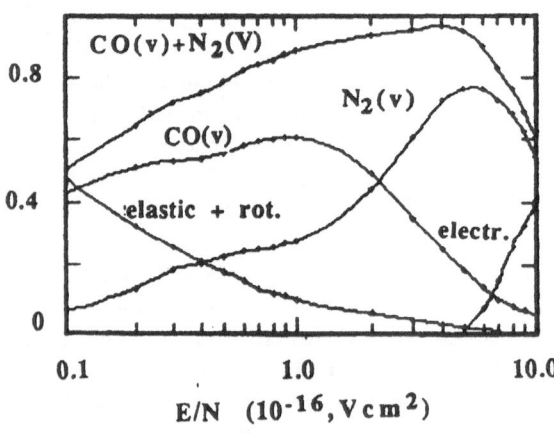

Figure 5. Branching of electron energy input vs E/N

Fig. 5 illustrates the branching of the electric power dissipated in the discharge in a $CO:N_2$ = 1:10 mixture as a function of the reduced electric field strength E/N, E is the electric field strength, N is the gas density. The excitation of the molecular vibrations is quite effective (≥ 0.8) for $1 \leq E/N$ (10^{16} V cm^2) ≤ 7. Using of a beam of fast electrons with the energy ≥ 100 keV to ionize the gas allows sustaining the discharge at an E/N-values close to optimum. One should keep in mind however, that in Fig. 5 results of calculations are shown which were made for an unexcited gas. Actually, the concentration of vibrationally excited molecules in CO laser mixtures are as a rule greater than in mixtures typical for the CO_2 laser. It means that the role of plasma electron collisions with vibrationally excited molecules can not be neglected. In particular during pumping in the discharge, in a mixture $CO:N_2$ the energy branching between the N_2 and CO vibrational level alters in the manner that the laser efficiency is decreasing [11]. Moreover, the explicit correlation between the discharge glow and the laser power noticed in several publications (see, for example [12, 13]) highlights the great role which most probably play molecular vibrations in the processes of electronic level excitation and ionization. The most adequate kinetic model of the CO laser should include a self-consistent description of the vibrational and electron energy distributions [14]. An indication that not everything is okay with the numerical modeling of CO laser kinetics is the fact that a direct gas heating rate calculated from

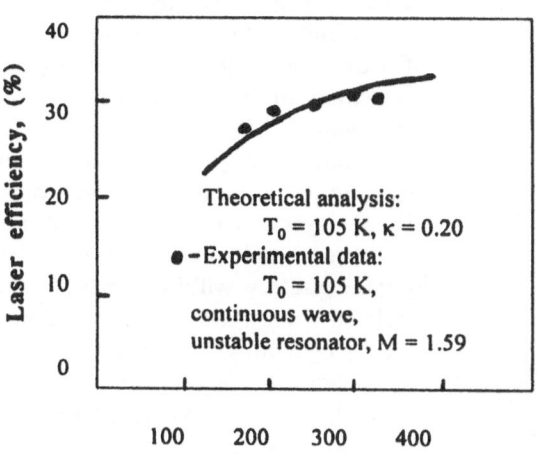

the electron Boltzmann equation seems to be lower than the one used in the overall CO laser models [14]. The direct gas heating rate is defined by excluding the processes of vibrational and electronic relaxation from the total discharge power. In calculations this quantity is taken usually to be proportional to the discharge power and its share κ serves as a fitting parameter found when the computed results agree with the experiment. The result of such a procedure for the conditions of the experiments [10] is shown in Fig. 6, where the dependence of the

Figure 6. Laser efficiency vs the reduced input energy

laser efficiency on the reduced energy input in J/g is shown. The magnitude of κ for which the calculations agree with the experiments does not coincide with the electron kinetics calculations.

4. 2-D DIFFRACTION MODEL

To have a radiation beam of high quality an unstable cavity usually is employed. It allows one to realize a single-mode operation with a high power output beam. The gas flow across the cavity axis induces a regular nonuniformity in gain and refraction index along the flow (see Fig. 7). These nonuniformities influence the optical mode pattern and may initiate multi-mode lasing. The higher the gas density the stronger the nonuniformities affect the beam quality. A high power cryogenic CO laser (a photograph is presented in Fig. 8) was pumped by an e-beam sustained discharge [10]. While a high laser efficiency (≥ 30 %) was achieved in this device, there is almost no information about the laser beam quality. Experimental estimates show that the output beam quality is far from of the diffraction limit.

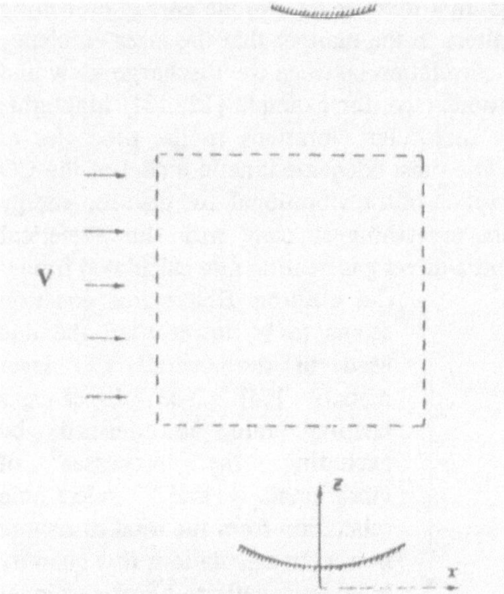

Figure 7. Schematics of CO laser with a mixture flow excited by the discharge (dashed line shows the discharge boundary)

The 2-dimensional diffraction model of the CO laser with a confocal unstable cavity developed in [15] allows one to study theoretically the effects of regular nonuniformities of the refraction and the gain in directions along the gas flow and laser axis. The model includes a detailed kinetic description of the vibrational distribution functions for CO and N_2 molecules, a calculation of the electron kinetic data by solving the electron Boltzmann equation and a 2-D diffraction description of the radiation propagation in the active medium for more than 10 laser transitions. The gasdynamic flow was considered as consisting of a number of transverse uniform layers. The evolution of the gasdynamic and kinetic quantities' along the gas flow within each layer was calculated from a system of vibrational kinetic equations and gasdynamic equations. The number of layers along the laser axis was varied up to 10.

Theoretical predictions for the conditions of the experiment [10] are illustrated in Fig. 9, where the far-field intensity distributions are presented, neglecting the gas refraction (a), and according to the full model (b). As one can see, the gas refraction nonuniformity induces the laser beam drift against the flow and its widening. Moreover, the far-field distribution can be viewed as a result of an almost independent summing of two beams diffracted over the left

Figure 8. Cryogenic cw CO laser excited by e-beam sustained discharge : "COL-2". Lasing mixture : CO:N_2=1:9, gas flow rate : 1kg/s, precooled gas temperature 90 K, total power 85 kW, efficiency (25-30)%, beam size 10x5 cm^2, operating time up to 5 s

and right edges of the output mirror. It should be emphasized that Fig 9 is drawn for a total intensity of multifrequency radiation (calculations predicted over 10 lasing transitions). The effect of the laser beam drift against the gas flow depends on the magnification number M of the cavity. Fig. 10 demonstrates that the greater M is the stronger is the beam drift. This drift originates in the refraction index decrease along the flow associated with the gas heating. For larger M the laser beam took less passes in the medium, resulting in smaller beam deflection.

If somebody is interested only in calculations of laser power and spectrum, a more simple model may be used where instead of the description of the radiation propagation one puts the gain on the lasing transitions to be equal to the threshold value at each laser beam position. Comparison between the diffraction model and this simplified one shown in Fig. 11, demonstrates that the differences in the calculated laser power and spectrum are insignificant. However, the distributions of the near-field intensity computed in the frames of these models may differ very strongly. Fig. 12 illustrates such differences for conditions of a supersonic CO laser [16]. The intensity spike found in the

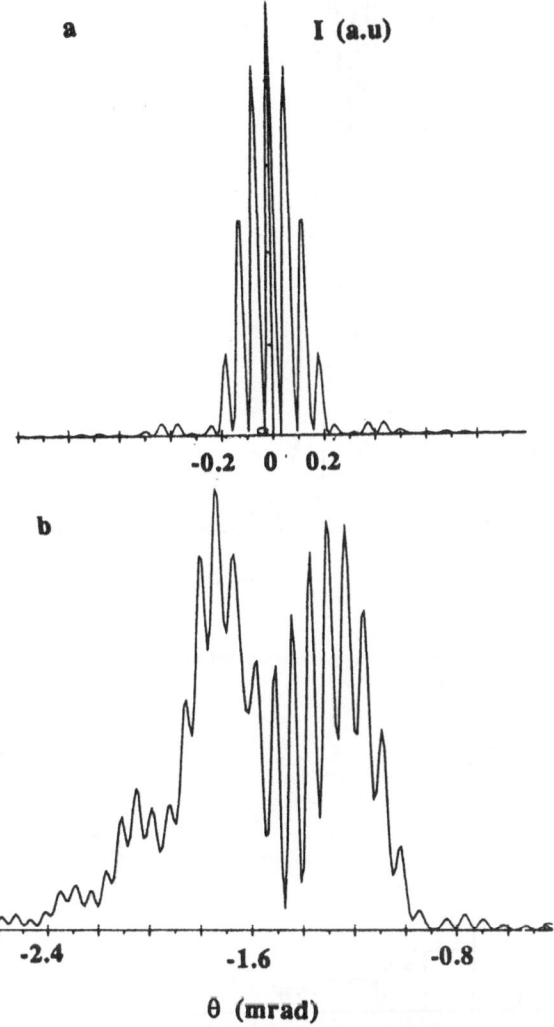

Figure 9. Calculated far-field intensity distributions, M=1.56
a : Refraction index variations neglected
b : Full model, the same conditions

constant-gain model is an artifact disappearing in the more correct 2-D diffraction model. Nevertheless, the output power in the left beam is evidently greater than in the right one. The physical reason for this lack of symmetry is due to gain and gas temperature variations along the gas flow.

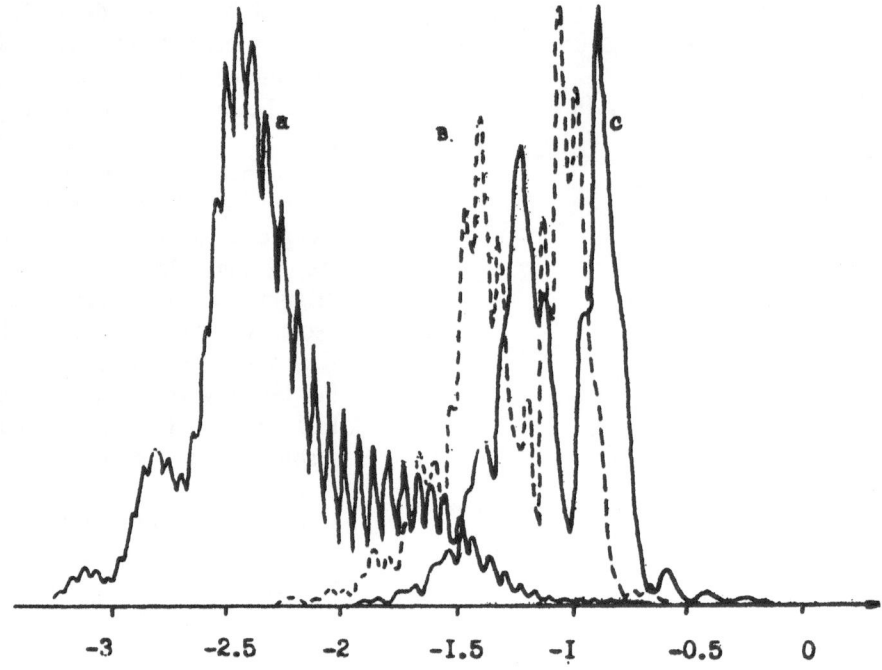

Figure 10. The far-field intensity distribution for different magnification numbers.
a-M=1.3; b-M=2; c-M=3.

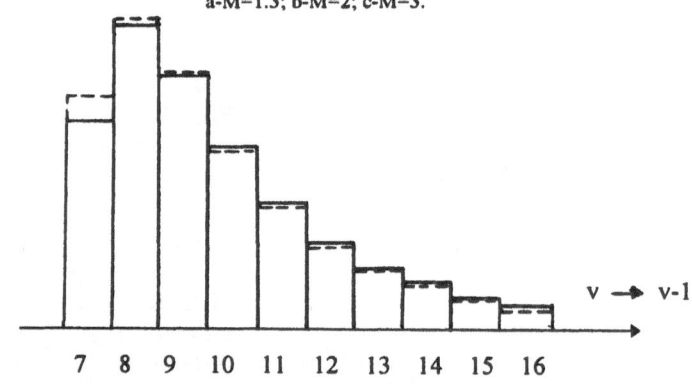

Figure 11. Laser spectrum calculated in a diffraction model (solid line) and a
constant-gain model (dashed line), T=120 K, $CO:N_2$=1:2, v=50 m/s,
Q_{input}=280 J/g, M=1.57 cylindrical resonator

5. DESCRIPTION OF ACTIVE MEDIUM NONUNIFORMITIES

In reality, the ionizing electron beam behind the electron-gun foil is highly divergent
resulting in a nonuniform distribution of the ionization rate in the laser aperture. It

means that the discharge parameters, the gain and the refraction index are nonuniform along the gas flow and along the discharge current.

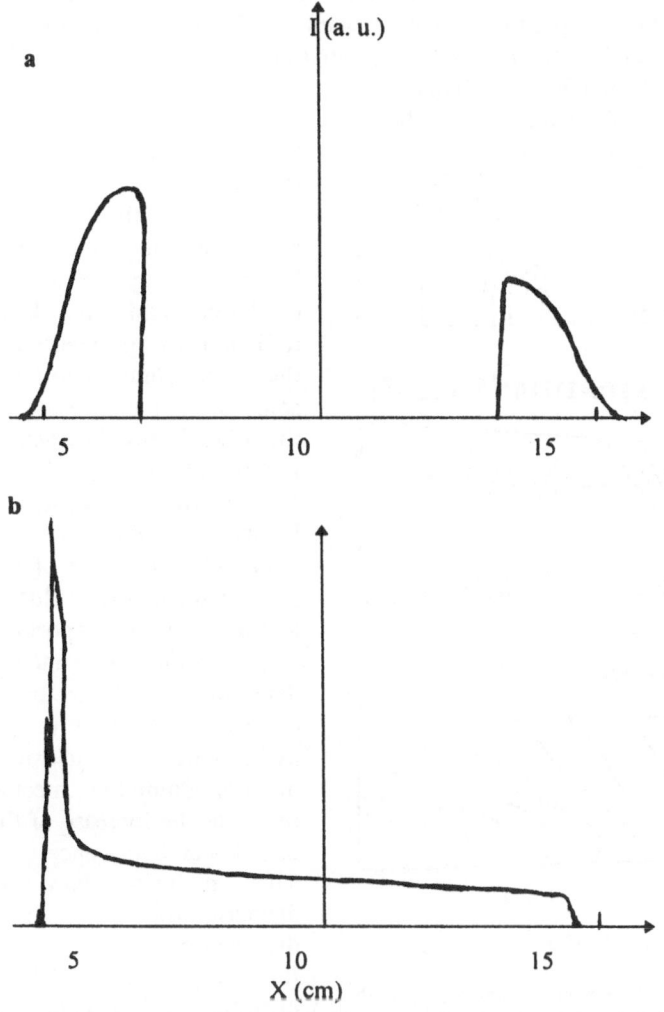

To describe these effects a 2-D model was developed where in contrast to the above mentioned 2-D diffraction model (coordinates along the gas flow and optical beam, respectively) the distributions of the plasma and optical parameters were calculated as functions of coordinates along the flow and the discharge current. In this model the distribution of the ionization rate by the e-beam was calculated according to the interpolation formula proposed by the authors of [17].

The electric field distribution and the plasma conductivity were found by solving the electric current continuation equation and kinetic equations for charged particle concentrations. For a description of the vibrational kinetics and lasing a simplified

Figure 12. Calculated near-field intensity distributions for conditions of the experiment [16]: a. 2-D diffraction model; b. Model g=const

model was formulated where the large set of kinetic equations was replaced by several analytical relationships, derived earlier in [18]. To find the laser radiation spectrum and its distribution over the laser aperture the constant-gain model was employed. Some results of the numerical simulations for the conditions of the experiment [10] are presented in Fig. 13, where a contour plot is drawn describing the spatial distributions of the physical quanities in a discharge channel of which the height was growing along the gas flow at an angle of 6°. One can see that the peaks in the reduced electric field strength (E/N) and in the laser power density (Q) appear in the region at the e-beam boundaries near the foil separating the electron gun and discharge volumes. The resulting nonuniform gas heating produces a complicated gas density distribution (ρ). Estimates made on the ground of the calculated distributions of the refraction index predict the beam quality to be much worse than the DL.

The theoretical models allow for estimating the laser beam divergence connected with the regular nonuniformities in the gain and index, induced by the complex self-consistent interaction between the gas flow, the discharge and the laser radiation. In an experiment there are some additional sources for the increase of the laser beam divergence, in particular connected with the gasdynamic turbulence. Because of the low gas temperature the role of the gas temperature fluctuations is very important. The existing experimental data on the laser beam quality do not allow one to make definite conclusions on the dominant mechanisms of the increase of the laser beam diver-gence.

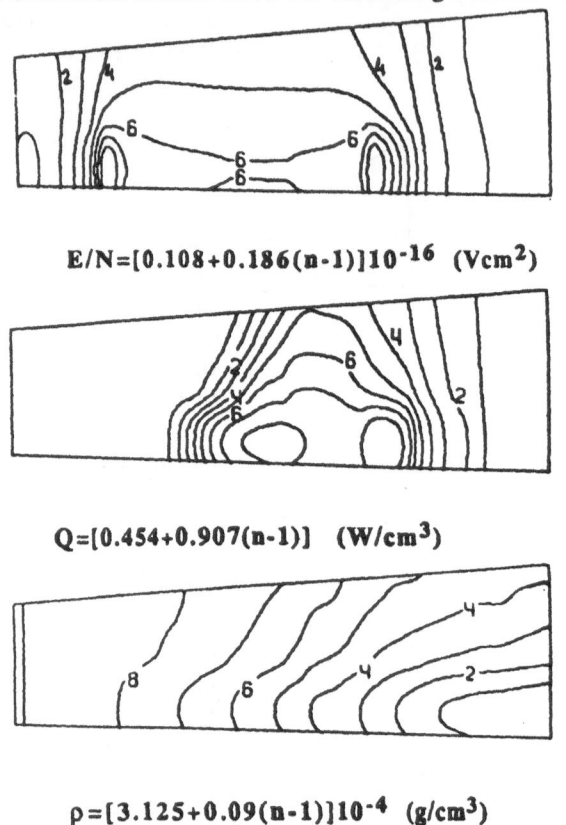

$$E/N = [0.108 + 0.186(n-1)]10^{-16} \quad (Vcm^2)$$

$$Q = [0.454 + 0.907(n-1)] \quad (W/cm^3)$$

$$\rho = [3.125 + 0.09(n-1)]10^{-4} \quad (g/cm^3)$$

Figure 13. Contour plots for the reduced electric field strength, E/N, laser power density, Q, and gas density, ρ for cryogenic CO laser shown in Fig. 8

The problem how to decrease the laser beam divergence is very serious for a high power CO laser operating with a cryogenic high-density gas mixture.

One possible way to solve this problem was pointed out recently in [19], where the phase conjugation effect in the active medium of the CO laser was observed in processes of 4-wave mixing. As was proven in Ref. 20 the processes of 4-wave mixing in this case involve the formation of frequency selective gain gratings. The cascading nature of the lasing transitions is responsible for the mutual increase of the gain modulation depth on adjacent transitions. These results [19,20] show that it is possible to improve the beam quality by virtue of the application of the phase conjugation effect in the active medium of the CO laser.

6. CONCLUSIONS

High power CO lasers are still being developed, and quite a few number of lasers with output powers in the range of tens of kilowatts are available for industrial applications. The straight way to construct lasers with powers in the range of hundred of kilowatts is to use a cryogenic e-beam sustained discharge with a subsonic gas mixture flow. The device will be compact and simpler (less expensive) in expoitation for a high gas density ($\geq$ 1/3 of the normal density). Then of great importance become the problems with sustaining the stable gas discharge at e-beam currents which do not destroy the foil separating the electron gun and the active volume. It was shown that this problem is solvable by virtue of a very careful gas cleanup. The next problem arising because of the high gas density is connected with refractive index nonuniformities induced by gas heating in the discharge. Modeling of the CO laser including a self-consistent description of the discharge and radiation propagation effects made it possible to estimate the output laser beam drift and divergence. A simplified description of the vibrational kinetics allowed one to estimate the effects of index nonuniformities connected with the ionizing e-beam spread in the discharge volume induced by electron scattering in the foil and gas. The theory predicts a strong impact of the spatial index variations on the laser beam quality. As a possible way to neutralize the negative influence of the induced index nonuniformities the phase conjugation effect in the active medium may be considered.

7. ACKNOWLEDGEMENTS

It is my pleasure to acknowledge the help of Dr. Kochetov in the preparation of this article. Thanks to Dr. Kuz'min for the photograph of the laser. I am very grateful to Professors Ochkin and Witteman who invited me to give this talk.

8. REFERENCES

1. Baranov, V.Yu., Bakhtadze, A.B., Belykh, A.D., et al. (1986) *Kvantovaya Elektronika*, **13**, 206
2. Maisenhaelder, F. (1988) *SPIE Proceedings* **1031**, 98
3. Ionin, A.A. this book
4. Averin, A.P., Basov, N.G., Vasil'ev, L.A., et al. (1982) *Kvantovaya Elektronika*, **9**, 2357
5. Mann, M.M. (1976) *AIAA J.* **14**, 549
6. Zhdanok, S.A., Napartovich, A.P., Starostin, A.N. (1979) *Sov. Phys. JETP* **49**, 66
7. Treanor, C.A., Rich, I.W., Rehm, R.G. (1968) *J. Chem. Phys.* **48**, 1798
8. Borodin, A.M., Vysikailo, F.I., Gurashvili, V.A. et al. (1989) *Doklady Akademii Nauk*, **306**, 1397
9. Gurashvili, V.A., Izyumov, S.V., Turkin, N.G. et al. (1989) *Zhurnal Tekhnicheskoy Fiziki*, **59**, 177
10. Gurashvili, V.A., Kuz'min, V.N., et al. (1991) *Soviet-American Symposium on Research, Technology and Trade*, San Francisco, USA

11. Islamov, R.Sh., Konev, Yu.B., Kochetov, I.V., Kurnosov, A.K. (1984) *Kvantovaya Elektronika*, **11**, 142
12. Grigoryan, G.M., Dymshits, B.M., Ionikh, Yu.Z. (1988) *Optika i Spektroskopiya*, **65**, 766
13. Margolin, A.K., Shmelyev, V.M. (1990) *Khimicheskaya Fizika*, **8**, 794
14. Konev, Yu.B., Kochetov, I.V., Kurnosov, A.K., Mirzakarimov, B.A. (1994) *J. Phys. D: Appl. Phys.*, **27**, 2054
15. Elkin, N.N., Kochetov, I.V., Kurnosov, A.K., Napartovich, A.P. (1992) *J. Sov. Laser Res.* **13**, 46
16. von Buelov, H., Schellhorn, M. (1993) *Appl. Phys. Lett.*, **63**, 287
17. Cason, C., Perkins, J.F., Werkheiser, A.H., Duderstadt, (1977) *AIAA Paper #77-65*
18. Berdyshev, A.V., Kochetov, I.V., Napartovich, A.P. (1988) *Khimicheskaya Fizika* **7**, 470
19. Belousov, D.V., Borodin, A.M., Bunkina, M.B. et al. (1991) *Abstracts of papers presented at All-Union Conference on Coherent and Nonlinear Optics*, Leningrad, **1**, 177
20. Berdyshev, A.V., Kurnosov, A.K., Napartovich, A.P. (1994) *Quantum Electronics*, **24**, 87

GENERATION OF VUV RADIATION WITH IONIC AND F_2 EXCIMER LASERS

P.J.M. PETERS and H.M.J. BASTIAENS
University of Twente
Department of Applied Physics
P.O. Box 217
7500 AE Enschede
The Netherlands
Tel. x-31-53-89(4)3967
Fax x-31-53-338065

1. Abstract

Light sources and especially coherent light sources in the VUV are extremely important for a large number of applications. Because of the fact that only a very limited number of optical crystals are transparent in this wavelength region several promising new gaseous media have been investigated recently. We will describe in this paper the results of our research on gaseous VUV emitting media. As a typical example of a class of totally new excimer molecules $XeRb^+$ was studied in detail. It is a combination of an alkali-ion with a rare gas atom resulting in an ionic complex isoelectronic to the well known rare gas excimers. Experimental fluorescence signals at $\lambda = 164.5$ nm will be compared with predicted values from a kinetic model. We also will report on the molecular $F_2{}^*$ laser emitting at $\lambda = 157$ nm. For the first time we were able to produce a laser pulse of 1,8 J with a pulse length of 160 ns. Model calculations also showed that although the finite lifetime of the lower laser level reduces the extraction efficiency of the laser to approximately 15% it is possible to realise long optical laser pulses.

2. Introduction

Coherent UV or VUV optical radiation at high power levels can only be generated by a special type of gas lasers namely by excimer lasers. Excimers (from excited dimers) are molecules which are only bound in the excited states but are unstable in their electronic ground states. Examples are diatomic molecules like $Xe_2{}^*$ (which was the first lasing excimer[1]) $Ar_2{}^*$ and $Kr_2{}^*$, or rare gas halide molecules like for example ArF^*, KrF^* or $XeCl^*$. They all have identical potential diagrams. A typical diagram is sketched in Fig. 1. One can see the bound excited state which acts as the upper laser level and the ground state which has only a very small minimum in the potential. The depth of this minimum is small compared to kT at room temperature so there are no stable molecules in the ground state which means that the lower laser level is always empty.

23

W. J. Witteman and V. N. Ochkin (eds.), Gas Lasers - Recent Developments and Future Prospects, 23–36.
© 1996 *Kluwer Academic Publishers.*

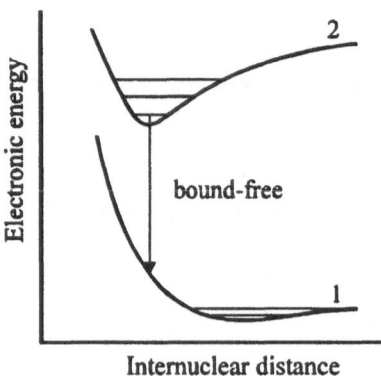

Fig. 1 A bound-free transition for a diatomic molecule, as represented in a potential diagram. The vertical transition is according to the Franck-Condon principle.

Almost 15 years after the discovery of the first excimer laser by Basov a new class of excimers were proposed at the same time by Sauerbrey[2] and Basov[3]. They predicted that the combination of an alkali ion (A^+) and a rare gas atom (Rg) or a halogen atom (X) leads to the formation of ionic excimers which are iso-electronic to the rare gas dimers and the rare gas halogen excimers respectively. By calculating their potential diagrams they were able to predict their emission wavelengths. All ionic excimers appeared to emit in the VUV region of the spectrum. The experimentally observed wavelengths can be seen in table 1 for the A^+Rg ionic excimers[4,5].

	He^+	Ne^+	Ar^+	Kr^+	Xe^+
Li	66.78	80.70	124.54	149.75	188.9
Na	65.76	79.40	121.30	144.95	182.9
K	63.82	77.18	115.04	135.70	167.38
Rb		76.66	113.70	133.85	164.1
Cs			112.26	131.58	160.50

Table 1 Experimentally observed spontaneous emission wavelengths (in nm) of ionic excimers, resulting from the combination of a rare-gas ion and an alkali atom.

Next to the A^+Rg and A^+X ionic excimers a number of other types of ionic excimers were proposed (A^{2+}A, Rg^+X, Rg^+ + atom of the oxygen group) with transitions in the wavelength region from the visible extending to the VUV. Until now only theoretical work has been done on these very special ionic excimers.
In this paper we will report on one special ionic excimer namely $XeRb^+$ emitting near 164.5 nm[6,7,8]. Apart from this study we also will report on the molecular F_2 laser emitting at $\lambda = 157$ nm in the VUV region[9]. For both systems a computer model was developed and the predicted results were compared with the experimentally obtained values.

3. Experimental setup

The XeRb$^+$ ionic excimer molecule is a combination of a rare gas atom (Xe) and a alkali metal ion (Rb$^+$) or a combination of Xe$^+$ and Rb. Such an ionic molecule can be produced by irradiation a closed volume containing the two gas species (Xe and Rb) by high energy photons (X-rays) or by high energy particles like relativistic electrons. Both methods are well know techniques for exciting laser gas mixtures. However an additional constraint is the number density of the Rb atoms. In order to achieve a partial pressure of several mbar Rb necessary for the XeRb$^+$ formation the active volume containing the Rb metal has to be heated up to about 400°C.

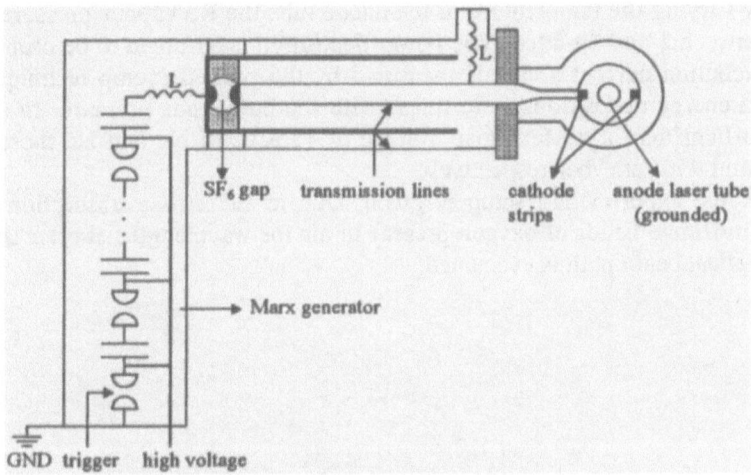

Fig. 2 Schematic view of the electron beam generator

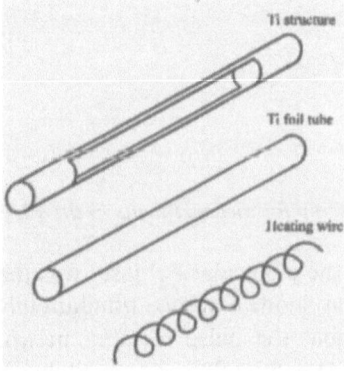

Fig. 3 Basic components of the heated anode tube

The electron beam generator (see Fig. 2) which was used as pumping source for the XeRb$^+$ excimer consists of a 10 stage Marx generator (10,8 nF/stage, maximum load voltage 45 kV each), a pulse forming network (PFN) which compresses the pulse from the Marx generator down to about 8 ns and a coaxial shaped field emission diode which contains a heated anode tube.

The heated anode tube configuration which contains the gas mixture is sketched in Fig. 3. A titanium foil with a wall thickness of 25 μm and an inner diameter of 10 mm is slid over a supporting titanium tube (with a wall thickness of 1 mm) and soldered at the ends. Two windows of 24 cm length are cut out at each side of the supporting tube to allow the electron beam to enter the gas volume inside. The heating is performed by a heating wire wound as a spiral around the tube. At the same time this wire reinforces the strength of the tube so that it can withstand pressures up to 10 bar at a working temperature of 700ºC for several hundreds of shots. By varying the temperature of the anode tube the Rb vapour pressure can be varied between 2 and 20 mbar. The power deposition is assumed to be proportional to the excitation current and it is measured by the pressure jump technique. The measured energy depositions were linear with the buffer gas pressure. In argon it was 33 mJ/cm^3/bar at a Marx load voltage of 45 KV. In Ne and He these values were 19 and 9 mJ/cm^3/bar respectively.

In Fig. 4 the experimental setup is given. Due to the strong absorption on the Schumann-Runge bands of oxygen present in air for wavelengths shorter than 190 nm the optical beam path is evacuated.

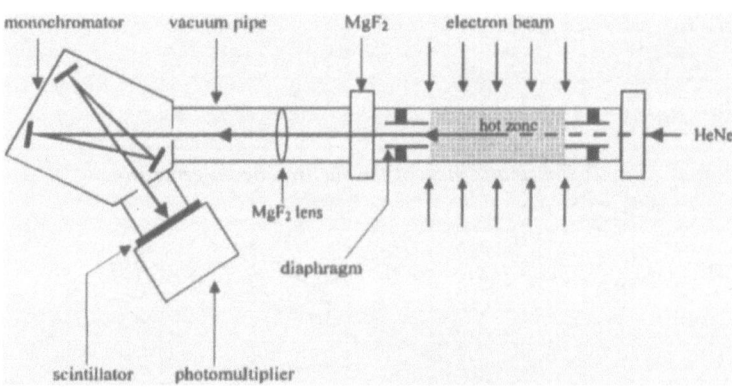

Fig. 4 The experimental setup for measurements in the VUV region of the spectrum

The experimental setup for the molecular F_2^* laser was the same as described above but with the exception that as anode a normal titanium tube with a wall thickness of 25 μm was used and without the pulse forming network (PFN). In this way a current pulse with a (FWHM) width of 25 ns is available as pumping pulse with the same energy deposition characteristics.

At the end of this paper we will also report on long pulse experiments with the F_2^* laser. For these experiments we used a larger coaxial e-beam pumping system working along the same principles but with an active laser gas volume of 157 cm^3 and with a pumping pulse length of about 160 ns (FWHM).

4. Gas kinetic model

Within the class of ionic excimers $XeRb^+$ is a typical representative which is relatively easy to produce experimentally. Therefore a kinetic model was created for this ionic excimer for pumping the medium with an e-beam. The high energy electrons from the e-beam deposit their energy in the laser gas mixture by means of ionising and exciting the atoms in the gas. Because the main component of the gas mixture are buffer gas atoms ($\approx 95\%$) the energy mainly will be deposited in the buffer gas with an efficieny $\eta_{i,e} = E_{i,e} / W_{i,e}$ where i stands for ions and e for excited atoms. $W_{i,e}$ is value in eV is necessary to create the ionic or excited state of the atom. The deposited energy is then transferred by association reactions and charge transfer reactions to finally $XeRb^{+*}$ as can be seen in Fig.5. Also shown in the figure are the quenching reactions which we took into account in our model.

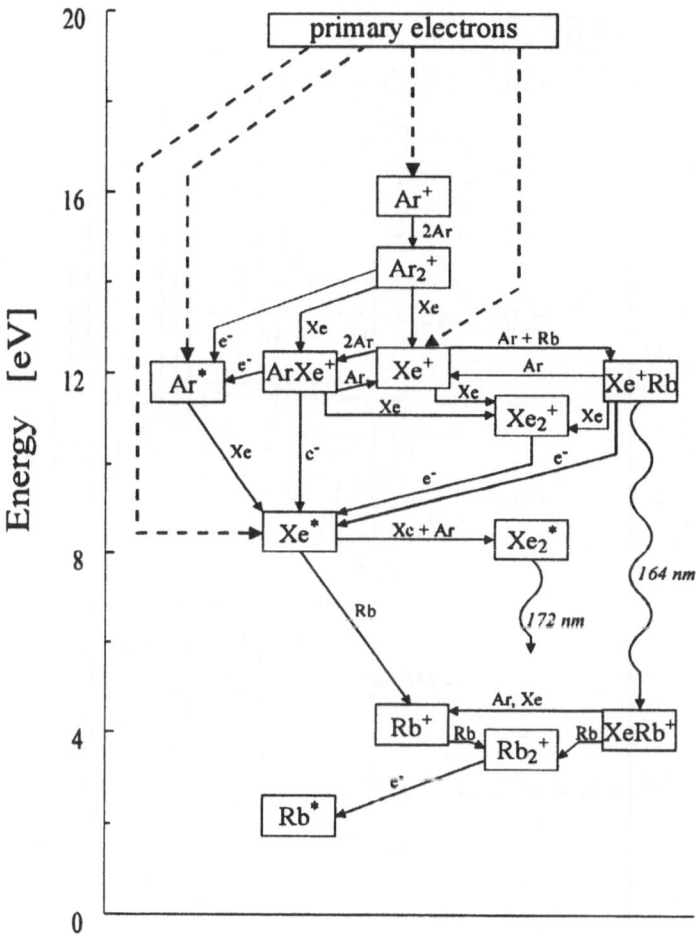

Fig. 5 *Flowchart indicating the most important kinetic paths in the electron beam pumped Ar/Xe/Rb gas mixture. The energy levels of the several species are indicated. Primary electron energies are in the order of 200 keV.*

28

A serious bottleneck in the attempt to obtain gain could be the absorption of XeRb$^+$ radiation by photo-ionisation of the Rb vapour. The cross-section for this process is about 0,15 10^{-18} cm^2 at λ = 164 nm. At a gas temperature of 380^{0}C the Rb density is 1,3 10^{17} cm^{-3} ($\approx$12 mbar) so at least an absorption threshold of 1.9 %/cm has to overcome under these conditions. Other absorbing species at this wavelength are the metastables of all rare gases used but their cross-section are relatively low.

The stimulated emission cross-section of the strongest $\Omega = 0^+ \rightarrow 0^+$ transition of XeRb$^+$was estimated to be $\sigma_{es} \approx 2.9 \ 10^{-16}$ cm^2. With this value it is possible to predict the time dependency of the gain or the XeRb$^+$ density provided all data for the electron beam are known by means of solving the set of rate equations.

For the molecular F_2^* laser basically the same kinetic model but with different gas species was used (see Fig. 6) and coupled to optical resonator equations based on a Rigrod analysis.

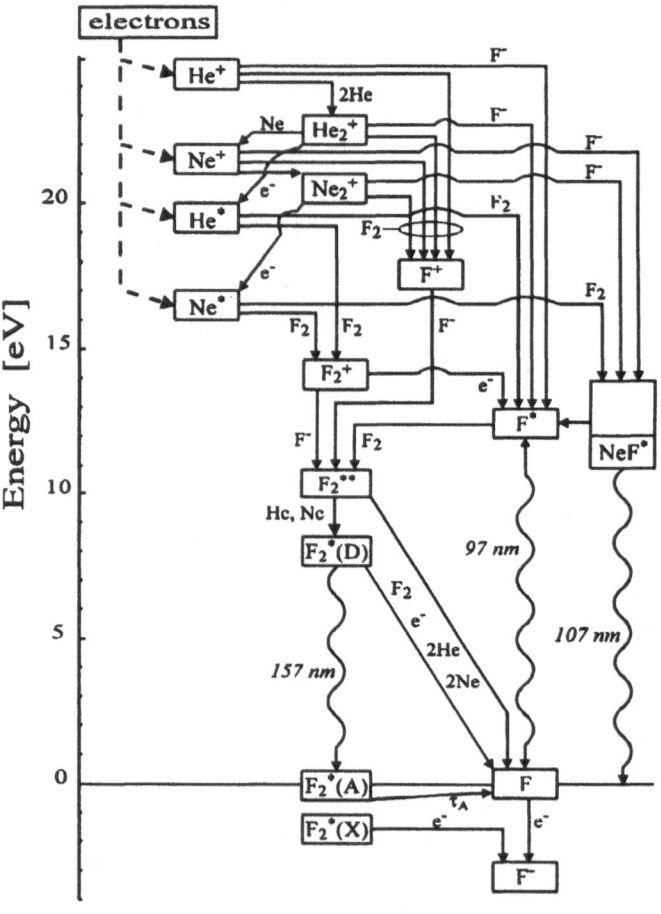

Fig. 6 Flowchart indicating the main reaction channels in an electron beam pumped He/Ne/F$_2$ gas mixture. Energy levels of the species are indicated. Primary electrons: $\approx$ 250 keV.

Results of the model calculations will be compared with the experimental results in the next paragraph for both the XeRb⁺ ionic excimer and the molecular F_2^* laser.

5. Experiments

5.1 XeRb⁺ measurements

From the kinetic model we expect that the fluorescence signals of XeRb⁺ reflect the time dependency of its precursor Xe⁺. In Fig. 7 typical traces are shown of a fluorescence signal and of the excitation current pulse as measured inside the tube. The rise time of the fluorescence pulse coincides with the duration of the excitation pulse. The pulse also exhibits a single exponential decay which is predicted also by our kinetic model.

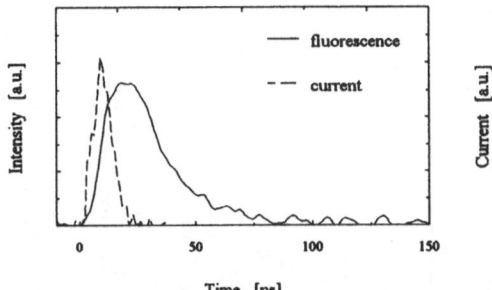

Fig. 7 *Typical traces of the (XeRb⁺) fluorescence and of the excitation current pulse as measured inside the tube. (gas mixture : 94.7% Ar, 5% Xe and 0.3% Rb (12 mbar) at a total gas pressure of 4 bar)*

We measured the fluorescence peak intensity and the decay frequency of the signal as a function of the Rb density and Xe concentration for different total gas pressures with various buffer gases like He, Ne and Ar.

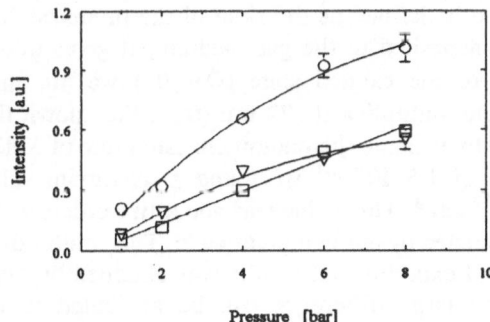

Fig. 8 *Fluorescence peak intensity of XeRb⁺ as function of total gas pressure for several buffer gases, with [Rb] = 12.7·10¹⁶ cm⁻³, Xe = 5%, (O: Ar ; ∇: Ne ; □: He).*

In Fig. 8 the fluorescence peak intensity of Xe Rb$^+$ is shown as a function of the total gas pressure with He, Ne, Ar as a buffergas. The increase of the peak intensity with the total gas pressure for the different buffergases can be explained by the enlarged power deposition in the gas and the increasing rate of the XeRb$^+$ formation reaction.

At higher pressures the intensity seems to saturate for the Ar and Ne gas mixtures while for He the peak intensity still increases linearly. In Fig. 9 the decay frequency of XeRb$^+$ (determined by the Xe$^+$ decay) is shown as a function of the same parameters. The increase of this decay frequency with the total gas pressure is due to the increasing formation rate of XeRb$^+$ while the saturation is attributed to the collision induced dissociation of XeRb$^+$ into Xe$^+$ and Rb.

The fluorescence pulses appeared to be almost independent on the energy deposition. This means that electron quenching plays an important role in the kinetics of the XeRb$^+$ ionic excimer which was also predicted by the model calculations. The temporal behaviour of the XeRb$^+$ density also was predicted reasonably by the model.

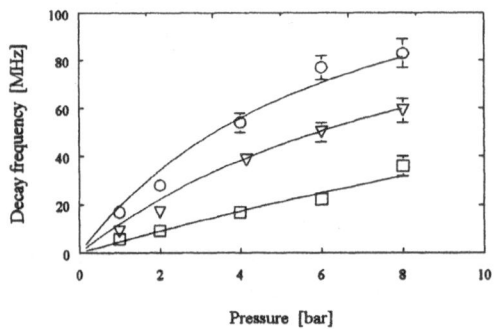

Fig. 9 Fluorescence decay frequency of XeRb$^+$ as function of total gas pressure for several buffer gases, with [Rb] = 12.7·10^{16} cm^{-3}, Xe = 5%, (O: Ar ; V: Ne ; □: He). The solid lines represent the calculated decay frequencies.

The fluorescence yield is defined as the ratio of the produced XeRb$^+$ fluorescence energy to the energy deposited in the gas medium. It gives information about the formation efficiency of the excited state ($\Omega = 0^+$) we are interested in. After calibration of the photo-multiplier at 172 nm (from the known fluorescence source Xe$_2$*) it was possible to estimate the photon emission rate of XeRb$^+$ and by using a transition probability of 1.5 10^8 s^{-1} we found a maximum value for the XeRb$^+$ density of about 2 10^{12} cm^{-3}. This value was about two orders of magnitude smaller than the densities predicted by our computer code. The similar discrepancy between model calculations and experimental results was obtained by mantel for the KrK$^+$ ionic excimer[10]. The large difference can be attributed to the rapid thermal relaxation within the upper level of the ionic excimer. Due to this effect the population probability of the state with the highest transition probability ($\Omega = 0^+$) at the moment of the radiactive decay was estimated to be about 1 : 65.

6. F_2* measurements

With the small coaxial e-beam system the gain and the dependance of the F_2* laser performance on gas composition, total gas pressure pumping power and cavity configuration were investigated and compared with the simulated results. In Fig. 10 the output energy of the F_2* laser from an extractable volume of 12.7 cm^3 is plotted as a function of the F_2 concentration at different Marx load voltages. The total gas pressure was kept constant at 12 bar Ne/F_2.

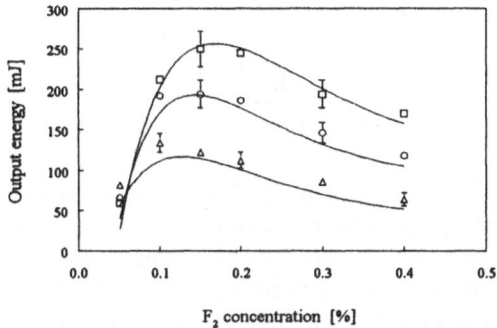

Fig. 10 Measured (markers; □: 60, O: 50, Δ: 40 kV) and simulated (solid curves) output energy as function of the F_2 concentration at different Marx load voltages. Gas mixture: 12 bar Ne/F_2. (The simulated lower level lifetime is 2.3 ns).

One can see that the laser energy increases with increasing power deposition except for low F_2 cocentrations where F_2 depletion becomes important and the laser performance improves at lower energy depositions. The solid lines are the simulated curves. The trends in this curves are the same as the experimentally observed results. The optimum intrinsic efficiency of 4.0 % was obtained with a He/Ne/F_2 = 39.9/60/0.1 gas mixture at a total gas pressure of 12 bar. The decrease of the efficiency at higher pressures is due to quenching reactions of the upper laser level and/or quenching of the precursors of the F_2* upper laser level. Measurements of the gain in this system already showed values a high as 63 %/cm[11)] obtained at a total gas pressure of 8 bar. Such high gain coefficients will drive the laser into saturation in typically 1 or 2 round trips, therefore quasi stationary laser operation may be assumed and it also was used in the model.

The fluorescence yield (efficiency) η_{fl} is defined as the ratio of emitted fluorescence energy to the deposited energy E_{dep}. Assuming a pressure independent formation efficiency η_f of the molecular D state η_{fl} scales like:

$$\eta_{fl} = \eta_f \frac{1/\tau_r}{1/\tau_r + 1/\tau_q} \tag{1}$$

where τ_r denotes the radiactive lifetime of the D state and τ_q denotes the finite lifetime induced by quenching processes mainly caused by F_2 molecules and electrons:

$$1/\tau_q = k_e[e^-] + k_{F_2}[F_2] \qquad (2)$$

Equation (1) can be rewritten to:

$$\frac{1}{\eta_{fl}} = \frac{1}{\eta_f}\left(1+\tau_r\left(k_e\left[e^-\right]+k_{F_2}\left[F_2\right]\right)\right) \qquad (3)$$

The quantum efficiency is given by:

$$\eta_{F_{l\,max}} = \frac{h\nu}{Q_{cost}} \qquad (4)$$

where Q_{cost} is the quantum cost for producing either a He(Ne) ion or an excited He(Ne) atom as defined earlier. The maximum fluorescence yield is then calculated to be 28 % in He/F_2 and 31 % in Ne/F_2 gas mixtures. The fluorescence yield is measured by recording the fluorescence signals of an excited 0.3 cm long cylindrical volume element (area = 2.82 cm^2) with a calibrated solar blind vacuum photodiode (ITT F4115) as shown in Fig. 11.

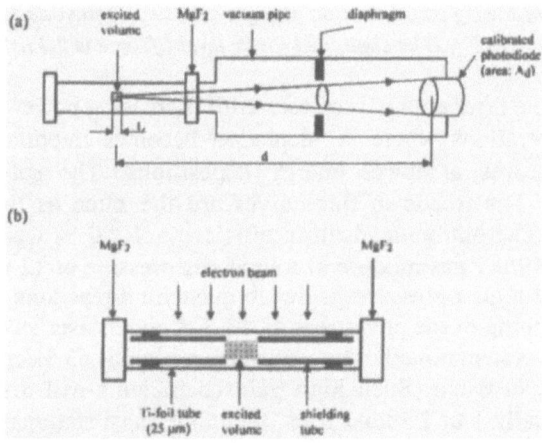

Fig. 11 The experimental set-up used for the determination of the fluorescence yield (a) and enlarged view of the electron beam excited volume in the cavity (b). Parameters: L=0.3 cm, d=116 cm, and A_d=2.8 cm^2. The solid angle $d\Omega$ is defined by $d\Omega=A_d/d^2$.

Fig. 12 shows the inverse fluorescence yield for He/F_2 and Ne/F_2 gas mixtures as a function of the total gas pressure at different fluorine concentrations. As expected from equation (3) $1/\eta_{fl}$ shows a linear dependency on the pressure. From the slopes the quenching parameter of F_2 can be determined to be:

$$k_{F_2} = (7 \pm 4)\ 10^{-10} \quad \left[cm^3 / s \right] \tag{5}$$

After we showed the realisation of a short pulse, high power F_2^* laser with a high efficiency the question arose if it would be possible to realise also a F_2^* laser with a long optical output pulse. In literature it was suggested that due to the strong bottlenecking in the lower laser level the laser always will show a self-terminating behaviour[12,13].

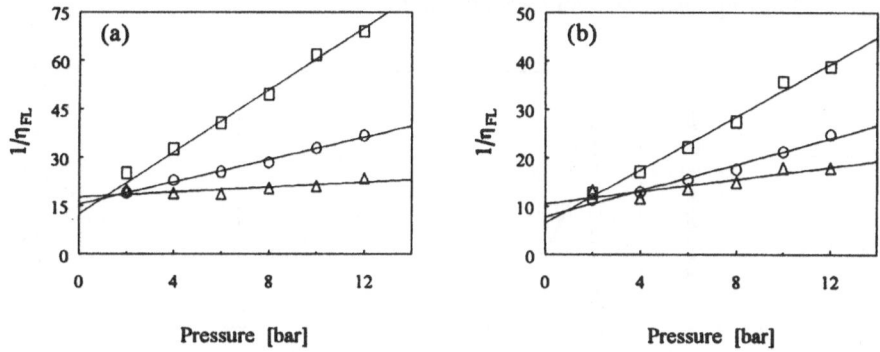

Fig. 12 *Inverse fluorescence yield of He/F$_2$ mixtures (a) and Ne/F$_2$ mixtures (b) as function of total gas pressure. (□: 0.4% F$_2$; O: 0.2% F$_2$; △: 0.1% F$_2$). 60 kV.*

With the larger coaxial electron beam machine (for details see ref. 9) it was possible to generate a pumping pulse length of about 200 ns duration. Using basically the same gas mixtures as in the smaller system described earlier, it appeared that the bottlenecking effect was not that large in order to prevent oscillation. We achieved a maximum laser pulse duration of approximately 160 ns and more than 1,8 J per pulse corresponding to a specific output energy of 11.4 J/l and an intrinsic efficiency of 2,0 % in a He/Ne/F$_2$ (2.85/97/0,15) gas mixture at a total gas pressure of 8 bar. A typical plot is given in Fig. 13.

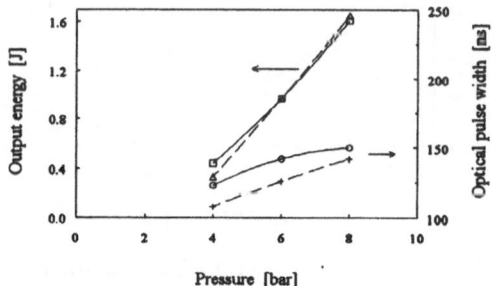

Fig. 13 *Measured total output energy and laser pulse width as function of total gas pressure for a He/Ne/F$_2$ = 1.9/98/0.1 gas mixture in a full reflector-MgF$_2$ (solid lines) and MgF$_2$-MgF$_2$ (dashed lines) cavity configuration. (output energy: □: FR-MgF$_2$; △: MgF$_2$-MgF$_2$; pulse width: O: FR-MgF$_2$; +: MgF$_2$-MgF$_2$).*

Also these long pulse behaviour was predicted well by our kinetic model. In Fig. 14 the temporal behaviour of the measured laser intensity is given. Although the simulated and measured pulse shape are not fully overlapping, the simulations give a reasonable estimation of the actual pulse width.

If according to the simulations, the lower lifetime is shorter than the radiative lifetime of the upper laser level then the laser is not self terminating as long as the gain is high enough to overcome the threshold. We also can say that the removal rate of the lower laser level should be able to keep up with the total population rate of this level during lasing to avoid accumulation.

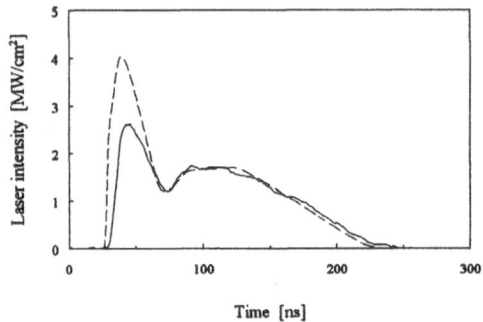

Fig. 14 Measured (solid line) and simulated (dashed line) laser pulse for a 8 bar He/Ne/F₂ (3.8/96/0.2) gas mixture pumped at 100 kV. The curves represent the emitted laser beam intensity from one side of a MgF₂-MgF₂ resonator.

However, bottlenecking in the lower laser level decreases dramatically the extraction efficiency and consequently the output power of the laser as can be seen from Fig. 15.

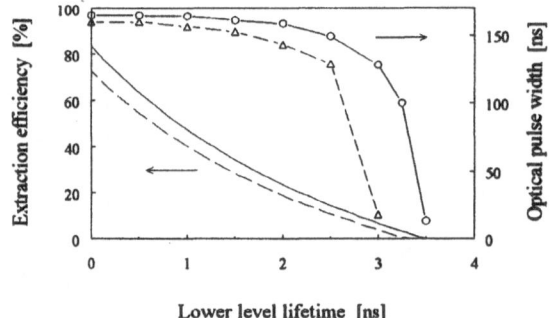

Fig. 15 Simulated pulse width (markers) and extraction efficiency in the large system as function of the lower level lifetime for a typical gas mixture (He/Ne/F₂ = 2.85/97/0.15). (solid lines: 8 bar; dashed lines: 4 bar).

Under the given experimental conditions no signs of self-terminating optical pulses are observed for stimulated radiation production up to 100 kW/cm³ and additional fluorescence radiation production of about 200-300 kW/cm³. Kinetic modelling showed that the maximum extraction efficiency of a F_2^* laser is probably limited by the non-negligible lifetime of the lower laser level. Our observation of a long optical laser pulse does not support the suggestion that the optical pulse width of the F_2^* laser is restricted by the process of self terminating.

7. Conclusions

The simulation results showed a maximum $XeRb^+$ density of about 2.10^{14} cm^{-3}. With the assumed cross-section for stimulated emission $\sigma_{se} = 2.9.10^{-16}$ cm^2 and a negligible lifetime of the lower laser level which means that the upper laser level reflects the actual inversion a gain coefficient of about 6% cm^{-1} can be expected. At 380 °C the photoionisation absorption coefficient of Rb on the λ = 164.1 nm transition is about 1,9 % cm^{-1} So a net gain of about 4 % cm^{-1} remains during a period of time of 25 ns yielding a gain length product of about 30 which should be enough to drive an active medium into saturation. However the formation probability for several upper level states decreases the gain of this particular $\Omega = 0^+$ level to a value below the absorption coefficient of the Rb vapour which means that there are hardly lasing prospects in our configuration.

The molecular F_2^* laser showed very good and promising results. A high specific output energy was obtained and it was shown that the optical pulse width was not restricted by self-terminating effects.

References

1. N.G. Basov, V.A. Danilychev and Y.M. Popov,: "Stimulated emission in the vacuum ultraviolet region", Sov. J. Quant. El., 1 18-22 (1971).
2. R. Sauerbrey and H. Langhoff, "Excimer ions as possible candidates for VUV and XUV lasers",: IEEE J. Quant. El., QE-21, 179-181 (1985).
3. N.G. Basov, M.G. Voitik, V.S. Zuev and V.P. Kutakhov,: "Feasibility of stimulated emission of radiation from ionic heteronuclear molecules", Sov. J. Quant. El., 15, 1455-1469 (1985).
4. M. Mantel, G. Herre, H. Langhoff, K. Petkau and W. Hammer,: "The fine structure of the first excited states in the rare-gas-alkali ions", J. Phys. B., 23, 4111-4117 (1990).
5. K. Petkau, W. Hammer, G. Herre, M. Mantel and H. Langhoff, J. Chem. Phys. 94, 7769 (1991).
6. H.M.J. Bastiaens, F.T.J.L. Lankhorst, P.J.M. Peters and W.J. Witteman,: Vacuum ultraviolet fluorescence of $XeRb^+$ produced in an electron beam pumped gas mixture", Appl. Phys. Lett., 60, 2834-2836 (1992).
7. F.T.J.L. Lankhorst, H.M.J. Bastiaens, P.M.J. Peters and W.J. Witteman,: "High specific laser output energy at 157 nm from an electron beam pumped He/Ne/F_2 gas mixture", Appl. Phys. Lett., 63, 2869-2871, (1993).
8. F.T.J.L. Lankhorst, H.M.J. Bastiaens, P.J.M. Peters and W.J. Witteman: "Formation and quenching mechanisms of the electron beam pumped $XeRb^+$ ionic excimer in different buffer gases", Appl. Phys. Lett., 64, 2471-2473 (1994).

36

9. F.T.J.L. Lankhorst, H.M.J. Bastiaens, H. Botma, P.J.M. Peters and W.J. Witteman,: Long pulse electron beam pumped molecular F_2^* laser", J. Appl. Phys., **77**, 399-401 (1995).
10. M. Mantel, M. Schuman, A. Giez, H. Langhoff, W. Hammer and K. Petkau : "Investigations on the production kinetics of ionic alkali-rare gas excimers", J. Chem. Phys., **97**, 3325-3332, (1992).
11. H.M.J. Bastiaens, B.M.V. Van Dam, P.J.M. Peters and W.J. Witteman,: "Small-signal gain measurements in an electron beam pumped F_2 laser", Appl. Phys. Lett., **63**, 438-440 (1993).
12. M. Ohwa and M. Obara,: "Theoretical evaluation of high-efficiency operation of discharge pumped vacuum ultraviolet F_2 lasers", Appl. Phys. Lett., **51**, 958-960 (1987).
13. M.Kakehata, F. Uematsu, F. Kannari and M. Obara,: "Efficiency characterisation of the VUV molecular F_2^* laser excited by an intense electric discharge", IEEE. J. Quant. EL., **QE-27**, 2456-2464, (1991).

RADIO-FREQUENCY CAPACITIVE DISCHARGES
AND GAS LASERS WITH RF EXCITATION

Yu.P.RAIZER, and N.A.YATSENKO
Institute for Problems in Mechanics
Russian Academy of Sciences
101, prosp. Vernadskogo, Moscow,
117526, Russia

1. Introduction

Radio-frequency capacitive discharges (RFCD) in the frequency range (1-150) MHz and gas pressures from to hundreds of Torrs are widely used at the present time for gas laser pumping. In spite of some drawbacks, essential for practical application (complicated and expensive power sources, difficulties with their matching with the discharge, small total efficiency, etc.) typical for this method of pumping, this discharge technique is used for excitation of high-power CO_2 lasers with both a transverse [1-5] and axial [2,5-7] fast gas flow of the active medium, for CO_2 lasers with a diffusion cooled capillary [8-10] for slabs [1,5,11-15] or multichannel systems [15-18] and for lasers with other active media [19-23]. It is due to the fact that RF excited gas lasers (in spite of the above mentioned drawbacks) with the appropriate choice of parameters, may considerably improve their emitter quality, in particular to decrease the emitter size and weight, to simplify their design, control and maintenance, to increase their durability and reliability, and to achieve an easier adjustment when changing from one laser operation mode to another.

However, to make the best use of the opportunities of a RFCD as the gas laser active medium source, it is necessary to choose the appropriate operation mode and RF frequency [24]. Difficulties arise from the fact that even under the same external conditions - (gas composition, gap geometry, field frequency and even electrode voltage a moderate-pressure RF discharge can operate in two strongly different modes: the low-current mode and the high-current one [25] (these modes are often called "α-" and "γ-discharge" respectively). The modes differ in the distribution of the basic characteristics (the plasma density and the RF field amplitude) and in the direction along the current flow which influence the gas laser parameters. These differences cause the various performance characteristics of RF-excited gas lasers [12,26].

That's why it is easy to explain the great attention for studying the different RF discharge modes, the conditions of their occurrence and existence and the transition of one mode into another. It is one of the most exciting areas in the fundamental and applied physics of gas discharge and gaseous electronics. There is a large amount of publications devoted to this field. The first book on this subject, appeared recently [27]. In this paper the general properties of the RFCD, important for laser operation, are discussed.

W. J. Witteman and V. N. Ochkin (eds.), Gas Lasers - Recent Developments and Future Prospects, 37–54.
© 1996 *Kluwer Academic Publishers.*

2. RF capacitive discharge among other types of the gas discharges

It is well known that the two most abundant types of DC self-sustained discharges, which have different processes in the cathode space are the glow discharge and the arc discharge [28-30]. The dark Townsend's discharge is rare in occurrence. However looking to the physical phenomena, the dark discharge is close to the glow discharge since the fulfillment of the Townsend criterion is the condition for self-supporting as in the glow discharge

$$exp \int_0^l \alpha(E \, / \, N)dx - 1 = \frac{1}{\gamma} \tag{1}$$

Here γ is the secondary emission coefficient of the electrode (serving as a cathode); α is the volume ionization coefficient dependent on the local value of the electric field E; N is the gas density; l is a characteristic size (for a dark discharge $l = h$ - size of interelectrode gap, for a glow discharge $l = d_c$ - the cathode area thickness and , as a rule, $d_c << h$).

It is known, that (1) is executed at a minimal potential difference U_l, its value is more than hundreds of volts [28-30]. The typical discharge current densities in the glow discharge are in the range from one up to hundreds of milliamperes per square centimeter, depending on the gas composition and pressure p. It must be noticed that the functions U_l (j) and U_l (pl) have minima, it is the reason of a normal density current effect on the cathode.

Increase of the current discharge density on the cathode results in the transformation of the glow discharge into an arc discharge. But this type of the discharge is not considered in this work.

Interesting and important practical features of a gas discharge appear at excitation by an alternating electric field, as a rule sinusoidally variation. The whole range of frequencies is divided usually into four areas: 1) low-frequency (AC) : $f < 10^5$ Hz, 2) radio-frequency (RF) : $10^5 < f < 10^9$ Hz, 3) microwave : $10^9 < f < 10^{11}$ Hz and 4) optical. In this work only RF capacitive discharges are considered. The range of frequencies, relating to the RF - area is conventional. The lowest limit of the frequency is defined by the characteristic time τ_l of charge losses in the discharge gap $(f > 1 \, / \, \tau_l)$ and the highest one is due to the occurrence of potential non uniformities along the electrodes. As a rule a frequency range of 1-150 Mhz is used.

The most characteristic and important feature of capacitive RF-discharges is their stationary glow with from a practical point of view interesting current density which is impossible for a DC discharge. It is important to emphasize that the RF discharge current density j depends upon the frequency of the RF field under other wise identical conditions (see below). I.e. an additional opportunity of the plasma parameters control appears in a RF discharge in comparison with a DC discharge.

The electrodes can also be located outside the discharge chamber in a specified range of frequencies which simplifies the design of a plasma generator and excludes the contact of the metal electrodes surface with the discharge plasma. It is especially important for the discharge excitation in harsh media, when the presence of metal electrodes is impossible.

One more feature of the RF capacitive discharge in which it distinguishes from other types of discharges   the opportunity of the RFCD perpendicular to the current direction. This process is made difficult due to plasma contraction [30,31] in DC discharges and in many cases the contraction is initiated by instabilities arising in the nearelectrode areas and usually in the anode area [31]. This compels us to sectionize the electrodes and to include ballast elements in each section, which complicates the design of the discharge device and reduces its reliability. In a RF capacitive discharge, if $f > 1 / \tau I$, the discharge plasma has no time to decay during the polarity reversal of the electrodes, and the anode area is excluded from the discharge structure [25] which increases its stability to contraction [1].

There are other practical features of an RFCD which will be considered below. Here we shall note, that the RFCD (as a method of gas laser active medium excitation) has large advantages compared to DC discharges, due to the opportunity of a more adaptable control of its parameters.

3. Peculiarities of the RFCD space structure

The most well known characteristic feature of an RFCD is the opportunity of placing the electrodes outside the discharge chamber. In this case the active current loop through the plasma is closed by displacement currents. The maximum value of the RF-current density j_{max}, which can be passed through the dielectric, not destroying it, is determined by the expression:

$$ j_{max} \le \varepsilon_d \varepsilon_0 \frac{\partial E_{br}}{\partial t} = \varepsilon_d \varepsilon_0 \omega E_{br} \qquad (2) $$

where $\varepsilon_0 = 8{,}85 \ 10^{-12}$ F/m, ε_d is a relative dielectric permittivity, E_{br} is the electric field in the dielectric at which breakdown occurs, $\omega = 2\pi f$ the cyclic frequency of the RF-field. For typical dielectric materials $E_{br} \sim 10^5$ V/cm and $\varepsilon_d \sim 5$. Therefore according to (2), the limiting current densities may vary from 0.5 to 50 A/cm^2 in the frequency range $f = 1\text{-}100$ Mhz accordingly. It is enough for a wide range of applications as far as the RF-discharges with current densities from one to hundreds of milliamperes per square centimeter [1-31] are used more frequently.

The opportunity to close the discharge current by displacement currents in the dielectric covering the electrodes, implies that the processes on the electrodes and in the nearelectrode areas in RF discharges are not important [28,29]. Therefore the space structure of the RFCD, especially in the range of pressures p >> 1 Torr, was considered to be very simple [29] for a long time. It was supposed, that the stationary maintenance of the plasma in a RFCD is provided in each point of the discharge by the local ionization balance according to equation:

$$ v_i (E_{pl} / N) = vD + \beta n_e + v_a \qquad (3) $$

where n_i (E_{pl}/N) is the ionization frequency, and ν_D, bn_e and ν_a are the frequencies of the losses (diffusion, recombination and attachment) of the charged particles, E_{pl} - the electric field in the plasma and β the recombination factor.

From equation (3), having set the value of the electron concentration n_e, it is possible to define E_p if the particular mechanism of charge losses is known. Thus it is important to emphasize that the significant n_e and value of the discharge current density j (dependent from n_e) depend from external circuit parameters, assuming:

$$j = e n_e \, \mu_e \, E_{pl} = I/S_{el} \qquad (4)$$

where I is the discharge current set by the external circuit, S_{el} is the area of the electrode, μ_e , e are the mobility and electron charge respectively. The contribution of the displacement current in the plasma was neglected in (4) which is true for the majority of all practically important cases.

However expressions (3) and (4) are in contrast with following experimental facts [25,40]:

1) At the minimum values of the discharge current I the electrode area, filled by discharge plasma S_{pl} may be smaller than S_{el} . When I increased S_{pl} grew also according to the equation:

$$j_{n1} = I / S_{pl} = const \qquad (5)$$

Not only can the RFCD fills just a part of the interelectrode gap in a direction perpendicular to the current, but the effect of the normal current density can take place in it, i.e. the effect at which the value of the current density has a minimal value and does not depend upon the total discharge current. Particular values of j_{n1} are defined by the gas pressure p and its composition, by the frequency ω and by the interelectrode gap size h [32-34] and it practically does not depend upon the electrode material.

2) There is a threshold current density j_{tr} (ω) dependent upon the frequency at which for $j > j_{tr}$ another RFCD form appears. The normal current density effect with a greater current density j_{n2} ($j_{n2} > j_{tr} > j_{n1}$ under other identical conditions) is also observed under these conditions. An influence of the electrode material or its dielectric cover on j_{tr} and j_{n2} was discovered.

3) There are zones (for $j_{rf} > j_{tr}$) with a very low E_{pl} in the interelectrode gap and filled by an RFCD plasma for which (3) is not valid.

4) Moreover nearelectrode space charge sheaths (NESCS) separating the plasma from the electrode surfaces in the RF-discharge are always present. The processes in NESCS are not described by (3) and can have qualitatively a various nature, depending upon the RF-voltage (current density) value in the sheaths.

Let us mention following [25], the RFCD with $j < j_{tr}$ the low-current discharge, and the RF-discharge with the current density $j > j_{tr}$ the high-current one.

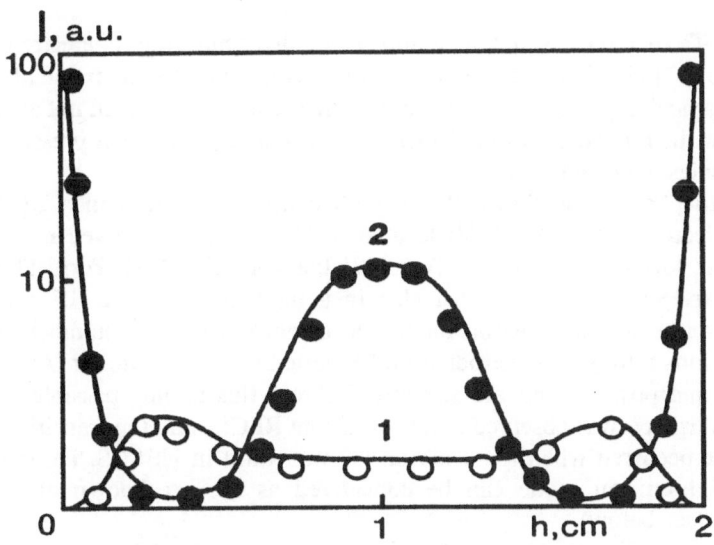

Figure 1. Glow intensity distribution in the gap of low (1) and high-current (2) discharges with air, p = 10 Torr, f = 13.56 MHz, and h = 2 cm between brass electrodes. U_{rf}=300 V in both modes.

The difference between these forms of the RFCD is, first of all, the qualitatively different distributions of the charged particles concentration and the electric field along the direction of the current [12,25,26]. Visually it is displayed in the transformation of the distribution character of the integrated RFCD radiation intensity (fig.1) at the variation of the discharge mode. One should mention the sharp increase of the discharge current density (from 7 up to 120 mA/cm2 under the conditions of fig.1), the change of the light intensity distribution I(x) and especially the occurrence of the dark areas which separate the plasma column in the discharge center from the bright nearelectrode areas (fig.1). Also a variation of the spectral structure of the radiation in the nearelectrode areas was observed. It appeared that in a high-current discharge the spectrum of the radiation is close to that of a negative cathode glow of the DC discharge.

The similarity of the light distribution in a high-current RFCD and in a DC glow discharge is known for a long time [35]. Nevertheless the processes on the cathode are not necessary the same for a RF discharge, since the phenomena near both RFCD electrodes in this case are symmetric. The necessary number of electrons near each of the electrodes is formed by ionization of gas particles by collisions with plasma electrons [28].

Another point of view about the nature of the two discharge forms [36] exists. The assumption about the different quenching mechanisms of the electrons in a discharge interval is the basis for it. It was proposed that the quenching plasma electrons by the electrodes is due to diffusion in the RF discharge with a small input energy. On the contrary electron quenching due to their drift in the RF field in RF discharges with large energy input becomes dominant. The discharge of the first type was called the a discharge and the second one is the g discharge. In the second mode of the RF discharge a high constant plasma potential relative to the electrodes U_0 was observed ($U_0 \gg kT_e / e$, where k is the Boltzmann constant, e is the charge of the electron and Te is the electron temperature) and also sputtering of electrode material. According to [37] U_0 should decrease with the increase of pressure p and at $p > 1$ Torr its value became comparable with kT_e / e, i.e. the second form should not appear.

Now the knowledge of the RFCD nature in a large range of pressures and RF-field frequencies is much better.

1) In [38] it was shown that a high constant potential in a RFCD plasma relative to electrodes ($U_0 > 100$ V) located in NESCS was observed not only at the comparatively low pressure $p < 1$ Torr [37] but also at $p \gg 1$ Torr. Therefore the powerful nearelectrode sheaths exist also in a moderate pressure RF discharge. At pressure increase the main reason for the occurrence is that the number of electrons and positive ions falling on electrode for a RF period should be similar. As the mobility of electrons and positive ions are strongly different this is only possible at large U_0. Moreover a large U_0 was observed in a low-current RFCD, and the transition to a high-current mode occurred when the criterion (1) was valid in NESCS, i.e. the transition into the high-current mode can be considered as the breakdown of a capacitive NESCS [25] (see below).

2) It was discovered [25] that the existence of two RF discharge modes and transitions between them was not dependent upon whether the electrodes are covered by dielectric or not. However, the dielectric cover of the electrodes influences greatly the parameters of the RF discharge [39].

3) Direct experiments [25,33] showed (see Fig. 2,3) that the active conductivity of the NESCS was increased significantly comparable with the conductivity of the cathode sheath in a DC glow discharge under identical experimental conditions at the transition into a high-current mode. Then areas with abnormal low electric fields and maxima of electron concentration between the NESCS and the plasma column appear.

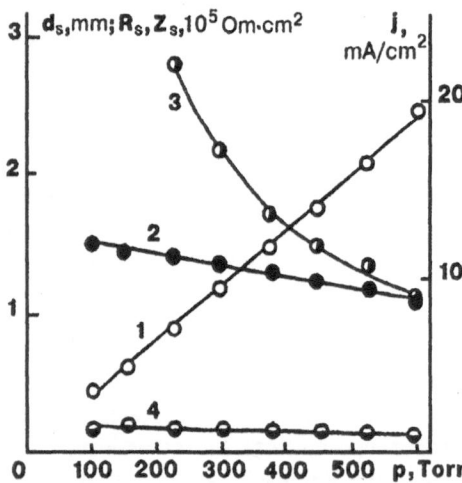
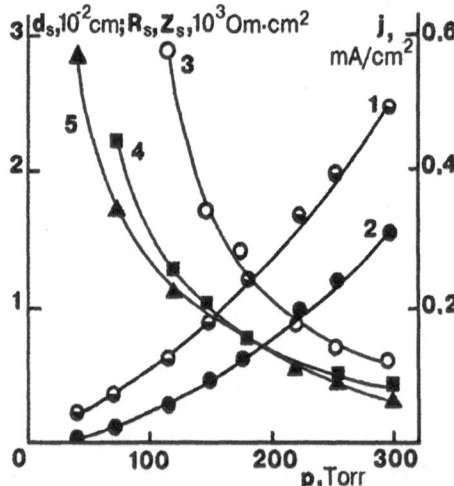

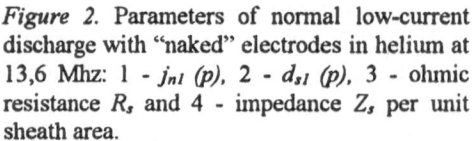

Figure 2. Parameters of normal low-current discharge with "naked" electrodes in helium at 13,6 Mhz: 1 - j_{n1} (p), 2 - d_{s1} (p), 3 - ohmic resistance R_s and 4 - impedance Z_s per unit sheath area.

Figure 3. Parameters of normal high-current discharge with "naked" electrodes in helium at 13,6 Mhz: 1 - j_{n2} (p), 2 - j_a (p), 3 - ohmic resistance R_s and 4 - impedance Z_s per unit sheath area, 5 - d_{s2} (p).

4) One more important feature of a RF capacitive discharge discovered in [25], is the maximum pressure p and interelectrode gap h; beyond this values a RFCD with current density j satisfying the expression $j_{nl} < j <_{jtr} < j_{n2}$ does not exist (Fig.4) (see below).

It is especially important for practical applications, that in a high-current RFCD the space structure becomes qualitatively different, the active component in a NESCS j_a sharply grows and at pressure tens of Torrs, as a rule, it exceeds the capacitive component [40] which results in a sharp growth of the ion current to the electrode surfaces, their destruction and release of high power in the NESCS.

Neglecting the real situation in the nearelectrode layers can result in unexpected details. For example if one places RF-electrodes outside the discharge chamber, it should be possible to create a pure plasma. But this specified statement is true only for a low-current RFCD. If one chooses the pressure, the interelectrode gap and RF field frequency not carefully enough (Fig.4), only a high-current RFCD can exist in the discharge chamber and then the discharge plasma can be contaminated by sputtered products of the chamber wall, placed under the electrodes.

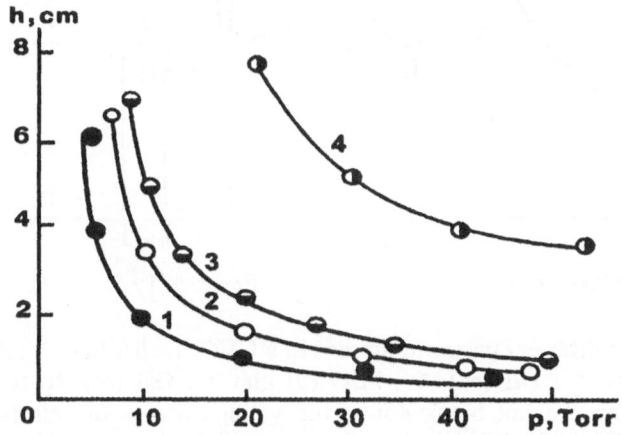

Figure 4. The boundaries of low-current regions at f = 13.56 Mhz with the high-current mode only above and on the right of the curves: nitrogen (curve 1), air (curve 2), CO_2 (curve 3), and helium (curve 4).

Another example is borrowed from a laser engineering experiment. In 1974 there was an attempt to use the RF capacitive discharge, perpendicular to the optical axis of a resonator [41] for pumping a CW CO_2 laser. The result was negative. Nevertheless, at present time CO_2 lasers with RF-excitation have the best specific characteristics [5,9,10,27], due to rational use of the RFCD considering the particular character of its space structure, though the main elements of the design remained old.

The received experimental data about the RFCD structure and the forms of its appearance are confirmed by numerical modeling (see references in [27]).

4. The current-voltage RFCD characteristics

The integrated current-voltage characteristics (CVC) of the RF capacitive discharge i.e. the dependence of the RF-voltage across the electrodes U_{rf} from the discharge current I or its density j^p permits us to receive important qualitative knowledge about the structural features of the RFCD.

Let's consider the static CVC of the RFCD at a moderate pressure together with the light distribution in the direction of the field. In fig.5,6 the most typical dependences $U_{rf}=F_1(I)$, $U_{rf}(j)$ are plotted.

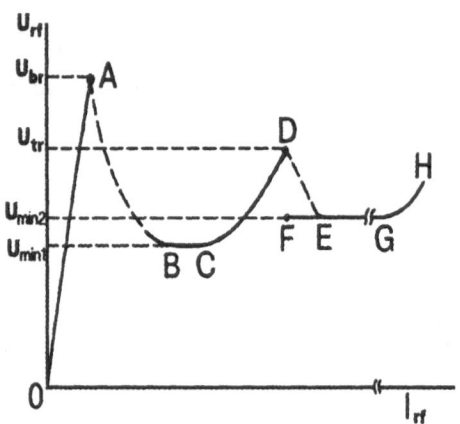

Figure 5. Typical CRC: $U_{rf}(I_{rf})$ Figure 6. Typical CRC: $U_{rf}(j_{rf})$

In general seven different parts are present in a CVC: 1 : 0A (0a), 2 : AB (ab), 3 : BC (b), 4 : CD (bc), 5 : DE (cd), 6 : FEG (d) and 7 : GH (df). In the first part the interelectrode gap did not break down, the voltage across the electrodes increased linearly with I. When U_{rf} becomes equal to the breakdown voltage U_{br} (point A,a) the RF discharge starts however, the plasma of the discharge can not fill the interelectrode gap completely in the direction perpendicular to the current. The voltage across the electrodes decreases at the start of the discharge (portions AB, ab). The growth of I results only due to an increase of the plasma cross section. It is important to note that the structure of the light in the direction of the current and the value j does not change (portion BC in fig. 5).

The fourth part of the CVC has a positive derivative with respect to I, j. The gap is then completely filled by the discharge plasma in the direction perpendicular to the current. The character and distribution of the light along the current lines here is the same as in the third part of the CVC. Only an increase of the total intensity of the light radiation especially at the ends of the plasma column, is observed. The discharge light in this case is uniform along the field, except from some bright near electrode areas separated from the electrodes by dark zones with a characteristic size dependent upon the nature of the gas and frequency of the RF field.

Measurements of the NESCS conductivity [33] have shown (Fig.2,3) that the active conductivity of the RFCD sheaths in the different parts of the CVC (BCD, bc) is small and it is comparable to the conductivity defined by the ion concentration $n_+ = n_e$, where n_e is the electron concentration in the discharge plasma.

One should notice that under real conditions after the onset of the discharge the value U_{rf} will not necessarily accept the minimum value U_{min1}. It is defined by the parameters of the external circuit and matching of RF-generator with the discharge since when the plasma starts, the complete impedance of the discharge gap changes strongly. As a result the matching of the RF-generator with the discharge can either improve, or fail. Is not excluded that after breakdown of the interelectrode gap an abnormal low-current RFCD mode (portion CD, bc) could be established in it. A normal mode of the RFCD is obtained in this case by an appropriate adjustment of the external circuit, causing an U_{rf} reduction.

When the voltage across the electrodes achieves the value U_{tr} (points D,c) a jump or smooth transition is observed. The jump takes place when the gas pressure in the discharge was higher than p_{cr}. The value p_{cr} is determined by the nature of the gas and the frequency of the RF field (see below). For example, the value p_{cr} at f=13.6 Mhz is about 1 Torr for molecular gases and tens of Torrs for inert gases. The jump in the CVC is accompanied by a redistribution of light along the direction of the current (fig.1). There is also an opportunity to get a high-current mode immediately after appearing of the RFCD.

The most important features of the RFCD mode with $j > j_{tr}$ are the following:
1) The sudden increase of the conductivity of the nearelectrode sheaths up to the characteristic values corresponding to the conductivity of the cathode area in a DC glow discharge at the same conditions [33].
2) A sharp increase of the discharge current density (in air at $p = 30$ Torr and frequency 13,6 Mhz it changes from 12 up to 240 mA / cm^2 at the transition into a new mode). The jump of j is connected not only with the occurrence of the active component of the current density j_a, but also with a sharp reduction of the NESCS thickness d_s (from d_{s1} in the low-current mode to d_{s2} in the high-current one) especially in molecular gases. The thickness of the NESCS defines the value of the current density displacement j_d at a fixed voltage on the sheath U_s

$$j_d = \varepsilon\varepsilon_0\omega E = \varepsilon\varepsilon_0\omega U_s / d_s \qquad (6)$$

For example, according to [25], the value d_{s1} was equal to 4 mm in a low-current RFCD, while in a high-current mode d_{s2} was not more than 0,3 mm at p=15 Torr in air. I.e., according to (6) after the jump in the CVC j_d increased 13 times. It is important to notice that the thickness of the sheath at $j > j_{tr}$ is almost indepent of the frequency of the RF-field (fig.7) but has a strong dependence on the pressure (fig.8) which can be approximated by the expression

$$d_{s2} = C_1 / N = C_1 / \{p(T/_0/T)\} \qquad (7)$$

where the C_1 is a constant determined by the type of gas and electrode material at T_0 =300 K, T is the temperature of the gas in the NESCS. On the contrary d_{s1} dependeds

46

slightly upon the pressure but very strongly upon the frequency of the RF-field (fig.7). Its behavior is described by the expression [25]:

$$d_{sl} = V_d \left(E_{pl} / N \right) / \omega \qquad (8)$$

where V_d is the electron drift velocity near the plasma sheath boundary.

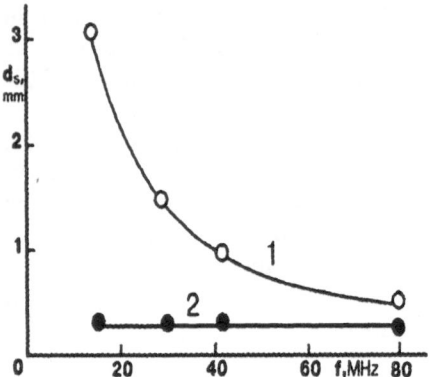 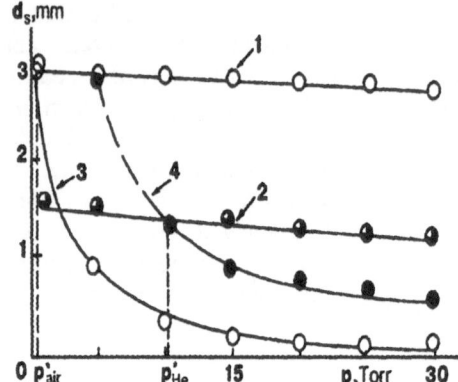

Figure 7. Dependence of NESCS thickness d_s from frequency of RF field: 1 - in low-current RFCD and 2 - in high-current one, in air at pressure p =10 Torr.

Figure 8. Experimental plots of the sheath thickness in the low- (1,2) and high-current RFCD (3,4) in air (1,3) and He (2,4), f=13.56 Mhz

5. Transition of the RFCD into a high-current mode due to a breakdown of the NESCS

A correlation between the NESCS parameters at the left and at the right side of the jump in the CVC permits us to conclude that we can consider the transition into a high-current mode as a breakdown of the NESCS involving secondary - emission electrons. I.e. at $U_{rf} = U_{tr}$ in the nearelectrode sheaths the RFCD (one or both) the electric field E has such a value that the Townsend condition (1) is satisfied at the characteristic thickness $d_{sl} \sim 1/\omega$. In other words when (1) is valid the balance of charged particles in the NESCS of the RF-discharge is provided due to secondary-emission γ-electrons inside the NESCS, whereas in a low-current discharge (left part of (1) is less than 1) the ionization is essential only in the plasma phase.

Special experiments were conducted in order to check the formulated criterion of the RFCD transition from one mode to another. In particular, the influence of the electrode material and gas type upon the value of the transition voltage in the high-current mode [25] was studied. The value of the constant plasma potential U_{0tr} at the moment of the jump in the CVC was measured simultaneously with U_{tr}. The obtained results were in good agreement with similar ones for the case of a breakdown in a DC field [29,30]. As well as in DC, U_{tr} and U_{0tr} were defined by the combination: gas composition - electrode material. The higher the breakdown DC voltage for a particular gas-electrode system was, the larger U_{tr} and U_{0tr} were required for the RFCD transition into a high-current mode.

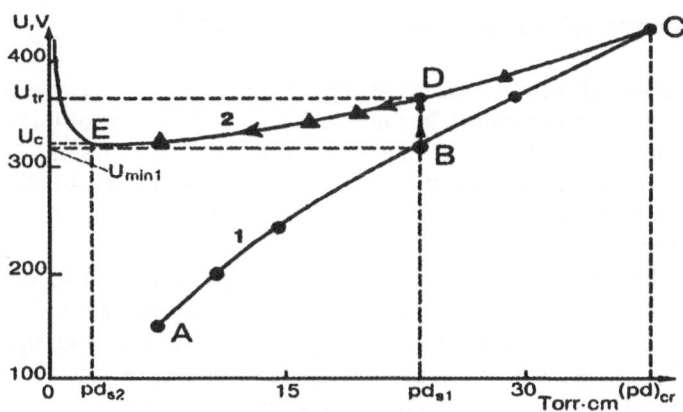

Figure 9. Dependencies of U_{min1} - 1 and U_{tr} - 2 in air at f = 13,56 MHz, h = 0.7 cm.

The experimental data in fig.9 [12,42] confirm the specific character of a RF discharge and the mechanism of its transition into a high-current mode with the breakdown of the capacitive NESCS. Curve 1 (ABC) represents the minimum voltage across the electrodes U_{min1} at which a low-current RFCD can exist (see. Fig.5,6). Curve 2 (CDE) gives the RF voltage across the electrodes U_{tr} at which the transition into a high-current mode (breakdown of NESCS) occurs (the analogue of Paschen curve) when (1) is valid.

Thus, below curve ABC the RF discharge does not exist (the curve itself is defined by the normal current density mode in the low-current discharge see below). An abnormal low-current RFCD is realized between ABC and CDE.

If we choose the interelectrode gap size so, that it should be possible to neglect the voltage of the plasma column U_{pl} in comparison with the voltage over the sheaths U_S, then $U_{rf} = U_S$. Under such conditions the data in Fig.9 were obtained. Therefore it can be considered as an illustration of the changes occurring in RFCD sheaths.

Let's consider a discharge with parameters, appropriate to point B. In this case we deal with a normal low-current RFCD. When U_{rf} increases d_{s1}, determined by (8), remains practically constant (see. Fig.10 [34]) and at $U_{rf} = U_{tr}$ (point D in Fig.9) equation (1) is valid in the sheath. But this condition is unstable as far as the Townsend criterion is satisfied, the CVC of the NESCS becomes "falling" on the right branch of the Paschen curve.

It is explained by the fact that the optimum value of the ionization factor $\alpha(E_S / N)$ is reached at the value E_S / N, larger than the one at which a breakdown on the right branch of the Paschen curve occurs. As far as the NESCS border opposite from the electrode is free, this value d_S is fixed, and the $\alpha(E_S / N)$ becomes an optimum. I.e. the thickness of the sheath will decrease until a new steady state condition appears. It will take place at point E at thickness d_{s2} of the sheath determined by (7).

6. Nature of the normal current density effect in a RFCD.

It was marked above that in both forms of the RFCD at moderate pressures the normal current density effect can be seen. The understanding of the nature of this phenomenon is important for a correct use of this type of discharge in lasers. For example, phenomena as the restriction of the electron concentration in the RFCD plasma column the minimum value of the RF voltage across the electrodes etc. are defined by the normal current density.

In the high-current mode of the discharge the nature of this effect does not differ from the one of the cathode area of a normal glow DC discharge [40,25,12]. The quantitative difference in current densities of DC and high-current RF discharges is due to the contribution of a capacitive component in the NESCS (6). As a result, the normal (minimum) current density in a high-current discharge can be evaluated from expression [40]

$$j_{n2} = (j_a^2 + j_d^2)^{1/2} = C_2(T_0/T)^2 p^2 \{1 + [\varepsilon\varepsilon_0\omega U_c/C_1C_2(T_0/T)p]^2\}^{1/2} \qquad (9)$$

Here $j_a = C_2 p^2 (T_0/T)^2$ is the active component of the current in the NESCS, C_2 is a tabulated constant determined by the gas composition and electrode material [28-30]. (The other parameters are mentioned above).

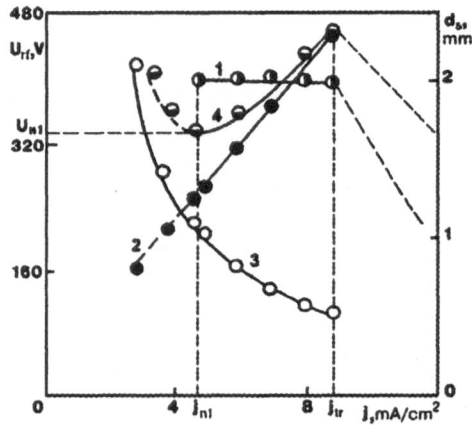

Figure 10. 1: d_{s1} (j), 2: U_{s1} (j),3: U_{pl} (j),4: U_{rf} (j).
RF discharge in air, at p = 30 Torr

The explanation of the low-current mode was found recently [34]. It appeared that the low-current RFCD is a consequence of the existence of the NESCS with a positive CVC $U_{s1}(j)$ in interelectrode gap and in the plasma column a negative CVC $U_{pl}(j)$. As a result a minimum in the CVC $U_{rf}(j)$ of the discharge appears which is realized at $j = j_{n1}$ (see Fig.10).

The particular form of $U_{pl}(j)$ is defined by the ionization balance equation in the plasma [30], but for simplicity we shall take the experimental value $U_{pl}(j)$ from [34]:

$$U_{pl(j)} = C p d_1 / j \qquad (10)$$

where $d_{pl} = h - 2 d_{s1}$ is the thickness of the plasma column, C, $\alpha \sim 1$ are values dependent on the gas composition. Neglecting the active conductivity of the NESCS in a low-current RFCD we receive an expression for the RF voltage across the electrodes $U_{rf}(j)$:

$$U_{rf}(j) = [U_{pl}^2(j) + U_s^2(j)]^{1/2} \qquad (11)$$

if we substitute $U_s(j) = 2 d_{s1} j / \varepsilon \varepsilon_0 \omega$ and $U_{pl}(j)$ from (10) in (11) and find a minimum of this expression, we receive, that $U_{rf} = U_{min1}$ at the following j_{n1}:

$$j_{n1} = [(\varepsilon_0 \omega p d_{pl} C \alpha^{1/2})/(2d_s /\varepsilon + 2\delta/\varepsilon_d)]^{1/(\alpha + 1)} \qquad (12)$$

For the derivation of (12) a dielectric cover of the electrodes with a thickness d and dielectric permittivity E_d was assumed.

The characteristic peculiarity of (12) is the growth of j_{n1} and hence, the minimum voltage across the sheath U_{s1} at an increase of p or h. It is confirmed by experiments (see Fig.9) and results for a limited number of values of pressure and interelectrode gap sizes exist for a low-current RFCD mode . (Fig.4 [25]).

7. Influence of a dielectric cover of the electrodes on the parameters of the active medium of the gas laser

The influence of a dielectric cover of the electrodes on the characteristics of the RFCD plasma, being the active medium of a gas laser is determined by the following reasons:

1)If the thickness of the dielectric cover satisfies the expression $\delta > \varepsilon_d d_{s1}$, there is an possibility to expand the range of change of j_{n2} in the low-current RFCD to a smaller area at $\omega = const$, (see (12)).

2)Choosing a dielectric cover with low emitting characteristics one can increase the value j_{tr} to expand the range of change of j in a low-current RFCD to a larger area at $\omega = const$. Especially one should note, that in this case the electron beam energy formed in the NESCS increases, because of the increase of the NESCS voltage U_{str}. The increase of the beam energy which dissipates basically outside the NESCS, is able to influence the ionization balance in the plasma column and its parameters.

3) Because a dielectric cover of the electrodes does not exclude the breakdown of the NESCS in a low-current RFCD and its transition into a high-current mode [25], it creates the possibility to realize RFCD modes with current densities intermediate between j_{tr} and j_{n2} [12,39]. The last is impossible on uncovered electrodes or covered by thin dielectric layers. A dielectric is thin if its thickness $\delta << \varepsilon_d d_{s1}$.

4) The dielectric cover of electrodes can be used for space modulation of the discharge parameters. (For example, to change the discharge current density along the electrodes according to the law given in [27]).

8. Influence of the RF excitation frequency on the characteristics of the active medium of a gas laser.

Using the knowledge about the sheath structure of the RF capacitive discharge one can easily determine the influence of the frequency on the characteristics of such discharge. The low-current RFCD mode is used most frequently in lasers since in this case the losses in near electrode sheaths are minimal, especially in a normal mode, because of the small active conductivity of NESCS. Therefore we can consider, that the active current in the plasma in the NESCS are displacement currents:

$$\sigma_{pl} E_{pl}(x = h / 2) = \varepsilon \varepsilon_0 E_s(x = 0, h) \tag{13}$$

here σ_{pl} is the conductivity of plasma and x is the coordinate in the direction of the current. It follows from (13):

$$\sigma_{pl} E_{pl}(x = h/2) = \varepsilon \varepsilon_0 \omega E_s(x = 0, h) / E_{pl}(x = h/2) \tag{14}$$

$$n_e(\omega) = \varepsilon \varepsilon_0 \omega E_s(x = 0, h) / e \mu_e E_{pl}(x = h/2) \tag{15}$$

We can see from (14) and (15) that σ_{pl} and n_e are proportional to the RF-field frequency at otherwise similar conditions. At $\omega = const$ n_e is defined by E_{s1} / E_{pl}. But at given p, h, ω and gas composition, E_{s1} is limited. According to (14-15) the change of σ_{pl} and N_e are proportional to the RF field frequency under identical E_{tr} at which the NESCS breakdown occurs from above, and by E_{s1n} from below. The last is determined by the effect of the normal current density in the low-current RFCD. We can find E_{pl} from the ionization balance equation. Therefore at a constant RF frequency there is an possibility to adjust N_e in a low-current RFCD plasma (by the RF voltage across the electrodes) in the range:

$$\varepsilon \, \varepsilon_0 \; E_{sn1} \; \omega / e \mu_e \; E_{pl1} \; < n_e \; < \; \varepsilon \, \varepsilon_0 \; \omega \, E_{t\,r} / e \; \mu_e \; E_{pl2} \qquad (16)$$

In the same way it is easy to receive an expression explaining the reduction of the minimum voltage across the electrodes with the increase of frequency [24].

9. Transition of the RFCD into a high-current mode with a jump in the CVC

Let's compare the expressions for d_{s1} (8) and d_{s2} (7) and their behavior depending upon the frequency and pressure (Fig.7, Fig.8). In general d_{s1} is not equal to d_{s2}. However for any sort of gas it is possible to choose such values for pressure and frequency, that $d_{s1} = d_{s2}$, i.e.

$$V_d \, / \, \omega = C_1 \, T / p \, T_0 \qquad (17)$$

Or

$$p^* = C_1 \, T \, \omega / V_d \, T_0 \qquad (18)$$

At $p = p *$ there wouldn't be any jump of the RF voltage U_s across the NESCS since the breakdown voltage across the sheath U_{str} corresponds to the minimum of the Paschen curve, when (17,18) are true. A jump of the capacitive component of the current in the NESCS would not take place as well, as far as neither d_s, nor U_s did not change. For the active current component j_a, it is easy to show that in this case $j_d \gg j_a$.

Thus, at $p < p *$ a transition into a high-current mode will occur without any jump. A jump will take place only at $p > p *$.

10. Modern developments of gas lasers with RF excitation

The comparison of the modern development of gas lasers with RF excitation which is partially described in [1-27] (though it is only a small part of all known publications in this field), with the achieved progress in the understanding of the RF capacitive discharge physics shows, that not all of its possibilities are realized now. Therefore it is possible to expect the occurrence of new and more advanced gas lasers with RF pumping.

It is obvious that due to the drawbacks marked in the introduction, it is profitable to use RF-pumping only when the methods of pumping by DC discharge do not work. The best example is the slab CO_2-laser with diffusion cooling. The idea of such laser is very simple and is known since the moment of the invention of the CO_2 laser in 1965. One of the sizes of the discharge chamber was made small for effective cooling of the active medium and the two others were chosen in order to generate the necessary laser power. However for the first time in 1981 the creation of a CO_2 slab laser with diffusion cooling was successfully [1]. It was a result of the investigation of the RFCD space structure [25]. The cross section of this laser is represented in Fig.11 (a).

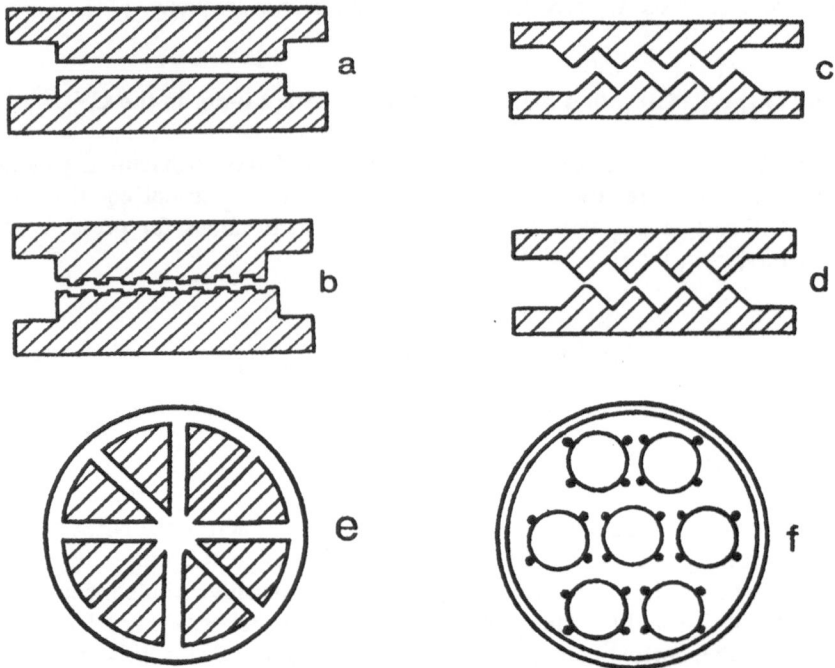

Figure 11. Perspective schemes of gas lasers with RF excitation

Further investigations of the low-current mode of the transverse RFCD in small interelectrode gaps showed that the plasma of such discharge can fill discharge gaps of complex cross sections (Fig.11, b-e). Some of the shown sections are successfully realized in working models of CO_2-lasers. For example the design of a discharge chamber according to Fig.11 (e) gives good results.

The use of a transverse RFCD for the creation of slab gas lasers in other active media than CO_2 is perspective. Such attempts are already known [21-23].

The application of a high-current RF discharge and in particular a "combined" one (when both RFCD modes exist in one discharge volume [24,27]), for pumping of gas lasers is not enough investigated. In this case the possibilities to control the parameters of the laser active medium are expanded. On the basis of such discharge a multi-channel laser with RF pumping at frequency 13.56 MHz (Fig.11, f) [24] was realized.

11. References

1.   Myshenkov V.I., and. Yatsenko N.A. (1981) Prospects for using high-frequency capacitative discharges in lasers, *Sov. J. Quantum Electronics* , **11**, 1297-1301.
2.   Hügel, H.E. (1986) RF-excitation of high power CO_2 lasers, *SPIE* **650**, 2-9.
3.   Schock, W., Walz, B., Wessel, K., and Wildermuth, E. (1990) Characteristics of a compact RF excited 12 kW CO_2 - laser, *SPIE* **1276**, 41-48.
4.   Offenhauser, F. (1988) Theory and experiments on the power modulation of CO_2 lasers, *IEEE Journal of Quantum Electronics* **24**, 1289-1296.

5. Yatsenko N.A. (1989) *Gas Laser with Radio-Frequency Excitation*, Preprint #381, Institute for Problems in Mechanics, Russian Academy of Sciences, Moscow (in Russian).

6. Wester, R., Seiwert, S., and Wagner, R. (1991) Theoretical and experimental investigation of the filamentation of high-frequency excited CO_2 laser discharges, *J. Phys. D: Appl. Phys.* **24**, 17961802.

7. Ehrlichmann, D., Habich, U., and Plum, H.-D. (1993) Diffusion-cooled CO_2 laser with coaxial high-frequency excitation and internal axicon. *J. Phys. D: Appl. Phys.* **26**, 183-191.

8. Lachambre, J.-L., Macfarlane, J., Otis, G., and Lavigne, P. (1978) A transversely RF-excited CO_2 waveguide laser, *Appl. Phys. Lett.* **32**, 652-653.

9. He D., and Hall D.R. (1983) A 30-W RF excited waveguide CO_2 laser, *Appl. Phys. Lett.* **43**, 723-728.

10. Heeman-Ilieva M.V., Udalov Yu.B., Witteman W.J., Peters P.J.M., Hoen K., and Ochkin V.N. (1993) RF excited 1.1 W/cm waveguide CO_2 -laser, *J. Appl. Phys.* **74**, 4786-4788.

11. Yatsiv, S. (1987) Conductively cooled, capacitevely coupled RF excited, CO_2 lasers, in S. Rosenwaks (eds.), *Proceedings of Gas Flow and Chemical Laser Symposium*, Springer, Berlin, pp.252-257.

12. Yatsenko N.A.(1988) *Space Structure of RF Capacitive Discharge and Prospects for its Using in Lasers*, Preprint #338 Institute for Problems in Mechanics RAS, Moscow, 1988.

13. Abramski, K.M., Colly, A.D., Baker, H.J., and Hall, D.R. (1989) Power scaling of large-area transverse radio-frequency discharge CO_2 lasers, *Appl. Phys. Lett.* **54**, 1833-1835.

14. Nowack, R., Opower, H., Schaefer, V., Wessel, K., and Hall, Th. (1990) High power CO_2 waveguide laser of the 1 kW category, *SPIE*, **1276**, 18-28.

15. Yelden, E.F., Seguin, H.J.J., Capjack, C.E., and Nikumb, S.K (1991) Multichannel slab discharge for CO_2 laser excitation, *Appl. Phys. Lett.* **58**, 693-695.

16. L. A. Newman, R. A. Hart J. T. Kennedy, A. J. DeMaria, "High-power-coupled CO_2 waveguide laser array", *Conf. on Lasers and Electro-Optics (CLEO '86)*, San Francisco, pp. 162-163,1986.

17. Abramski, K.M., Colly, A.D., Baker, H.J., and Hall, D.R. (1990) Offset frequency stabilization of RF excited waveguide CO_2 laser arrays, *IEEE Journal of Quantum Electronics* **26**, 711-717.

18. Yatsenko, N.A. (1992) Peculiarities of application of RF capacitive discharge in gas laser engineering, in S.Z. Zynabidinov (eds.), *Proc. Int. Conf. RF Discharge in Wave Fields and Pumping of Gas Laser*, University, Tashkent, pp. 3-6 (in Russian).

19. Pearson, G.N., and Hall, D.R., (1987) Carbon monoxide laser excited by radio-frequency discharge, *Appl. Phys. Lett.* **50**, 1222-1224.

20. Ivanov, I.G., Latush, E.L., and Sam, M.F. (1990) *Ion Lasers on Metal Vapor*, Energoatomizdat, Moscow (in Russian)

21. Yatsenko, N.A. (1991) Slab gas lasers, in *Proc. XIV Int. Conf. on Coherent and Nonlinear Optics*, Leningrad, **2**, 52-53 (in Russian)

22. Yu, G., Baker, H.J., Rodrigues, N.A.S., and Hall, D.R. (1994) Compact high-efficiency carbon monoxide laser at 1 kW, *Appl. Phys. Lett.* **65**, 2904-2906.

23. Kukhlevsky, S.V., and Kozma, L. (1994) Area scaling in N_2 waveguide lasers, *IEEE Journal of Quantum Electronics*, **30**, 759-762.

24. Yatsenko, N.A.(1993) Choice of the combustion regime and the frequency of radio-frequency capacitive discharge for the pump of the gas lasers. *Bulletin of the Russian Academy of Sciences. Physics* **57**, 2156-2163.

25. Yatsenko, N.A. (1981) Relationship between the high constant plasma potential and the conditions in an intermediate-pressure RF capacitive discharge, *Sov. Phys. Tech. Phys.* **26**, 678-683.

26. Yatsenko, N.A. (1992) Slot gas lasers, *Bull. Russ. Acad. Sci., Phys.* **56**, 1901-1907.

27. Raizer, Yu.P., Shneider, M.N., and Yatsenko, N.A. (1995) *Radio-Frequency Capacitive Discharges*, CRC Press, Boca Raton, Ann Arbor, Tokyo, London.

28. Kaptsov, N.A. (1950) *Electrical Phenomena in Gases and Vacuum*, GITTL, Moscow, Leningrad (in Russian).

29. Brown, S.C. (1959) *Basic Data of Plasma Physics*, Technology Press Willey, New York.

30. Raizer Yu.P. (1991) *Gas Discharge Physics*, Springer-Verlag, Berlin, New York.

31. Golubev, V.S. and Pashkin S.V. (1990) *Glow Discharge in Elevated Pressure Molecular Gases*, Nauka, Moscow (in Russian).

32. Yatsenko N.A. (1982) The normal current-density effect in moderate-pressure, capacitive HF discharge. *Sov. Phys. - Tech. Phys.* **27**, 741-743.

33. Yatsenko N.A. (1982) Integral characteristics of electrode layers in capacitive medium pressure HF discharge, *High. Temp.* **20**, 820-826.

34. Yatsenko, N.A. (1988) Mechanism of the formation of the spatial structure of high-frequency volume discharge, *Sov. Phys. - Tech. Phys.* **33**, 180-184.

35. Banerji, D., and Ganguli, R. (1932) On the distribution of space potential in striated and other forms of high-frequency discharge, *Phil. Mag.* **13**, 494-501.

36. Levitskii, S.M. (1957) Investigation of ignition potential high frequency discharge in as in transition field of frequencies and pressures, *Zh. Tekh. Fiz.* **27**, 970-977.

54

37. Levitskii, S.M. (1957) Space-potential and sputtering of electrodes in RF discharge, *Zh. Tekh. Fiz.* **27**, 1001-1009.
38. Yatsenko, N.A. (1978) About high constant potential of RF discharge plasma, in *Proc. General and Molecular Physics*, Moscow Institute for Physics and Technology, Moscow, **10**, 226-229 (in Russian).
39. Yatsenko N.A. Influence of dielectric coating of electrodes on characteristics of gas lasers with radio-frequency excitation, *Bull. of the Russian Acad. Scie., Physics*, **58**, 939-947.
40. Yatsenko N.A. (1980) medium-pressure high-current high-frequency capacitative discharge, *Sov. Phys. - Tech. Phys.* **25**, 1454-1455.
41. Goikhman V. Kh., and Goldfarb V. M. (1974) Stationary CO_2-laser with capacitive discharge in gas flow, *Zh. Prikl. Spektrosk.* **21**, 456-459 (*J. Appl. Spectrosc.*).
42. Yatsenko, N.A. (1991) Evolution of the RF capacitive discharge during the process of transition to high-current regime, in *Proc. XX Intern. Conf. on Phenomena in Ionized Gases*, Pisa, **5**, 1159-1160.

CAPACITIVELY COUPLED RF EXCITATION OF CW GAS LASERS AND ITS COMPARISON WITH HOLLOW CATHODE LASERS

J. Mentel, N. Reich, J. Mizeraczyk*, M. Grozeva**, N. Sabotinov**

Allgemeine Elektrotechnik und Elektrooptik
Ruhr-Universität Bochum
D-44780 Bochum, Germany

* Polish Academy of Sciences, Institute of Fluid Flow Machinery
Fiszera 14, PL-80-952 Gdansk, Poland

** Bulgarian Academy of Sciences, Institute of Solid State Physics
Tzarigradsko Chaussee 72, BG-1784 Sofia, Bulgaria

Abstract

We report the designs and performances of He-Kr$^+$, He-Ar$^+$, He-Cd$^+$, He-Cu$^+$ (CuBr) and Ne-Cu$^+$ (CuBr) cw lasers excited with a transverse capacitively coupled radio frequency discharge and compare their properties with those of the corresponding hollow cathode discharge lasers.

At similar laser output parameters the designs and operation of the radio frequency excited lasers are much simpler than those of the hollow cathode discharge lasers. A radio frequency excited He-Kr$^+$ laser with an active length of 40 cm delivered output powers of 22 mW at 469.4 nm and 11 mW at 431.8 nm. The output powers of a He-Cd$^+$ laser with an active length of 40 cm were 60 mW at 441.6 nm, 36 mW at both 533.7 nm and 537.8 nm and 14 mW at both 635.5 nm and 636 nm under single line operation. They were distinctly higher for the green and red lines under multiline operation. Using a He-CuBr gas mixture laser action on 4 infrared Cu II lines (740.4 nm, 766.5 nm, 780.8 nm and 782 nm) was achieved and with a Ne-CuBr gas mixture laser gains on 14 UV-lines between 240.3 nm and 272.2 nm were observed. As it was shown for different laser lines the rms noise-to-signal ratio was much lower for radio frequency excited lasers than for conventional positive-column lasers making them attractive for practical applications.

W. J. Witteman and V. N. Ochkin (eds.), Gas Lasers - Recent Developments and Future Prospects, 55–67.
© 1996 Kluwer Academic Publishers.

1. Introduction

Recently the technology of several gas ion lasers has been essentially improved in our laboratories by employing a transverse capacitively coupled radio-frequency (CCRF) discharge in combination with ceramic alumina oxide discharge tubes for the laser excitation. This concerns such lasers as $He-Kr^+$, $He-Ar^+$, $He-Cd^+$, $He-Cu^+$ (CuBr) and $Ne-Cu^+$ (CuBr) lasers. Until now successful cw operation of these lasers was only possible by using hollow cathode discharges (HCD) or positive column discharges.

The high ability of the HCD to generate laser oscillations in many mixtures of buffer gas (He or Ne) and a lasing additive (metal vapour or rare gas, usually much easier ionized than the buffer gas) is a result of two factors. First, a large number of high-energy electrons, needed for excitation of the laser transitions in the lasing additive ions exists in the HCD negative glow, used as an optically active volume [1]. These high-energy electrons are produced by the cathode-emitted electrons accelerating in the electric field of the cathode dark space. Second, in the HCD the number of high-energy electrons remains high even if a relatively large fraction of the lasing additive is present [2]. This is due to an increase in the operating voltage , caused by the increasing number of the low-mobility ions of the additive component. In contrast, in the positive-column of the glow discharge the presence of the more easily ionized additive decreases substantially the number of fast electrons [2]. The special excitation capability of the HCD is due to this high fraction of the lasing component, in the presence of the large number of the high-energy electrons in the laser mixture.

A similar mechanism of producing high-energy electrons exists in the so-called transverse capacitively coupled radio-frequency (CCRF) discharge, in which under certain conditions a high electric field occurs near the electrodes [3, 4]. This high electric field accelerates the ions towards the electrodes causing emission of electrons, even if the electrodes are covered with an insulator. This is important for the gas laser technology. The emitted electrons are accelerated in the near-electrode high electric field, becoming fast and capable of exciting the laser ionic transitions in the He- (or Ne-) metal vapour or the He- (or Ne-) rare gas mixtures, similarly as in the HCD. As in the HCD, a relatively high fraction of the low-ionization-potential lasing component can be introduced into the CCRF discharge, without decreasing the number of fast electrons, and therefore without reduction of the lasing capability of the discharge [5].

Despite the correspondence between the hollow-cathode and the CCRF-discharges, the CCRF-excited lasers show some superiority over the lasers excited by the HCDs. This superiority is due to the higher longitudinal homogeneity of the discharge, more efficient transforming of the input power into the energy of the high-energy electrons, the simplicity of the laser design, absence of arcing at higher input powers and the possibility of using external electrodes, which enable work with substances of high chemical activity.

These special properties of the CCRF discharges have been employed for generating of a considerable number of visible and infrared ionic laser transitions in Tl [5, 6], Cd [5, 6-12], Zn [5, 6, 13, 14], Hg [5, 6, 15], Se [5, 6, 16] and Cu [5, 17].

However, technological problems encountered in the CCRF discharges, such as overheating of the discharge tube, deterioration of the inner wall surface of the laser tube by the ion bombardment and the introduction of the wall-originated particles into the discharge, have limited the investigations mainly to a quasi-cw or pulsed regime. Actually, until now no cw CCRF-excited laser which assured good stability and long operating lifetime has been reported besides by the authors and their coworkers [18-22].

We present the results of our effort to develop simple CCRF-excited lasers, exhibiting long-life and stable generation in cw regime. The long-life and the stable operation of the CCRF-excited lasers were achieved mainly through the use of a ceramic alumina oxide discharge tube instead of the fused silica tube commonly employed in the past. The Al_2O_3 tube is less susceptible to damage by overheating and ion bombardment than that made of fused silica. Therefore less particle and gas impurities are released from the wall of the Al_2O_3 tube during operation of the discharge. More specific details on this subject, as well as on the discharge tube designs and operation of the CCRF-excited lasers are given in [22].

2. RF-excitation and matching

The RF power was capacitively coupled into the discharge established inside the Al_2O_3 capillary tube with 400 mm long and 4 mm wide transverse nickel plated copper electrodes mounted along the capillary tube with ceramic holders. The discharge was run by a 13.56 MHz RF generator with an output power up to 600 W.

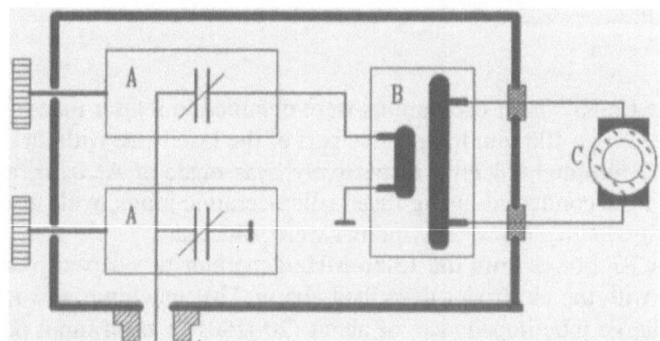

Fig. 1: Electrical circuit supplying RF power to the laser tube: A - capacitor, B - symmetrizing transformer, C - laser tube with electrodes

A special matching circuit (Fig. 1) was used to transform the laser discharge tube impedance Z_D to the 50 Ω output resistance of the RF generator. The matching circuit, consisting of two capacitors and a transformer, symmetrized the RF voltage and was essential to maintain a uniform discharge between the electrodes and to avoid strong RF interference. Both, a nonuniform discharge spreading outside the electrode gap and strong RF interference occurred when a nonsymmetric matching was used. The symmetrizing transformer consisted of two coils, the primary - e.g. of inductance $L_1 = 2.7$ µH and the secondary - e.g. of inductance $L_2 = 11.4$ µH. The capacitances could be varied from 45 pF to 650 pF to reach optimum matching. The order magnitude of the discharge current amounts to 5 A. Such a relatively high discharge current is typical of the so-called γ-type CCRF discharge [4].

3. He-Kr+ and He-Ar+ laser

Up to now laser oscillations on Kr^+ ion transitions have been obtained in different cw and pulsed HCDs [23] and in the positive column of a pulsed glow discharge [24]. Here we report cw laser oscillations on the Kr^+ ion transitions at $\lambda = 431.8$ nm, 438.7 nm, 458.3 nm and 469.4 nm excited with the CCRF discharge in He-Kr mixtures.

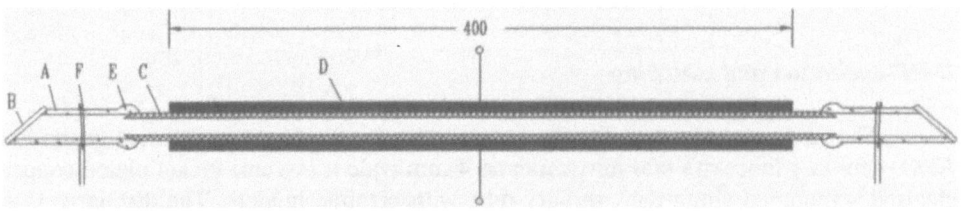

Fig. 2: Design of the CCRF-excited He-Kr+ laser tube with the Al_2O_3 ceramic active part: A- fused silica tube, B - Brewster windows, C - Al_2O_3 ceramic tube, D - RF electrodes, E - fused silica-ceramic joint, F - heating wire

The He-Kr+ laser oscillations were obtained in a laser tube of a very simple design (Fig. 2). The 400 mm long active part of the laser tube with an outer and inner diameter of 7 mm and 2.8 mm, respectively, was made of Al_2O_3 ceramic. The ceramic tube ends were connected, using fused silica-ceramic joints, with fused silica end stubs to which fused silica Brewster windows were soldered.

The RF power from the 13.56 MHz generator was capacitively coupled into the discharge with the electrodes described above. The matching network transformed the laser discharge tube impedance, of about $(20-j230)$ Ω at an input power of 600 W, to the 50 Ω output resistance of the RF generator.

At maximum laser output powers for all four lines the He pressure and the He to Kr partial pressure ratio were (13.0 ± 0.3) kPa and 1500 : 1, respectively. The laser oscillation at $\lambda = 469.4$ nm could be achieved in a He pressure range from 2.5 kPa to 26 kPa. For the corresponding HCD laser [23] the He to Kr partial pressure ratio extends from 200 : 1 to 1000 : 1.

The CCRF-excited He-Kr+ laser output powers and small-signal gains at $\lambda = 469.4$ nm and $\lambda = 431.8$ nm as a function of RF input power are shown in Fig. 3 [21]. It is seen from it that at 600 W RF input power the small-signal gains for these laser lines were about 7.5 %m⁻¹ and 5 %m⁻¹. The small-signal gains at λ = 469.4 nm and 431.8 nm of the corresponding HCD laser were 6.7 %m⁻¹ and 3.2 %m⁻¹, respectively [23]. At 600 W RF input power the laser output power and the small-signal gain at λ = 438.7 nm were 1.6 mW and 1.8 %m⁻¹, respectively. The laser oscillation at λ = 458.3 nm exceeded barely the threshold (gain was 0.5 %m⁻¹ at 600 W). No saturation of the laser output power with increasing RF input power up to 600 W was found. The rms noise-to-signal ratio of the He-Kr⁺ laser output power was lower than 0.6 %, and thus, similar to that of HCD lasers.

The investigations of the CCRF-excited He-Kr⁺ laser performance presented above were carried out during about 500 hours. Owing to the Al_2O_3 ceramic discharge tube the laser did not show any essential deterioration after that time. In contrast, cw operation of this laser in a fused silica tube is not possible because of severe deterioration of the capillary inner walls.

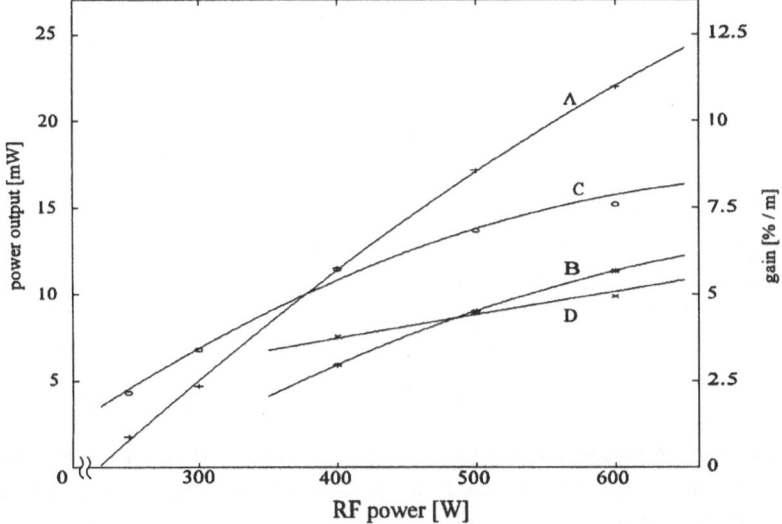

Fig. 3: The CCRF-excited He-Kr⁺ laser output powers and small-signal gains at λ = 469.4 nm (A and C) and λ = 431.8 nm (B and D), respectively.

To our knowledge there has been no report on laser oscillations on Ar⁺ ionic transitions excited in rare gas-Ar mixtures with the CCRF discharges. Using a CCRF discharge operated in the tube given in Fig. 2 we obtained cw laser oscillations on the Ar⁺ transitions at λ = 454.5 nm and λ = 476.5 nm in a He-Ar mixture. These lines were also emitted by the corresponding HCD laser [25]. The ratio of the partial pressures of the He : Ar mixture was 100 : 5. For the HCD laser it extends from 100 : 2 to 100 : 5 [25]. The total pressure of the operating mixture was 4.5 kPa. At 600 W of the RF input power the laser output power at λ = 476.5 nm was 10 mW. However, the operating conditions were not optimized.

4. He-Cd$^+$ laser

Preliminary experiments [5-12] have shown that for an efficient operation of a CCRF excited He-Cd$^+$ laser the Cd vapour density in the discharge zone has to be controlled independently from the RF input power. This can be realized by using an oven with Cd in it, heated separately. The manufacture of a laser tube with an oven attached to an Al$_2$O$_3$ capillary tube is possible but difficult since its design requires several ceramic-fused silica transitions. Therefore, a capillary tube made of Al$_2$O$_3$ ceramic (the length - 400 mm, the inner diameter - 4 mm) forming the active part of the laser tube was inserted symmetrically into a fused silica tube.

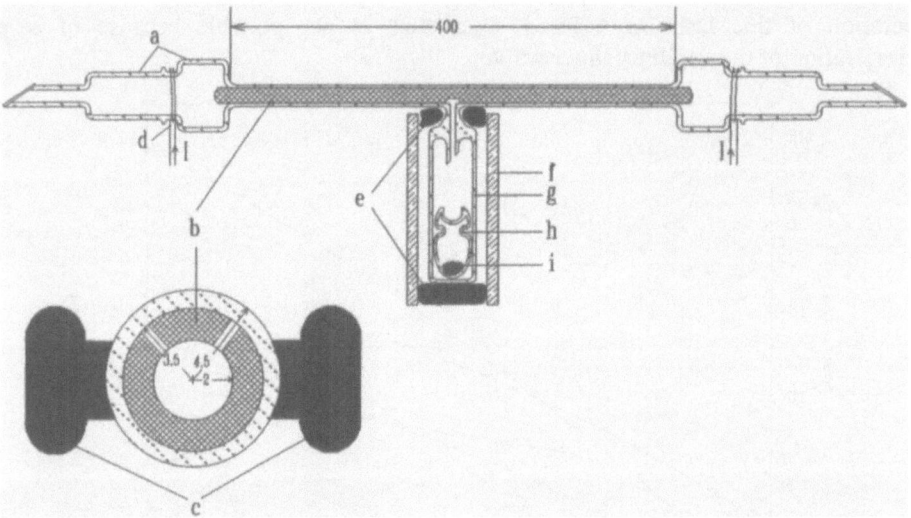

Fig. 4   Design of the CCRF-excited He-Cd$^+$ laser tube with Al$_2$O$_3$ capillary tube insert: a - fused silica tube, b - Al$_2$O$_3$ capillary tube, c - RF electrodes, d - heaters, e - insulating material, f - oven, g - Cd reservoir, h - ampoule, i - cadmium

In the design shown in Fig. 4 Cd vapour was supplied into the active part of the laser tube through a hole made in the Al$_2$O$_3$ capillary tube from a fused silica sidearm Cd reservoir with oven connected to the middle of the fused silica capillary tube. The Cd vapour diffused along the CCRF discharge in the Al$_2$O$_3$ capillary tube and deposited on the inner walls of the wide-diameter extension stubs, which served as condensation regions. Although we did not measure the distribution of the Cd vapour along the Al$_2$O$_3$ capillary tube, it seems to be axially uniform. This was checked by using five Cd reservoirs, distributed uniformly along the Al$_2$O$_3$ capillary tube and operated in an on- and off-mode. The higher number of the Cd reservoirs did not result in an increased output power of the laser [18]. Such a behaviour is in contrast with the theory of diffusion of Cd vapour in a laser tube [26]. During operation the temperature of the Cd reservoir was stabilised within ± 0.5 K.

To protect the Brewster windows from deposition of the particles sputtered from the Al_2O_3 tube, electrically heated wires were wound around the condensation stubs. The heated wires produced thermal buoyancy whirls of the operating gas within the condensation region. The convective gas whirls deflected the travelling particles onto the walls of the stubs, thus avoiding contamination of the Brewster windows.

The discharge was run by a RF generator operating at 13.56 MHz with an output power up to 600 W. The symmetric matching given in Fig. 1 was used to transform the laser discharge tube impedance (typically (19.5 - j 497 Ω) at a RF generator output power of 400 W and a He pressure of 3 kPa) to the 50 Ω output resistance of the RF generator.

The CCRF-excited He-Cd$^+$ laser exhibited cw single- or multi-line operation at seven wavelengths in the blue (λ = 441.6 nm), the green (λ = 533.7 nm and λ = 537.8 nm), the red (λ = 633.5 nm and λ = 636.0 nm) and the infrared (λ = 723.8 nm and λ = 728.4 nm) regions.

He-pressure-dependencies of the CCRF-excited He-Cd$^+$ laser intra-resonator powers of both red lines, the blue line and the green lines, when oscillating separately are shown in Fig. 5a. To separate each of the green lines a birefringent Lyot filter [27] was inserted in the resonator. It is seen that the laser intra-resonator powers of the red and blue lines exhibit narrow maxima at He pressures around 1.6 kPa and 2.1 kPa, respectively, while for each green line exists a broad maximum of the intra-resonator power ranging from 5 kPa to 15 kPa. Similar behaviours were observed for the intra-resonator powers of the HCD white-light He-Cd$^+$ laser operating in the single-line mode [28].

The curves presented in Fig. 5b show that the pressure dependencies of the output powers of the HCD He-Cd$^+$ laser are similar to those of the CCRF-excited He-Cd$^+$ laser, although some differences are present [28]. The absolute values of the He pressure at which the maxima occur are lower by a factor of 2 than those for the CCRF laser. This can be attributed to the lower gas temperature in the HCD. The two maxima of the laser output power at λ = 441 nm correspond presumably with two different operation modes of the HCD [28]. The maximum at lower He pressure is connected with the longitudinal HCD mode having similar properties as the positive column of the glow discharge while the maximum at higher pressure is related to the so-called transverse mode which resembles the negative glow.

The output powers and the small-signal gains of the separately oscillating blue line, both green lines, and both red lines increased with increasing RF input power and no saturation was reached up to 400 W. At 400 W the maximum output powers (and small-signal gains) were about 60 mW (11 %m^{-1}), 36 mW (16 %m^{-1}), and 14 mW (7 %m^{-1}) for the blue line, both green lines, and both red lines, respectively. These values are comparable to those obtained in a HCD He-Cd$^+$ lasers of the same active length being equal to 12 %m^{-1}, 15 %m^{-1}, and 4.5 %m^{-1} [28-32].

62

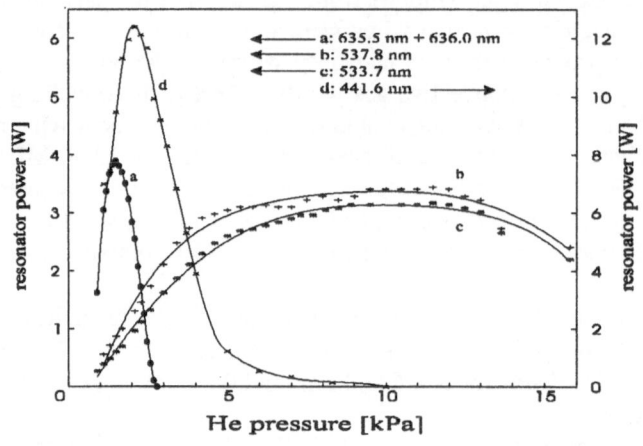

a)

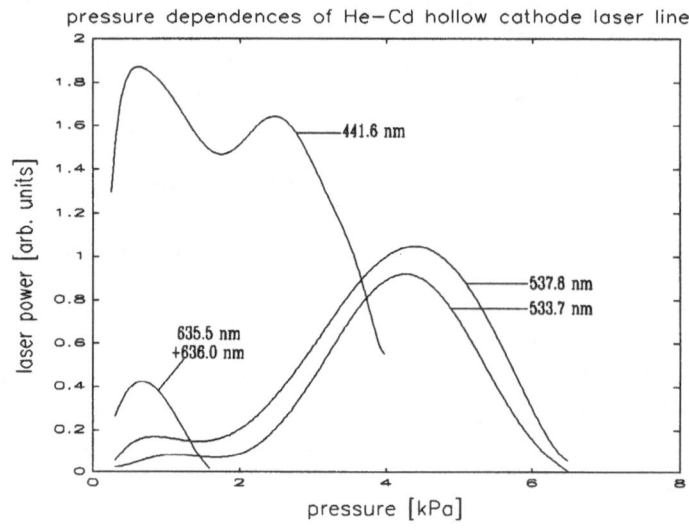

pressure dependences of He—Cd hollow cathode laser lines

b)

Fig. 5   He-Cd^{+} laser intra-resonator power at λ = 441.6 nm, λ = 533.7 nm, λ = 537.8 nm, and λ = 635.5 nm
and 636 nm as a function of helium pressure.
a)   CCRF discharge (RF input power - 400 W)
b)   Hollow cathode discharge

The observed increase in the laser output power of the red lines when they operated simultaneously with the green lines can be beneficial if an optimum white-light operation of CCRF-excited He-Cd$^+$ lasers is considered [19, 33].

The superiority of the CCRF-excited He-Cd$^+$ laser over a positive-column He-Cd$^+$ laser is seen from Fig. 6 which shows the noise spectra of the laser output powers at $\lambda = 441.6$ nm of the presented CCRF-excited He-Cd$^+$ laser and a typical positive-column He-Cd$^+$ laser. The noise-to-signal ratio of the laser output power was less than 0.4 % for the CCRF-excited He-Cd$^+$ laser and about 7 % for the positive-column one.

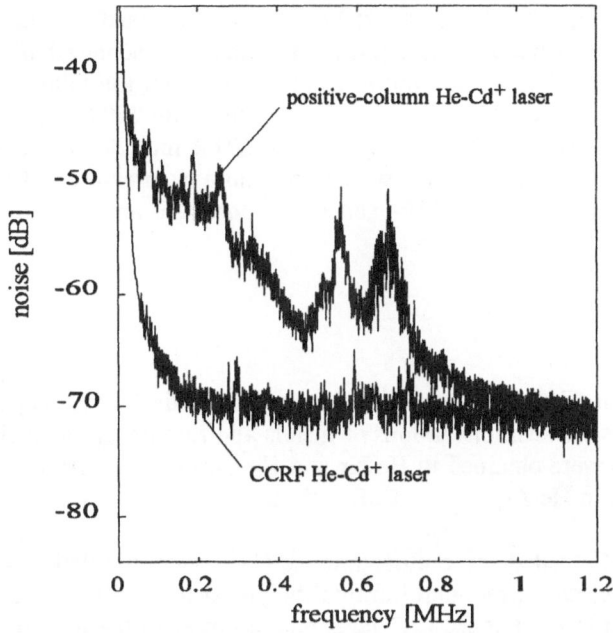

Fig. 6 Noise spectra of the CCRF-excited and positive column He-Cd$^+$ lasers

The investigations of the CCRF-excited He-Cd$^+$ laser have been performed during more than 400 hours. Any essential deterioration of neither the laser tube nor the CCRF discharge have been observed during that period. The above allows inferring that a long-life He-Cd$^+$ laser can be built when the CCRF discharge is used for its excitation.

5. He-Cu$^+$ (CuBr) and Ne-Cu$^+$ (CuBr) laser

Up to now CCRF-excitation of a He-Cu$^+$ laser was only achieved in a pulsed mode [17]. The Cu vapour was produced in an uncontrolled manner by sputtering from an internal Cu-electrode. On the other hand cw laser oscillation was demonstrated using a HCD which was seeded with Cu-halides [34]. Thus, it was obvious to transfer the technology already developed to operate a CCRF-excited He-Cd$^+$ laser to CCRF excitation of a Cu$^+$ laser by vaporising CuBr into a CCRF discharge in He or Ne. For realization of this idea a discharge tube similar to that shown in Fig. 4 was used. The CuBr vapour was produced by heating the sidearm container filled with CuBr powder up to a temperature of 723 K. From the sidearm container CuBr vapour diffused to the transverse CCRF discharge region where it was dissociated into Cu and Br atoms. The Cu atoms were ionized and excited by charge transfer reactions with the noble gas ions.

Filling the tube with He laser oscillations on the Cu$^+$ infrared transitions at $\lambda =$ 740.5 nm, 766.5 nm, 780.8 nm and 782.6 nm were obtained. At the 780.8 nm line which is the strongest line an output power of 10 mW was achieved at an optimum He pressure of 7 kPa. Using Ne as a buffer gas laser gains on the following 14 UV lines were observed: 240.3 nm, 242.4 nm, 247.3 nm, 248.6 nm, 250.6 nm, 252.9 nm, 254.5 nm, 259.1 nm, 260.0 nm, 270.1 nm, 270.3 nm, 271.4 nm, 271.9 nm, 272.2 nm. The strongest line was the 248.6 nm line with an optimum Ne pressure at 1 kPa.

Further details on the CCRF-excited Cu$^+$ ion laser are given in [35] published in this issue.

6. Conclusions

In this contribution it was shown that the CCRF discharge is capable of efficient exciting many laser transitions in ions of metals and rare gases. In particular, efficient laser oscillations were obtained in He-Kr and He-Cd mixtures. As it was shown, also laser generations in He-Ar, and He-CuBr mixtures are possible with the CCRF excitation.

The CCRF-excited cw multicolour He-Cd$^+$ laser, oscillating at seven wavelength in the blue, green, red, and infrared regions at tens miliwatts levels, is promising as a practical laser. The He-Cd$^+$ laser output power in the red (also in the green) can be increased by a factor of 3, if all lines operate simultaneously. Owing to the power interaction of the green and red lines, a white-light operation of the CCRF-excited He-Cd$^+$ laser (40 cm active length) with a total output power of about 60 mW is possible.

As far as the laser output power level is concerned the CCRF-excited lasers show no inferiority to the lasers excited by HCDs. However, output powers comparable to those of the HCD laser are obtainable using the CCRF-excited lasers of much simpler laser tube design. The similar laser capabilities of the HCD and CCRF discharges suggest correspondence between the plasma properties of both discharges when optimized for laser operation.

The other advantages of using the CCRF discharge for exciting gas ion lasers, noticed in this experiment, are:
- absence of arcing, which occurs very often in the case of the HCD excitation

- low degradation of the operating gas by impurity gases owing to the absence of the metal electrodes inside the discharge tube
- no cataphoretic effects along the tube axis
- shorter starting time compared to that of HCD lasers
- relatively low noise of the laser output power

Owing to the Al_2O_3 capillary tube the CCRF-excited He-Kr$^+$ and He-Cd$^+$lasers exhibited stable operation for more than 400 hours without any essential deterioration of the discharge and laser tube. This allows us to claim that the presented CCRF-excited He-Kr$^+$ and He-Cd$^+$ lasers should be useful as a simple, long-lived, continuously operating laser sources operating at tens miliwatts output power levels.

Moreover there is a good chance to realise by CCRF excitation reliable cw lasers in the far UV with copper ions and other metal ions.

Acknowledgement

M. Grozeva expresses her deep gratitude to NATO and J. Mizeraczyk his deep gratitude to the Alexander von Humboldt Foundation and the Heinrich Hertz Foundation for granting them a scholarship which enabled their research at the Department of Electrical Engineering, Ruhr-Universität Bochum.

This work was supported by the Bundesminister für Forschung und Technologie under Grant No. 13N5714/7 and by the European Commission Copernicus Programme CIPA-CT 93-0219.

References

[1] P. Gill and C.E. Webb (1977), 'Electron energy distribution in the negative glow and their relevance to hollow cathode lasers,' J. Phys. D: Appl. Phys. **10**, p. 299-311

[2] J. Mizeraczyk (1987), Discharge physics of positive-column and hollow-cathode discharge excited He-metal vapour lasers in Lasers and their applications, Proc. 4th Summer School on Quantum Electronics, Sunny Beach, Bulgaria, Ed.: A. Y. Spasov, Publ.: World Sci. Publ. Co Pty Ltd., p. 1-30

[3] V. Ya. Khasilev, V. S. Mikhalevskii, and G. N. Tolmachev (1980), Fast electrons in a transverse rf discharge, Fiz. Plazmy, **6**, 430-435, (also Sov. J. Plasma Phys., **6**, 2, p. 236-239 (1980))

[4] Yu.P. Raizer (1991), Gas discharge physics, Springer-Verlag, Berlin Heidelberg, p. 378-414

[5] M.F. Sem, V.Ya. Khasilev, V.S. Mikhalevskii and G.N. Tolmachev (1981), Metal ion lasers with excitation by transverse radio-frequency discharge, Proc. Int. Conf. Lasers '80, Ed.: C.B. Collins, p. 182-189, STS Press, McLean, VA, New Orleans

[6] Latush, V.S. Mikhalevskii, M.F. Sem, G.N. Tolmachev and V.Ya. Khasilev (1976), Metal-ion transition lasers with transverse HF excitation, Pis'ma Zh. Eksp.Teor. Fiz. (USSR) **24** (2), p. 81-83. Translation in JETP Lett. **24** (2), p. 69-71

[7] Mikhalevskii, G.N. Tolmachev and V.Ya .Khasilev (1980), Optimization of the excitation conditions of a transverse rf discharge He-Cd laser, Kvantovaya Elektron. (Moscow) **7** (7), p. 1537-1542. Translation in Sov. J. Quantum Electron. **10** (7), p. 884-887 (1980)

[8] Dyatlov, V.G. Kas'yan and V.G.Levin (1977), Frequency selection in a helium-cadmium laser with transverse rf excitation, Pis'ma Zh. Tekh. Fiz. (USSR) **3** (7), p. 644-646. Translation in Sov. Tech. Phys. Lett. **3** (7), p. 264-266 (1977)

[9] Alexandrov, V.V .Elagin and A.E. Fotiadi (1980), New laser transitions in a He-Cd laser with transverse rf pumping, Pis'ma Zh. Tekh. Fiz. (USSR) **6** (2), p. 160-161. Translation in Sov. Tech. Phys. Lett. **6** (2), p. 70-71 (1980)

[10] A.N.Korolkov and S.A. Rudelev (1981), Quasi-cw operation of the He-Cd laser with rf pumping, Zh. Prikl. Spektrosk. (USSR) **34** (1), p. 89-92

[11] V.G. Doronin, A.N. Korolkov, S.V. Pismennyi and S.A. Rudelev (1980), Investigation of cascade generation in He-Cd laser with transverse high-frequency excitation, Zh. Prikl. Spektrosk. (USSR) **35** (2), p. 252-256

[12] G.P. Strokan and G.N. Tolmachev (1984), Comparison of helium-cadmium laser designs with transverse high frequency discharges, Avtometriya (USSR) **1**, p. 62-65

[13] N. Sabotinov and P. Telbizov (1982), Laser oscillation on the 758.8 nm zinc line in a transverse high-frequency discharge, Bulg. J. Phys. vol. **9** (6), p. 667-672

[14] N. Sabotinov and P. Telbizov (1986), He-Zn laser with a transverse high-frequency excitation, Optics Commun. vol. **59** (4), p. 290-292

[15] N. Sabotinov and P. Telbizov (1990), Laser oscillations on HgII transitions in a transverse HF discharge, Optical and Quantum Electron. vol. **22**, p. 83-88

[16] V.S. Mikhalevskii, G.N. Tolmachev and V.Ya. Khasilev (1981), Lasing and excitation of levels in a Se-He discharge, Zh. Prikl. Spektrosk. (USSR) vol. **34** (4), p. 623-625

[17] V.S. Mikhalevskii, M.F. Sem, G.N. Tolmachev and V.Ya. Khasilev (1980), Laser action on the ionic transitions of copper in a rf discharge, Zh. Prikl. Spektrosk. (USSR) vol. **32** (4), p. 591-593

[18] N. Reich, J. Mentel and G. Jakob (1993), Capacitively coupled transverse rf-discharges for multiline lasers, Proc. XXI. Int. Conf. Phenomena in Ionized Gases, Ed.: G. Ecker, U. Arendt and J. Böseler, **vol. II**, p. 102-103, Arbeitsgemeinschaft Plasmaphysik (APP), Ruhr-Universität Bochum, Bochum

[19] N. Reich, J. Mentel, and J. Mizeraczyk (1994), Characteristics of a cw multiline He-Cd[+] laser with transverse radio frequency excitation, Conf. Lasers and Electro-Optics Europe, CLEO-E '94, Amsterdam, Ed.: IEEE, p. 144-145

[20] N. Reich, J. Mentel, and J. Mizeraczyk (1994), Correspondence between laser excitation with ccrf discharges and hollow cathode discharges, Bull. Am. Phys. Soc., 47th Gaseous Electronics Conference, vol. **39**, no.6, p. 1495

[21] N. Reich, J. Mentel, G. Jakob and J. Mizeraczyk (1994), cw He-Kr[+] laser with transverse radio-frequency excitation, Appl. Phys. Lett., vol. **64** (4), p. 397-399

[22] N. Reich (1994), Transversale Kapazitive Hochfrequenzanregung von Gasentladungslasern, Ph. D. dissertation, Ruhr-Universität Bochum, Germany

[23] M. Janossy, K. Rozsa, P. Apai, L. Csillag (1984), He-Kr Ion Laser in a DC hollow cathode discharge, Optics Comm., vol. **49**, p. 278-280

[24] L. Dana, P. Laurès (1965), Stimulated emission in krypton and xenon ions by collisions with metastable atoms, Proc. IEEE, vol **53**, p. 78

[25] M. Janossy , L. Csillag, K. Rozsa (1977), cw laser oscillation in a hollow cathode He-Ar discharge, Phys. Lett., **63A** , No. 2, p. 84

[26] T. P. Sosnowski (1969), Cataphoresis in the helium-cadmium laser discharge tube, J. Appl. Phys., vol. **40** (13), p. 5138-5144

[27] J. Mentel, E. Schmidt, and T. Mavrudis (1992), Birefringent filter with arbitrary orientation of the optic axis: An analysis of improved accuracy, Appl. Optics, vol. **31** (24), p. 5022-5029

[28] J. Mizeraczyk, N. Reich, J. Mentel, and E.Schmidt (1994), Performance of the hollow-cathode discharge cw multicolour He-Cd$^+$ laser, J. Phys. D: Appl.Phys., (in print)

[29] J. Mizeraczyk, J. Mentel, E. Schmidt, N. Reich, C. Carlsson, and S. Hård (1994), A hollow-cathode discharge cw multicolour He-Cd$^+$ laser module, Meas. Sci. Technol., **5**, p. 936-941

[30] A. Fuke, K. Masuda, and Y. Tokita (1986), Characteristics of He-Cd$^+$ white light laser, Trans. of the IECE of Japan, vol. **E 69**, p. 365-366

[31] A. Fuke, K. Masuda, and Y. Tokita (1988), High-power He-Cd$^+$ white- light laser, Electronics and Communications in Japan, Part 2, vol. **71**, p. 19-27

[32] A. Fuke, K. Masuda, and Y. Tokita (1988), Gain characteristics of hollow cathode He-Cd$^+$ white-light laser, Jap. J. Appl. Phys., vol. **27** (4), p. 602-606

[33] N. Reich, J. Mentel, J. Mizeraczyk, cw radio frequency excited white-light He-Cd$^+$ laser, IEEE J. Quantum Electron, accepted for publication

[34] J. A. Piper, D. F. Neely (1978), cw laser oscillation on transitions of Cu$^+$ in He-Cu halide gas discharges, Appl. Phys. Lett., **33**, No. 7, p. 621-623

[35] J. Schulze, C. Lücking, N. Reich, D. Teuner, J. Mentel, M. Grozeva, J. Mizeraczyk, CCRF excited copper-ion-laser, this issue

High power small-bore planar waveguide CO_2 laser

S.B. Chernikov, A.I. Karapusikov, S.A. Stojanov
Institute of Laser Physics SB RAS,
13/3, Lavrentyv st., Novosibirsk, Russia,
tel.: (383-2) 35-42-66, e-maile: bagaev@ilph.nsc.su

Abstract:The slab CO_2 -laser is presented. The output power of 46 W in continuous and 160 W in quasicontinuous regimes of single-mode are gained.

1 Introduction

Some technical, medical, physical applications of laser require the compact source of radiation in IR range and that stimulated the research developments in the creation of CO_2 lasers with the power up to 1 kW and more based on slab structures where the generation power increases in proportion to electrode square and one can reach 30 kW per m^2 that makes possible the creation of extremely small oscillators.

2 Experimental Setup

A laboratory sample of slab CO_2-laser was created by us to investigate the possibilities of using similar lasers. The laser was based on a metal/ceramic electrode block cooled by water and maintained in the stainless steel tube with diameter 100 mm. The tube serves both as a vacuum shell and the carrying construction of the resonator. The dimensions of RF-discharge slab zone were 2.5x27x270 mm^3. The electrode block was presented by a unified construction with the electrodes linked between themselves by cooling channels and plurality inductances. The side walls of slab waveguide were formed from ceramic plates (Al_2O_3 ceramics) and electrodes consisted of oxidated plates from aluminium alloy.

In this laser we used the unstable structure resonator. Its basic diagram is shown Fig.1 . This resonator consists of two mirrors: one is concave with the radius of curvature R1=6 m and other is convex with the radius of curvature R2=-5.42 m. To provide the rigidity both mirrors were maintained on spherical tuning heads. Such type of the resonator allowed us to get single-mode radiation. In order to examine the spatial structure of radiation we created a special experimental setup. Its scheme is shown in Fig.2. Results of this test are given in Fig.4

W. J. Witteman and V. N. Ochkin (eds.), Gas Lasers - Recent Developments and Future Prospects, 69–72.
© *1996 Kluwer Academic Publishers.*

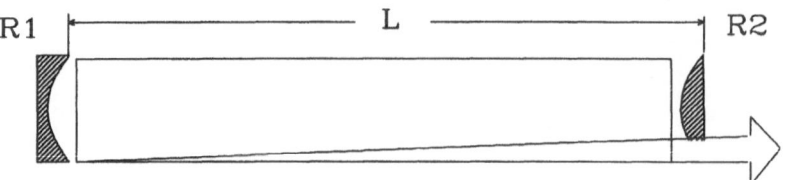

Figure 1: basic diagram of the unstable structures resonator

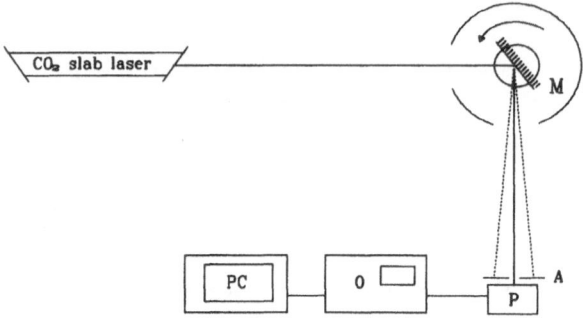

Figure 2: The experimantal scheme for measuring the spatial structure of radiation; M–rotated mirror; A–aperture; P–photodetector; O–digital oscillograph.

3 Experiment

Spatial and energetic characteristics of the laser were measured with operating mixture as follows: $CO_2 : N_2 : He : Xe$=1:1:3:0.5 at different pressures in continuous, impulse and quasicontinuous modes. The results of these experiments are shown in Fig.3.

In continuous mode the maximal output power of radiation was 45 W under pumping power equal to 360 W (this is the maximal output power of the RF-generator used). In quasicontinuous mode the generator allowed us to get the pumping impulses of the power equal to 1700 W lasting 200 μs. In this case the power of radiation impulses ranged up 160 W at the mixture's pressure 105 Torr and with the presence of a small constant generation in the intervals between the impulses.

4 Conclusions

The investigations of a spatial composition of the generation showed the possibility of realizing single-mode one. A spatial distribution diagram of the radiation in the laser beam appears to be Gauss one and has a comparatively small side difractional maximum. The resulting measurements of parameters allow us to suppose the possibility to obtain the output power about 120 W on this laser model providing the generator delivers the output rf power about 1 kW. The preliminary

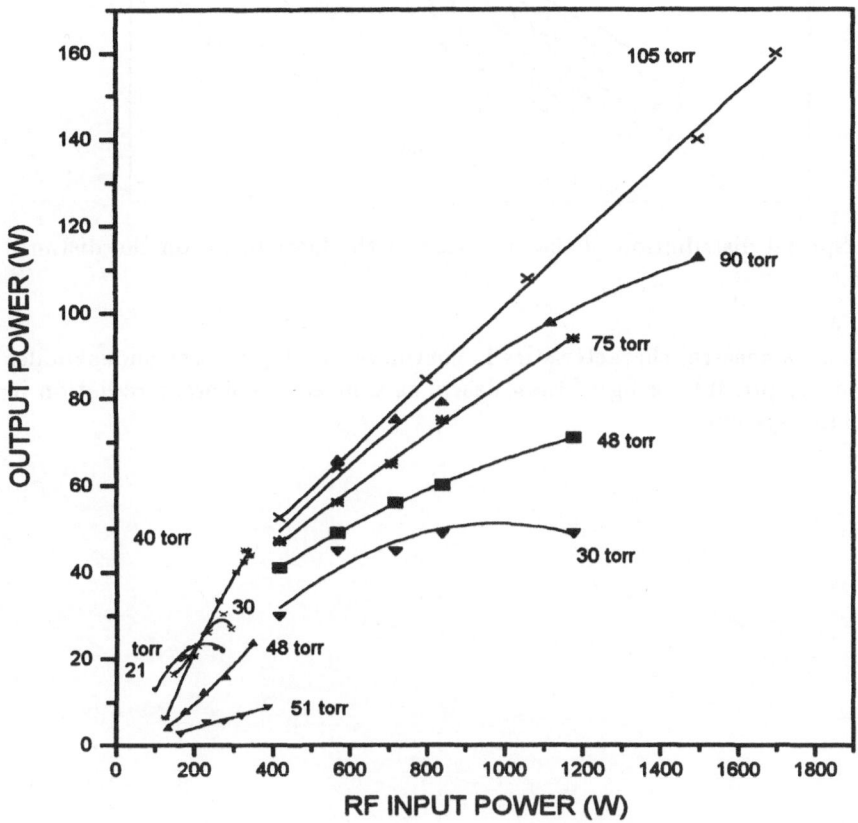

Figure 3: The dependence of output power of CO_2 slab laser on rf power input under different gas mixture pressures.

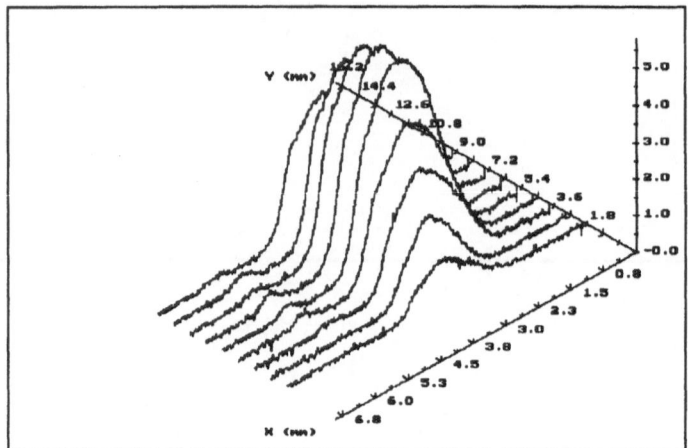

Figure 4: Spatial distribution of the radiation in the laser beam on the distance about 2 m.

investigations of spectral characteristics in continuous and quasicontinuous modes have shown the possible using of these lasers as sources of coherent radiation in geterodin laser systems.

RF EXCITED GAS LASERS -RECENT PROGRESS

**Y.B.UDALOV, S.N.TSKHAI, P.J.M.PETERS,
and W.J. WITTEMAN**
*University of Twente
Department of Applied Physics
P.O. Box 217
7500 AE Enschede
The Netherlands*

V.I.KOCHETOV and V.N.OCHKIN
*Low-Temperature Plasma Laboratory,
P.N.Lebedev Physics Institute
Leninsky Prosp., 53, 117924
Moscow, Russia*

1. Introduction

Already in the early stage of the laser research continuous wave (CW) radio-frequency (RF) discharges proved to be an effective mean for the excitation of gaseous active media. Lasing in most atomic and molecular gas lasers like the He-Ne laser [1], CO laser [2], CO_2 laser [3,4], He-Xe laser [5] for the first time was obtained in devices with RF excitation. Although being quite easy to operate, these constructions were soon abandoned because of the limited range of operating conditions. The laser parameters like the gas pressure and input power that were typical for DC excitation appeared to be much higher at that time than with an RF excitation. However lasers with RF excitation once more began to attract interest in the eighties. In a mean time it has been experimentally proven that the performance of the lasers with RF excitation can be significantly improved if a transverse rather than a longitudinal excitation geometry is chosen. Transverse RF excitation requires a lower operating voltage. Furthermore, it has the advantage that the gas pressure can be substantially increased, especially in waveguide laser configurations. RF excitation also appears to be more flexible than a DC one because an additional parameter for control of the characteristics of the active medium namely the electric field frequency is present.

A unique feature of transverse RF excitation is the possibility of CW homogeneous excitation of large volumes of active media and with large cross-sections This is very attractive for practical applications because thetput power per unit length increasesthen substantially.

All these as well as many other features make of these lasers with CW RF excitation an interesting object for scientific research as well as for technological and industrial applications.

W. J. Witteman and V. N. Ochkin (eds.), Gas Lasers - Recent Developments and Future Prospects, 73–88.
© 1996 *Kluwer Academic Publishers.*

2. Basic principles of RF excitation

Generally speaking there are two different ways for the excitation of gases by an RF electric discharge. A volumetric electrical discharge may be sustained by an inductively or capacitively coupled electric field. In the first case the discharge tube is placed inside a solenoid or a coil consisting of several turns. A high frequency current flowing through the coil creates an alternating magnetic field inside the tube, with the direction parallel to the tube axis. The alternating magnetic field in turn creates an electric field with the lines concentric to the tube axis. The inductively coupled RF discharge is basically electrodeless.

In the case of a capacitively coupled RF discharge an electric voltage is applied to the two parallel electrodes. The electric field lines start on one electrode and end on another. The electric field between the electrodes is to a large extent a potential field. A capacitively coupled discharge can be ignited in a discharge tube with internal electrodes or in the electrodeless configuration.

Application areas for inductively and capactively coupled RF discharges differ substantially. Inductively coupled discharges at high pressures have been successfully applied for the formation of a relatively low temperature (~10.000 °C) equilibrium plasma for material processing. These techniques were proposed a long time ago, with efficient technical solutions suggested in early sixties [8].

The main application area for capacitively coupled discharges is an excitation of active media for gas lasers. As a result of joined efforts of several research groups (for a review see , for example [9]), RF pumping became a common mean for the excitation of various gases , with the input power ranging from tens of Watts to tens of kiloWatts.

There are some special features typical for RF excited gas discharges that make them rather attractive for practical applications but also interesting from a physical point of view.

Some basic properties of RF excitation can be understood from a simple model [10] that links microscopie and microscopie parameters of the plasma.

Let us start with the analysis of the electron motion in plasma. The equation of motion for the plasma elections can be written as

$$m\frac{dv_d}{dt} + mv_d\, v_m = e\, E_o\, e^{iwt} \qquad (1)$$

where m and e are the mass and the charge of electron, v_d is the drift velocity of the electrons, v_m is the effective electron collision frequency, E_0 - an amplitude of the electric field with a frequency ω oscillating in a plasma.

By integrating equation (1) the drift velocity can be determined. This parameter, along with the electron density N_e gives a value of the discharge current

$$J(t) = \frac{e^2 N_e\, S\, E_0}{m\left(i\omega + v_m\right)} e^{i\omega t} \qquad (2)$$

where S is the cross-section of the discharge area in the direction perpendicular to the current flow. The macroscopic parameters of the discharge can be derived from the expression (2).

The impedance of the central part of the discharge, i.e. positive glow can be written as

$$Z_d = \frac{m_d}{e^2 N_e S} \left(\nu_m + i\omega \right) = R_0 + i\omega L_0 \tag{3}$$

The plasma impedance can thus be represented by an equivalent circuit, as a combination of lumped active and reactive components. The operating conditions typical for RF excited gas lasers at a pressure of about 100 Torr and an excitation frequency 40 - 200 Mhz which means $\nu_m \gg \omega$. Therefore the reactive part of the impedance can be neglected. That does not mean however that the impedance of the discharge as a whole is purely resistive. The impedance of the inter-electrode discharge gap has a substantial capacitive part. To understand its origins let us analyse the spatial structure of transverse RF discharge.

Already in the very early experiments with the transverse RF excitation of gas discharges it was observed that their spatial structure is fairly inhomogeneous. Along with a bright central part two more intensive layers, separated from the electrodes with thin dark regions, can be observed. The mechanisms of the formation of this spatial distribution can be easily understood qualitatively. Let us assume that after the breakdown of the gas the interelectrode gap is homogeneously filled with an electrically neutral plasma. Oscillations of the electric field applied to the electrodes result in a motion of charged particles. The mobility of ions compared to the electrones is much lower, so it is enough to consider only the process of the oscillatons of an electron gas. The electrons which are close to the electrodes (i.e. within the amplitude of an oscillation under the influence of an RF field) will disappear in metal. After several oscillation cycles thin positively charged layers will be formed near the electrodes. These layers, also called "sheaths", have a capacitive nature. In a macroscopic equivalent circuit of the discharge they can be represented as two equal capacitors connected in series with a resistor representing the bulk plasma (positive glow). Because of their low conductivity the sheaths play an important role in the stabilisation of the discharge current. The breakdown of the sheaths that takes place under certain circumstances, generally results in the transition from a low to a high current discharge mode (which is also often referred to in the literature as a α- to γ-discharge transition).

A qualitative approach presented above was developed in [12] long before the RF excitation has been successfully applied for the development of gas lasers. This approach is still widely used for the basic description of the processes taking place in the laser active media. For example, the α- to γ-transition proved to be an important factor limiting the power deposition in RF excited CO_2 lasers. However, as we shell see later, for the simulation and evaluation of real devices more sophisticated models are required.

3. Experimental and theoretical studies of waveguide CO_2 lasers with RF excitation.

The model presented below was developed for modeling and optimization of CO_2 waveguide laser with transverse RF excitation [13-15]. It consistw of several numeric codes. Each of them deals with certain specific aspects of laser operation, and uses output parameters of other codes for calculations. The main simulation blocks are: modeling of the microscopic behaviour of a gas dicharge; calculations of the macroscopic parameters of the discharge; calculation of the molecular kinetic processes that lead to the formation of inversion population; laser resonator modeling and finally the calculation of the laser output parameters.

In the microscopic model a transverse RF excited discharge between two electrodes separated at a distance 1 is considered. The motion of the charged particles is characterised by a set of equations: the continuity equations for the flux of charged particles, along with the electrostatic equation for the electric field:

$$\frac{\partial N_{e,p,n}}{\partial t} + \frac{\partial \Gamma_{e,p,n}}{\partial x} = T_{e,p,n}$$

$$\frac{\partial E_o}{\partial x} = 4\pi e \left(N_p - N_n - Ne \right)$$

(4)

where $N_{e,p,n}$ are the concentrations and $\Gamma_{e,p,\,n}$ the flux densities for electrons, positive and negative ions respectively, e is the electron charge, E_o is the electric field strength and $J_{e,p,n}$ are the particle source functions dependent on the plasma composition. The fluxes of the charged particles are created by the electric field as well as by diffusion.

The rate coefficients in (4) depend on the gas temperature and neutral particle density. In the case of gas lasers the pumping rates are usually high. This results in essential non-uniformity of the discharge plasma. To take this fact into account an equation describing the processes of heat conductivity

$$\frac{d}{dx} \left(\lambda(T(x)) \right) \frac{dT(x)}{dx} + \delta(x) q(x) = 0$$

(5)

was solved along with equations (1) and (2). Here q(x) is the local electric power dissipated in the discharge, δ is the fraction of power converted into heat and λ is the heat conductivity coefficient.

The rate constants for the elementary processes involving electrons were determined on the basis of the kinetic Boltzmann equation for the spherical symmetrical part of the electron energy distribution function [16]. Equations (4) and (5) were solved using an implicit difference iterative scheme (see, for example, [17]). Several approaches has been developed to shorten the computation time. The details can be found in [13].

The numerical code was developed to perform the modeling and optimisation of RF excited CO_2 waveguide lasers. Several important microscopic parameters, such as the E/N value, the spatial distribution of the particle densities, the production rates for charged particles, etc. were determined. They were then used to estimate the macroscopic parameters of the discharge, like the current density and voltage, the discharge impedance, the specific electrical input power, etc.

To check the validity of the simulation some macroscopic parameters were calculated and compared with the measured values. Good coincidence of the experimental and calculated data entitled us to perform a broad range of the simulations.

The numerical results confirmed the qualitative model discussed above. For example, in Fig. 1 the structure of the charged particles distribution across the inter-electrode gap is presented. The particle densities N_e, N_n and N_p for the electrons, negative and positive ions respectively are plotted for two different time moments which correspond to the voltage phase $\pi/2$ (curves with the index 1) and π (index 2). The simulation was performed for the following discharge conditions: interelectrode separation is 2.25 mm, gas mixture composition - CO_2:N_2:He = 1:1:8 at a pressure 100 Torr. An external electric field with the frequency of 125 MHz and a voltage $V_0 = 180$ V is applied to the electrodes.

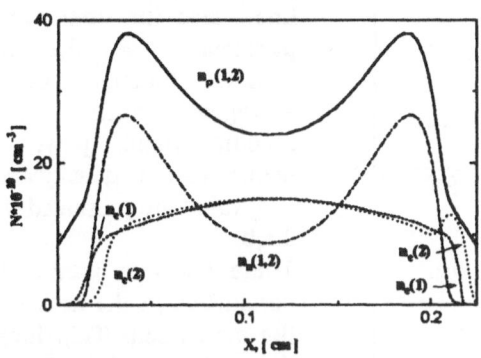

Fig. 1. The electron and ion density in the inter-electrode gap for the voltage phases $\pi/2$ (1) and π (2).

In accordance with the qualitative model described above, it is clearly seen that low conductivity regions exist near the surfaces of electrodes. The spatial profiles for the positive and negative ion densities remain practically unchanged and symmetric with respect to the gap centre. On the contrary the electrons follow the changes in external electric field and could oscillate between the electrodes. The central part of the discharge gap is filled with a quasi neutral plasma.

Fig. 2 illustrates the spatial distribution of the RF electric power deposited into the discharge. This structure is defined by both the external electric field and the internal distribution of the charged particles.

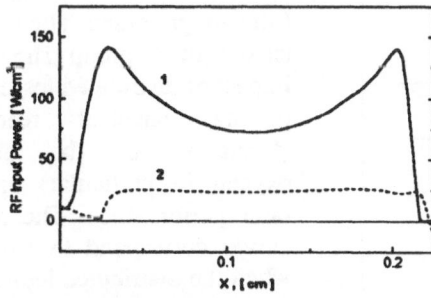

Fig. 2. Dynamics and spatial structure of local RF input power. $\pi/2$: (1); π : (2).

The structure of power input is again shown for two different field phases $\pi/2$ (1) and π (1), like in Fig. 1. It is interesting to mention that at some moments a narrow spatial region exist where the input power is negative. Of course this effect is not so large, as it is seen in the right side of the Fig. 2. It has to do with the specific features of the formation of the charged particles fluxes. Near the electrode surfaces a strong charge density gradient exists. As a result the electrons move due to diffusion in the direction opposite to the electric field. The kinetic energy of the electrons is thus transferred to the electric field. It results in some local cooling of the electron gas.

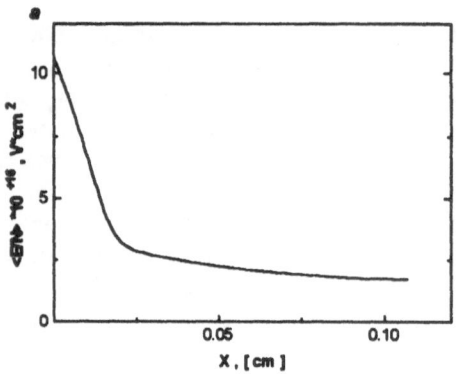

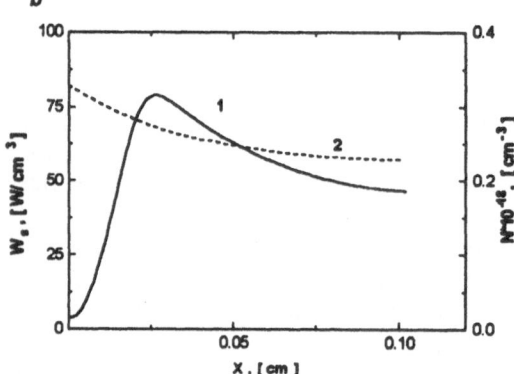

The results of the simulations shown above refer to the instant parameters of the discharge plasma. For the modelling of the laser performance it is necessary to know also the time averaged parameters of the active medium. Examples of time-averaged profiles of the specific input power, the neutral particle density and the E/N value are presented in fig. 3 a,b.

These data were used as input parameters in the next step of the simulation. This included the modelling of the CO_2 laser kinetics, the plasma chemical processes in discharge, the spatial structure of the laser radiation formed by resonator, etc. [13,14]. Some typical results of the simulations are presented in Fig. 4. The measurements were performed for a CO_2 laser system, with an active length of 37 cm and a square cross-section of 2.5 x 2.5 mm². A gas mixture consisting of CO_2:N_2:He =

Fig. 3 Mean square values of the reduced electric field strength as a function of the inter-electrode distance (a), and the spatial distribution of the specific input power W_s (1) and neutral particle density N (2) (b). 1:1:3+5% Xe at a pressure 100 Torr was used. The outcoupling mirror had 8% transmittance. Data for two different excitation frequencies of 85 MHz and 142 Mhz are presented. The various curves in a group show the impact of distributed losses and plasma chemical reactions (formation of the atomic oxygen in particular) on the laser performance. The dotted curves correspond to the case when the distributed losses and the oxygen atoms concentration are equal to zero. The dashed

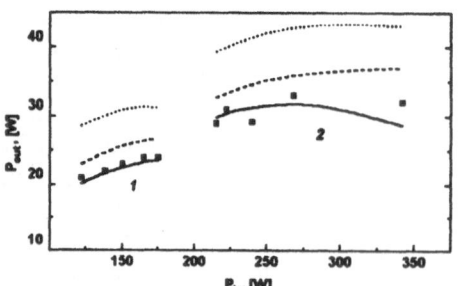

Fig. 4 CO_2 Laser output power as a function of the RF

curves correspond to the case when the distributed losses of $\alpha=10^{-4}$ cm^{-1} are taken into account. When the loss in inversion population due to depopulation of the upper laser level in collisions with the oxygen atoms ([O]=0.2%) was also accounted for (solid lines in Fig. 4), the calculated values of the output power decreased further. However this is a realistic approximation and it also gives a good agreement with the experimental points shown in the figure.

Along with the theoretical modelling an extensive parametrical study had been undertaken. Detailed information on the laser head design and experimental set-up can be found elsewhere [18-20]. The experimental procedure included a parametrical optimisation of the laser output with respect to the gas composition, gas pressure, electrodes material and excitation frequency.

A typical graph that characterises the behaviour of the laser output as a function of the RF input is shown in Fig. 5. The laser was filled with a mixture

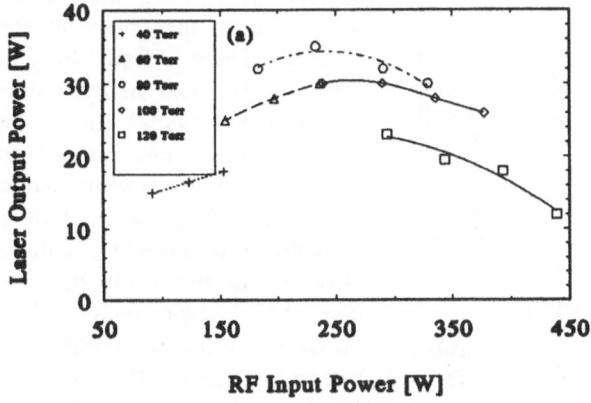

CO_2:N_2:He=1:1:5+5% Xe, at a pressure of 40 Torr. A maxumum laser power of 42.5 W was reached in our experiments. It corresponds to a specific output power of about 1.15 W/cm.

The laser output strongly depends on the excitation frequency. In helium rich mixtures the maximum output power is always obtained at frequencies of ~ 190 MHz, while in a

Fig. 5. Laser output power as a function of RF input power for different gas pressures. Excitation frequency 190 MHz.

"traditional" mixture with CO_2:N_2:He=1:1:3+5% the position of the maximum output is less pronounced. In the case of leaner mixtures the maximum output

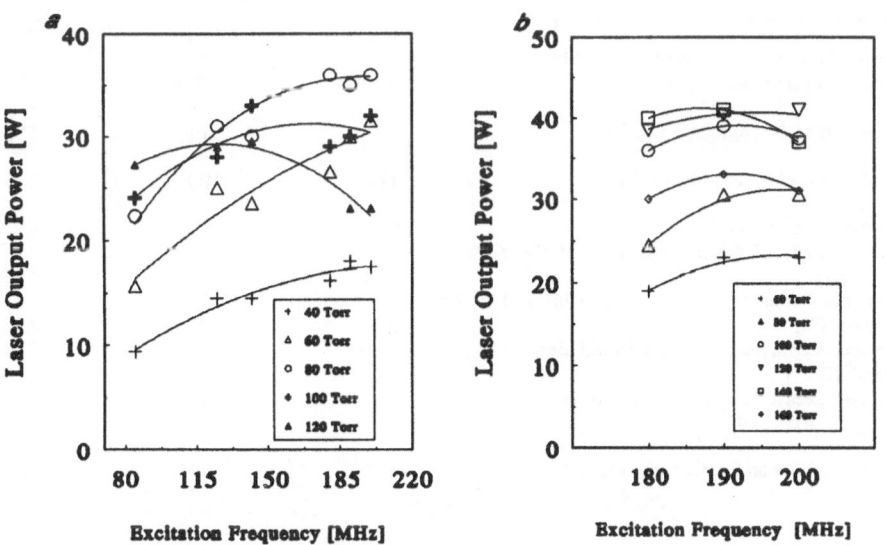

changes as the pressure changes. It moves from ~ 180 Mhz for the pressure of 40 Torr to ~130 MHz for the pressure of 120 Torr. It is clearly seen in Fig. 6 where the peak laser output power as a function of the excitation frequency is plotted for the gas mixtures $CO_2:N_2:He=1:1:3$ +5% Xe (Fig. 6a) and $CO_2:N_2:He=1:1:5$ (Fig. 6b). In the range of low pressures the laser output increases with the increase of both pressure and frequency. However in the higher pressure range from 100 to 120 Torr the pressure rise causes an increase in the laser output only at low frequencies, whereas at higher frequencies the output power drops more rapidly with pressure. This shifts the optimal frequency to still lower values.

The excitation frequency substantially effects the maximum RF input power. In Fig. 7 the variation of laser output as a function of input power is shown for different excitation frequencies. For each curve the gas pressure corresponds to the optimal

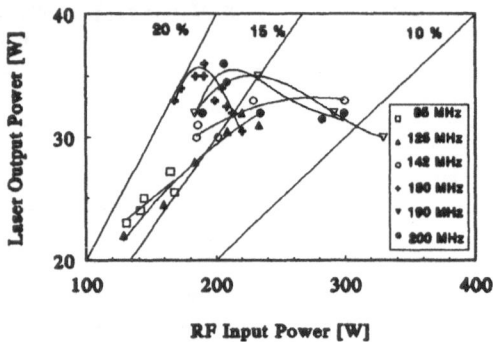

one for a particular frequency in accordance with Fig. 6a. Dashed lines indicate the values of the laser efficiency - from 20% to 10%. The first and the last points on the experimental curves for each frequency are determined by the minimum power needed to sustain the discharge and by a breakdown from the high voltage electrode to the walls of laser chamber, respectively. The power deposition into discharge can be increased as the excitation frequency rises.

Fig. 7. Variation of laser output power with RF inout power at the optimal gas pressure. The gas mixture is $CO_2:N_2:He:Xe=1:1:3+5\%$.

All these experimental facts clearly show that the excitation frequency has substantial impact on all laser characteristics. Numerical analysis gives better insight in the role of the excitation frequency in optimisation of the laser

Frequency f, [MHz]	62	125	223
Module of the discharge impedance Z, [Ω]	129	72.1	56.9
Cosine of phase angle between the first harmonic of the current and the voltage, cos Φ	0.508	0.719	0.828
Active part of the impedance R_g,[Ω]	65.6	51.8	47.1
Reactive part of the impedance X_g,[Ω]	111	50.1	31.8
Full current amplitude I_m,[A]	2.36	2.59	2.74
The amplitude of the first harmonic of the discharge current [A]	2.51	2.80	2.93
Difference between the full current value and its first harmonic [%]	0.6	0.8	0.8
Reduced electric field in a quasi neutral plasma E_{RMS}/N [V cm^2 10^{-16}]	1.86	1.99	1.99
Reduced electric field in the vicinity of the electrodes E_{RMS}/N [V cm^2 10^{-16}]	33.2	21.7	14.7
Gas temperature on the discharge axis T [K]	493	511	515

Table 1

performance. The computer model described above was used to calculate the parameters of gas discharges in the RF excited CO_2 lasers. Here we shall discuss the simulation results obtained for three different frequencies.

A discharge in a 37 cm long waveguide channel with a square cross-section of 2.25 x 2.25 mm^2 was considered in the simulations. The gas mixture was assumed to consist of CO_2:N_2:He = (1:1:8) at a pressure of 100 Torr. The discharge parameters were calculated for excitation frequencies of 62 MHz, 125 MHz and 223 MHz, and for an RF input power of 200 W. The results of the simulations are shown in table 1.

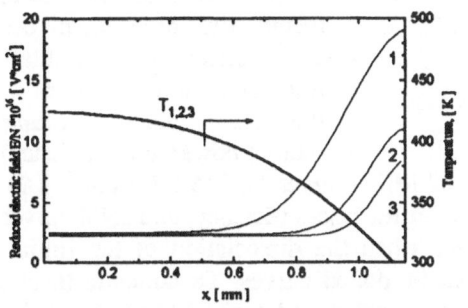

Fig. 8. Spatial distribution of reduced electric field and gas temperature for three different frequencies. Input power 100 W.

A significant variation of reduced electric field strength in the near-electrode regions is observed at different frequencies. It is noticeable however that the E_{rms}/N value remains almost constant and close to its optimum in the quasi neutral central part of the plasma. The gas temperature also remains practically unchanged which seems reasonable because a constant input power has been assumed. The spatial distribution of the reduced electric field strength and gas temperature is presented in Fig. 8. Here the RF input power is 100 W, and the curves marked 1, 2 and 3 correspond to excitation frequencies of 62 MHz, 125 MHz and 223 MHz respectively.

The internal parameters of the gas discharge undergo significant changes as the RF input power increases. Some typical simulation results for the case of an input power of 750 W are shown in Fig. 9. Apart from the spatial distribution of the gas temperature and the reduced electric field strength, the input power density is also plotted. The solid line 2 shows the total input power deposition while the dashed line 2 presents the part of RF input power which originates from the electronic component. The two curves coincide pretty well in the central discharge region filled with the quasi-neutral plasma. In the near electrode sheaths there are heavy particles that play an essential role in the processes of energy transfer and absorb substantially more energy than the electrons. When the absorbed input power reaches a certain level the breakdown of the sheaths will occur. This process is also refered to as a transition from low to high-current (or α- to γ-) mode and has been widely discussed in scientific literature (see for example article of S.A. Yatsenko and Yu.P. Raizer in this book). Generally speaking efficient operation of molecular

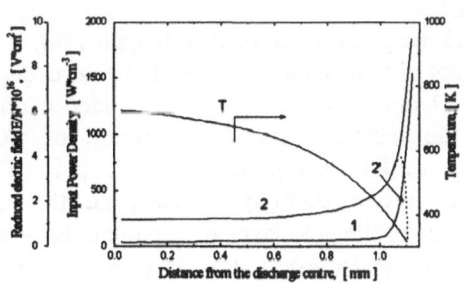

Fig. 9. Spatial distribution of reduced electric field, gas temperature and RF input power

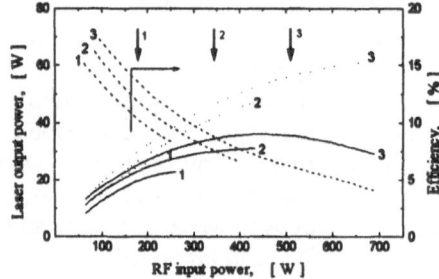

Fig. 10. Laser output power and efficiency as a function of RF input power

laser systems can be expected only in the case of a low-current discharge. So it is important to keep it in the α-mode. The higher is the excitation frequency the larger power can be pumped into the discharge before the breakdown occurs. This effect is quantitatively illustrated in Fig. 10 in which the laser output parameters (output power and efficiency) are plotted for three different frequencies: 62 MHz (curves 1), 125 MHz (curves 2) and 223 MHz (curves 3). Dashed curves are for the laser efficiency, and solid curves are for laser output power. The ideal case when the dissociation of the initial gas mixture can be neglected is illustrated by dotted curves. To compute these data separate codes for the laser kinetics and resonator model where used along with the basic program of the gas discharge simulation.

There are also arrows in the upper part of the figure which correspond to the input power levels at which the voltage-current characteristics reach their maximum. Further increase of the input power results in a transition to a part of the voltage-current curve with a negative slope. To achieve a stable operation in this power domain an extra ballast resistance should be connected in series with the discharge. This modification of the electric circuitry would be necessary only at lower frequencies because already for the curves 2 and 3 it would not result in the increase of output power. The maximum output for the frequencies of 125 and 223 MHz is reached before the system approaches the point of instability.

For the same parameters of the active medium it is possible to pump almost three times more RF power into the discharge excited with the frequency of 223 MHz instead of 62 MHz. Moreover our analysis of the spatial characteristics of the discharge, its inversion population and the laser mode structure shows [13] that higher excitation frequency results in larger volumes of active media where efficient lasing can take place. One can also see in Fig. 10 that a substantial improvement in laser performance can be achieved if the level of dissociation of CO_2 molecules will be decreased. That would be especially beneficial at higher specific pumping rates of 150-200 W/cm^3.

Recently we proposed a technique which helps to accomplish this goal [18]. The CO_2 concentration in the discharge is kept high by using a distributed gold catalyst that covers the surface of electrodes. Detailed study shows that under certain conditions this approach results in the increases the concentration of CO_2 molecules from 60 % to 85 % of the initial CO_2 partial pressure [21]. The catalytical activity of the gold plated surface strongly depends on the gas temperature [19]. The optimum temperature of the gold layer was found to be ~ 50° C. The changes in the catalytical activity are clearly seen in Fig. 11 where the small signal gain in a CO_2 gas mixture (CO_2:N_2:He:Xe = 1:1:5 +5%) is presented as a function of time for catalytically active gold coated electrodes and for regular aluminium electrodes which are supposed to provide the optimum performance [22]. Only one ground electrodes is cooled in both cases. The temperature of the

83

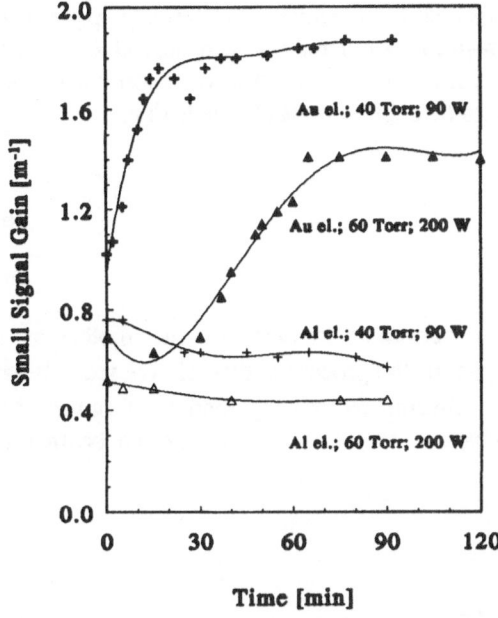

Fig 11 Small-signal gain as a function of time for
aluminium and gold plated electrodes.

high-voltage electrode increases with time until it reaches a steady state condition after ~ 30 minutes. A rise of the temperature in laser chamber results in a slight decrease of gain for the laser with aluminium electrodes. In the case of gold plated electrodes this minor negative effect is compensated by the increasing activity of the gold catalyst. A CO_2 laser with distributed catalyst can thus take the advantage of the possibilities of higher pumping rates offered by the operation at higher excitation frequencies. This technique is also very prospective for the future design of large laser systems in stable-unstable resonator geometry, with broad plasma electrodes where the effects of overheating of laser active medium become important [23,24].

4. Continuous wave noble gas lasers with RF excitation

Carbon dioxide lasers with RF excitation discussed in paragraph 3 proved to be the most advanced and widely used type of lasers. It is also confirmed by the fact that a few other articles in this volume deal with various aspects of RF excited CO_2 lasers. (see, for example, papers of D.R. Hall, C.A. Colley, N.A. Yatsenko, Yu.P. Raizer). There are, however, several other potentially interesting gas laser systems which have emerged recently. To our opinion an RF excited Ar-Xe laser belongs to the most prospective of its kind.

Lasing in Xe-based noble gas mixtures was observed in the infrared region of spectrum as early as in 1962 [25]. The researchers noted at that time that the laser has a remarkably high gain of 4.5 dB/m for the 2.026 μm transition [25] and 50 dB/m for the 3.51 μm transition [5]. However, a big disadvantage of this particular laser system (as well as of many others which ceased to exist soon after being developed) was the extremely low efficiency and output power. Several milliwats of CW output power was the maximum obtained with low pressure neutral Xe lasers [26].

The situation with the infrared xenon lasers has changed drastically after it was shown that a high pressure Ar-Xe laser can provide high energy efficient pulses in 1.73-3.5 μm range. The maximum energy of 600 J has been obtained with an e-beam pumped high pressure Ar-Xe laser. [27]Also, an efficiency of 6% was reported for long-pulse Ar-Xe laser [28].

There are some theoretical considerations which qualitatively explain high output energy and efficiency of high-pressure e-beam pumped neutral xenon lasers. In a typical Xe-Ar mixture pumped with an e-beam most of the energy is deposited in the Ar atoms. Ionised argon relaxes through various channels like.

$$Ar^+ + Xe = Xe^+ + Ar$$

$$Ar^+ + 2Ar = Ar_2^+ + Ar \tag{6}$$

$$Ar_2^+ + Xe = Xe^+ + 2A2$$

After a certain period of time a noticeable number of the molecular ions Xe_2^* and $Ar-Xe^+$ is present in the gas. In the process of dissociative recombination excited Xe^* atoms are produced. According to existing models of pulsed Ar-Xe laser operation this is the main channel of the relaxation for the beam energy deposited in the gas:

$$Xe_2^+ + e = Xe^6(6p) + Xe$$

$$ArXe^+ + e = Xe^*(5d) + Ar \tag{7}$$

$$Ar^* + Xe = Xe(5d) + Ar$$

The rate constants for the three above mentioned processes are $2.3 \cdot 10^{-6} \cdot \sqrt{T_e/300} \cdot cm^3/s$ [29], $1.0 \cdot 10^{-6} \sqrt{Te/300} \, cm^3/s$ [30] and $2.4 \cdot 10^{-10} \, cm^3/s$ [30] respectively.

As a result of the above mentioned processes an inversion population between the levels 5d and 6p occurs. After lasing the lower laser level relaxes into the meastable 6s level. This scheme is shown in Fig. 12. .

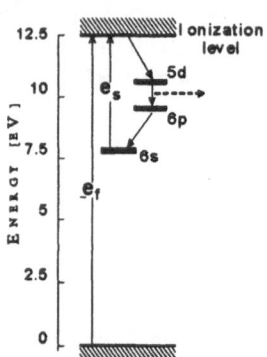

Fig. 12. Simplified 4 level scheme of lasing from Xe atoms.

Xe atoms from the metastable 6s level can be easily excited and ionised by low energy discharge electrons. This mechanism explains relatively high efficiency of pulsed Ar-Xe laser.

Recently we hawe shown that the potential capabilities of the CW Ar-(He)-Xe laser also have been underestimated [31]. It appears to be that transverse RF excitation of medium pressure atomic xenon lasers gives a substantial (hundreds of times) improvement over DC excitation under the same operational conditions [32]. A discharge was ignited in a quartz tube with 3 mm internal and 5 mm external diameter. The tube was mounted between the two electrodes which are used for the excitation of the transverse RF discharge through the dielectric. In the case of RF excitation the active length of the discharge was 25 cm. It was ~ 4 cm longer when the discharge was DC excited. Two platinum electrodes were mounted sideways. The laser was designed to operate in a DC, RF or combined mode so it was possible to make indeed the correct comparison between the different discharge regimes.

The results for RF and DC excitation of the discharge are shown in Fig.13 and Fig.14. The gas mixture is identical in both cases and consist of Ar:He:Xe=59:40:1.

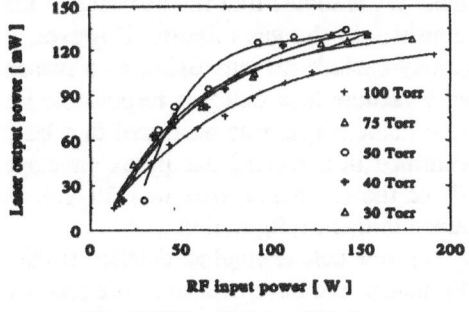

Fig 12.RF excitaion of Xe laser   Fig . 13 DC excitation of Xe laser

the input power was up to 2 times higher in the case of RF excitation, when it was also possible to operate the laser at higher gas pressures. There is a dramatic difference in output power for DC and RF excitation - more than 2 orders of magnitude. When DC and RF excitation were applied simultaneously the output power was comparable to the one observed in the DC mode. This clearly indicated that the high specific characteristics of the CW Ar-Xe laser can be only obtained in RF excited discharges. We assume that this due to some specific features of the the γ-mode of RF discharge, although to clarify the situation detailed experimental and theoretical studies on the kinetics of the Xe laser levels should be done.

First the laser output power was optimised. The highest output power obtained in

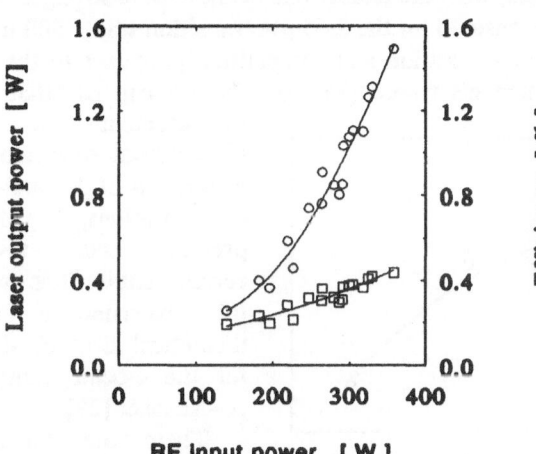

Fig. 15.Output power and efficiency of the slab Ar-Xe laser as a function of input power

our experiments was ~1.5 W [33] It corresponds to a specific output power density of 0.27 W/cm^3, with a power input value of ~ 65 W/cm^3, the gas pressure for the mixture consisting of Ar:He:Xe = 50:49:1 was 90 Torr. The size of the discharge volume formed by two metal electrodes and two dielectric sidewalls was 1.5 x 10 x 370 mm^3, and the excitation frequency was 125 MHz. The laser output power as a function of the input is shown in Fig.15. It is clearly seen that stronger pumping is preferable for the high output. Both the laser output power and efficiency increased as the input increased. It was also possible to change the dimensions of the discharge volume from 2 cm^3 to 10 cm^3 thus increasing the output power Higher specific input power also results in higher specific output, as it is shown in Fig. 15.

These experimental data look quite promising. The problem is however that the output remains stable for a relatively short time but then drops to a 2-3 times smaller value in a period of about 30 minutes. We assumed that the decrease in the laser output power originates from the impurities in the gas mixture. However, it was found that although we cannot completely exclude an outgassing as a reason for the poor output power stability, it is not a vacuum leak that it is responsible for it. With the same vacuum of 10^{-6} by far more stable output was measured in a laser system where the active gas media was confined in a rectangular pyrex envelope with 2×10 mm^2 cross-section. The length of the discharge area was 30 cm. A maximum output power of ~ 1 W was obtained in this configuration.

With this improvement of the time stability we were able to conduct detailed studies on various aspects of the Ar-Xe laser performance. We have measured the spectral composition of the laser output in the range of 1.73 μm - 3.51 μm. Totally 7 laser lines with the wavelengths of 1.73 μm, 2.03 μm, 2.48 μm, 2.63 μm, 2.65 μm, 3.11 μm, 3.37 μm and 3.51 μm were observed. The output spectrum of the CW RF excited Ar-Xe laser differs from the one observed in e-beam pumped Ar-Xe lasers. The transition $5d[^3/_2]_1 \rightarrow 6p[^5/_2]_3$ (wavelength 1.73 μm) which is dominant in the latter case, normally was not present in the output spectrum of the RF excited laser, irrespective of the gas mixture composition. Only in a frequency selective resonator with a diffraction grating replacing the totally reflecting mirror it became possible to observe lasing on the 1.73 μm transition. Again, its intensity was 10 times less than the intensity of the 2.03 μm line under similar experimental conditions. This ratio is close to the ratio of the transition probabilities of these lines. [34]

In a regular two-mirror configuration an output power was mostly concentrated in two lines 2.03 μm and 2.65 μm, with the former one being $\sim 10\%$ stronger. The maximum single-line intensity measured for the 2.65 μm transition was ~ 500 mW. Detailed studies on the spectral composition and competition processes in the RF excited Ar-Xe laser are presented elsewhere [35]. Briefly, we can say that the measurements of the laser output spectrum as a function of Xe and Ar concentrations, gas pressure, etc. reveal certain similarities with the experimental and theoretical data obtained for the e-beam pumped Ar-Xe laser [29].

There are however some substantial differences, like for example the poor performance of the 1.73

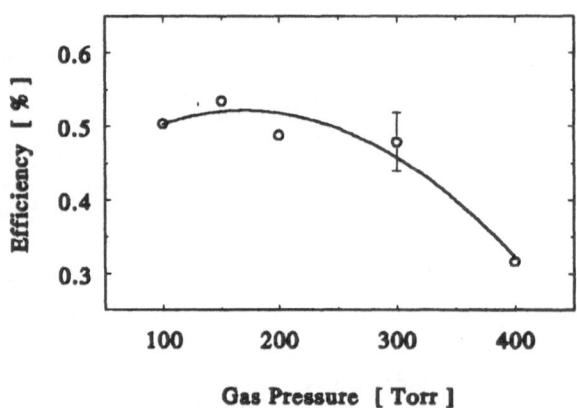

Fig. 16. Laser efficiency as a function of gas pressure

μm transition. Another example of the different behaviour is a decrease of laser efficiency with the increase of gas pressure. As we already mentioned one of the commonly accepted mechanisms of the population of upper laser level of Xe is recombination and dissociation of ArXe$^+$ ions. These ions are formed in three particle collisions in the plasma. The efficiency of this mechanism as well as the laser efficiency should increase as the gas pressure increases. However it is clearly seen in Fig.16 that initially the efficiency remains constant and then drops as the pressure continues to rise.

Of course this single experiment does not necessarily mean that the mechanisms which lead to the efficient operation of e-beam pumped pulsed Ar-Xe lasers should be completely excluded in the case of CW RF pumped lasers. Our experimental observation show that for the same input power the discharge can exist in two completely different forms. A bright and luminous discharge, with a transverse structure typical for the γ-mode results in high output power and efficient laser operation. A less intensive form of the discharge with thicker sheats typical for the α-mode also provides a substantially lower output [35, 36]. The most important fact is however that different discharge modes also have different discharge properties, especially with respect to the electron energy distribution. It was shown by J.P.Boeuf [36] that at certain condition two groups of high- and low-energy exist in RF excited gas discharge. As it was already mentioned above the similar situation exists in the e-beam pumped Ar-Xe lasers (Fig. 12). Further experiments which are now in progress will reveal whether the exitation mechanisms of various types of Ar-Xe lasers indeed have much in common.

Conclusion

In this article we discussed only a small part of the existing types of RF excited gas lasers. Metal vapour lasers, rare gas ion lasers - just to mention a few - have also recently made a substantial progress. Further experiments, detailed studies of the physics and special features of the gas discharges with RF excitation will certainly the performance of existing lasers, as well as increase the number of new prospective laser systems.

Literature

1. Javan A., Bennet, W.R., Herriot, D.R. (1961) Population inversion andcontinuous optical maser oscillation in a gas discharge containing a He-Ne mixture, *Phys. Rev. Lett.* **6**, 106-110.
2. Patel, C.K.N. (1964) CW laser on vibrational-rotational transitions in CO, *Appl. Phys. Lett.* **7**, 246-247.
3. Patel, C.K.N. (1964) Interpretation of CO_2 optical maser experiments, *Phys. Rev. Lett.* **12**, 588-590.
4. Patel, C.K.N.(1964) Selective excitation through vibrational energu transfer and optical maser action in N_2-CO_2, *Phys. Rev. Lett.* **13**, 617-619.
5. Paananen, R.A. and Bobroff, D.L. (1963) Very high gain gaseous (He-Xe) optical maser at 3.5 μm, *Appl. Phys. Lett.* **2**, 99-100.
6. Laakmann, K.D. (1979) Waveguide gas laser with high frequency transverse discharge ecitation, *U.S. Patent 4169251*, Sep. 25, 1979, 1-6.
7. Myshenkov, V.I., Yatsenko, N.A. (1981) Prospects for using high-frequency capacitive discharges in lasers, *Sov. J. Quanum. Electron.* **11**, 1297-1301.
8. Raizer, Yu.P. (1991) *Gas discharge physics*, Springer-Verlag, Berlin, etc.
9. Hall D.R. and Hill, C.(1987) RF discharge excited carbon dioxide lasers, in P.K.Cheo (ed.), *Handbook of Molecular Lasers*, Marecl Decker, New York, pp. 165-258.
10. Griffith, G.A. (1980) Transverse RF plasma discharge characterization for CO_2 waveguide lasers, *SPIE Proc* **227**, 6-11.
11. Kittel, C. (1986) *Introduction to solid state physics*, J.Wiley, NY etc.
12. Levitskii, S.M. (1957) An investigation of the breakdown potential of a high frequency plasma in the frequency and pressure transition regions, *Sov. Phys. - Tech. Phys.*, **2**, 887-889.

13. Ilukhin, B.I., Udalov, Yu.B., Kochetov, I.V., Ochkin, V.N., Heeman-Ilieva, M.B., Peters, P.J.M., and Witteman, W.J (1994), Research and optimization of a waveguide CO_2 laser with RF excitation, P.N.Lebedev Physical Institute, *Preprint*, N 38, 1-52.
14. Ilukhin, B.I., Udalov, Yu.B., Kochetov, I.V., Ochkin, V.N., Heeman-Ilieva, M.B., Peters, P.J.M. and M., Witteman, W.J (1995) A theoretical and experimental investigation of a waveguide CO_2 laser with r.f. excitation, *Appl. Phys. B*, 61,   .
15. Witteman, W.J., Ilukhin, B.I., Kochetov I.V., Ochkin, V.N., Peters, P.J.M., Udalov, Y.B., and Tskhai, S.N. (1994) On the impact of the excitation field frequency on the parameters of RF excited CO_2 laser, P.N.Lebedev Physical Institute, *Preprint* N 38, 1-28.
16. Witteman, W.J., (1987) *CO_2 laser*, Springer -Verlag, Berlin-Heidelberg, etc.
17. Boeuf, J.-P, (1987) Numerical model of rf glow discharge, *Phys. Rev.* A 36, 2782-2792
18. M.B.Heeman-Ilieva, Y.B.Udalov, W.J.Witteman, P.J.M.Peters, K. Hoen, and V.N.Ochkin (1992) RF excited 1.1 W/cm waveguide CO_2 laser, *J. Appl. Phys.* 74, 4786-4788.
19. V.N., Heeman-Ilieva,, Udalov, Yu.B.,Hoen, K., and Witteman, W.J., (1994) Enhanced gain and output power of a sealed-off rf-excited CO_2 waveguide laser with gold-plated electrodes, *Appl. Phys. Lett.* 64, 673-675.
20. Heeman-Ilieva, M.B., Udalov, Yu.B., and Witteman, W.J (1994) New trend in the technology of CW RF-excited CO_2 lasers, *SPIE Proc.* 2118, 230-237.
21. Tskhai, S.N., Udalov, Yu.B., Peters, P.J.M., Witteman, W.J., and Ochkin, V.N.Spectroscopic studies of the RF excited CO_2 lasers with distributed gold catalyst, (to be published).
22. Haas, W., Kushimoto, T. (1990), Investigation of the gas composition in sealed-off RF-excited CO_2 lasers, *SPIE Proc.* 1276, 49-57.
23. Abramski, K.M., Colley, A.D., Baker, H.J., Hall, D.R. (1989) Power scaling of large-area transverse radio frequence discharge CO_2 laser, *Appl. Phys. Lett* 54, 1833-1835.
24. Jackson, P.E., Baker, H.J., Hall, D.R. (1989) CO_2 large-area discharge laser using an unstable-waveguide hybrid resonator, *Appl. Phys. Lett* 54, 1950-1952.
25. Patel, C.K.N., Faust, W.L., and McFarlane, R.A. (1962) High gain gaseous (He-Xe) optical masers, *Appl. Phys. Lett.* 1, 84-85.
26. Laser Handbook
27. Litzenberger, L.N., Trainor, D.W., and McGeoch, M.W (1990) A 650 J e-beam-pumped atomic xenon laser, *IEEE J. Quantun Electron.* QE-26, 417-423.
28. Perkins, T.T. (1993) Steady-state gain and saturation flux measurements in a high efficiency, electron-beam-pumped, Ar-Xe laser, *J. Appl. Phys.* 74, 4860-4866.
29. Ohwa, M., Moratz, T.J., and Kushner M.J. (1989) Excitation mechanisms of the electron-beam-pumped atomic xenon (5d-6p) laser in Ar-Xe mixtures, *J. Appl. Phys.* 66, 5131 - 5145
30. Basov, N.G., Danilychev, V.A., Dudin, A. Yu., Zayarnyj, D.A., Ustinovskii, N.N., Kholin, I.V., and Chugunov, A.Yu. (1984) Electron-beam-controlled atomic Xe infrared laser, *Sov. J. Quantum Electron.* 14, 1158-1167.
31. Udalov, Y.B., Peters, P.J.M., Heeman-Ilieva, M.B., Ernst, F.H.J., Ochkin, V.N., and Witteman, W.J. (1993) New continuous wave infrared Ar-Xe laser at intermediate gas pressures pumped by a transverse radio frequency (RF) discharge, *Appl. Phys. Lett.* 63, 721-722.
32. Udalov, Yu.B., Peters, P.J.M, Heeman-Ilieva, M.B., Ochkin, V.N., and Witteman, W.J (1994) Continuous-wave high specific output power Ar-He-Xe laser with transverse RF excitation, *SPIE Proc.* 2118, 238-248.
33. Tskhai, S.N., Udalov, Yu.B., Peters, P.J.M., Witteman, W.J., and Ochkin, V.N. (1995) Continuous wave near-infrared atomic Xe laser excited by a radio frequency discharge in a slab geometry *Appl. Phys. Lett.* 66, 801-803.
34. Ayman, M. Coulombe, M.(1978) *Atomic data and nuclear data tables*, 21, p. 537-559.
35. Tskhai, S.N., Udalov, Yu.B., Peters, P.J.M., Witteman, W.J., and Ochkin, V.N. (1995) Spectral investigation of cw rf-pumpe atomic Xe laser with a slab geometry, *Appl. Phys. B* 61, ...
36. Belenguer, Ph. and Boeuf J.-P. (1992) Transition between different regimes of RF glow discharge, *Phys. Rev.* A 41, 4447-4459.

HIGH POWER CW MOLECULAR GAS LASERS USING NARROW GAP SLAB WAVEGUIDES

A D Colley, F Villarreal, A A Cameron, P P Vitruk, H J Baker, D R Hall

Department of Physics, Heriot-Watt University
Edinburgh EH14 4AS United Kingdom
Tel: +44 131 451 3081   Fax: +44 131 451 3088

1.   Introduction

The efficient extraction of high power outputs from carbon dioxide [1] and carbon monoxide [2] static-gas, sealed-off, compact lasers has been realised by exploiting the area-scaling concept in slab waveguide geometries. The gain volume is increased in size by extending the waveguide dimensions in one dimension only, while maintaining efficient cooling of the laser medium in the other. Since coupling losses between the waveguide and the mirrors are much reduced for wide aspect ratios in comparison with symmetric (e.g. square bore) waveguides, it is possible to reduce the electrode separation while maintaining efficient laser operation. From the discharge scaling laws it is apparent that small inter-electrode dimensions lead to increased gas pressures which in turn has a beneficial effect on the ability to modulate the laser output by discharge switching.

In this paper we indicate the factors which contribute to the stability of the discharge, particularly for small interelectrode separations and the types of resonators which may be employed with large aspect ratio gain media. We also provide results on a narrow gap slab-waveguide laser which operates close to atmospheric pressure, including modulation repetition rates for a variety of gas compositions and pressures.

2.   Laser construction

A schematic diagram showing the construction of the slab waveguide laser is shown in figure 1. Waveguiding is provided by the two electrodes and for some resonator configurations, by the ceramic sidewalls also, which are used as spacers to ensure uniform separation of the electrodes. The discharge is cooled by the close-proximity of the electrodes which are water cooled. In all cases the excitation frequency was

89

W. J. Witteman and V. N. Ochkin (eds.), Gas Lasers - Recent Developments and Future Prospects, 89–103.
© 1996 *Kluwer Academic Publishers.*

90

125MHz which was delivered by 50Ω cable from a remote generator and impedance matched to the laser medium by means of a two or three element LC impedance matching circuit.

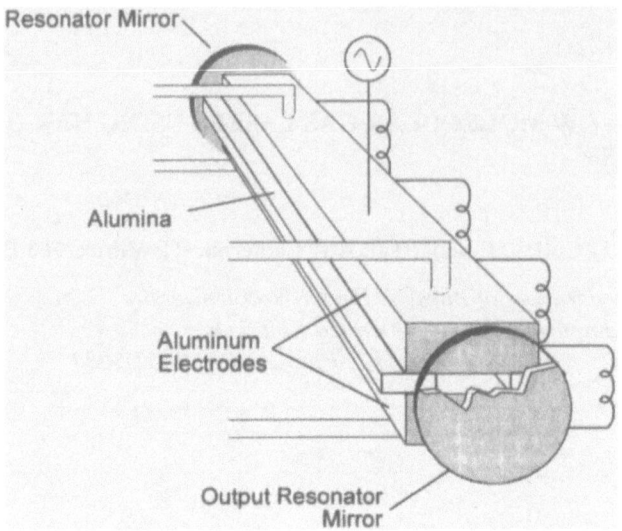

Figure 1   Schematic diagram of rf excited slab-waveguide laser

2.1.   RESONATOR

The invar-stabilised cavity is adjustable by means of mechanical mirror adjusters which exit the vacuum enclosure, and also by low-voltage piezoelectric actuators. In the transverse dimension, the small electrode separations invariably leads to single order mode operation which approximates closely to a gaussian profile in the far field. In the lateral dimension, two types of resonator design in particular have been employed.

For highly efficient exploitation of the gain medium the all-waveguide dual case I resonator has been used, where the sidewalls determine a high order lateral waveguide mode. In this configuration, a silicon back mirror with reflectivity of 99.7% and ZnSe output couplers have been employed. For the CO_2 laser an output coupling of 8% was used, and for the CO, 5% was more appropriate due to the reduced gain. The output beam from this resonator is highly multimode in the lateral dimension and not suitable for most applications [3]. However it does conveniently provide a suitable vehicle for characterisation of the gain medium and waveguiding properties of the resonator.

To achieve high beam quality, a cavity which controls the lateral mode is required and this may be implemented using a 1-D positive branch confocal unstable resonator [4], which may be implemented using cylindrical optics. However, as is usually the case, the magnification determined by the small output coupling leads to large radii of curvature mirrors. Hence spherical optics are employed which approximate closely to plane mirrors in the waveguide dimension.

2.2.   VOLTAGE UNIFORMITY

The use of high frequency excitation required to produce a stable discharge and to optimise power output and the capacitative nature of the electrode structure lead to transmission line effects along the laser electrodes. These effects can produce non-uniform excitation of the gain medium or worse still, to discharge instability and/or localisation into areas of high/low power density. The use of the distributed parallel resonance technique has provided adequate voltage uniformity for square bore waveguide lasers and has also been successfully applied to slab waveguide devices of modest size. Figure 2(a) shows a small sub-section of the electrodes with the parallel inductors in position. For large areas of discharge associated, for example, with large slab lasers, two effects can prove problematic.

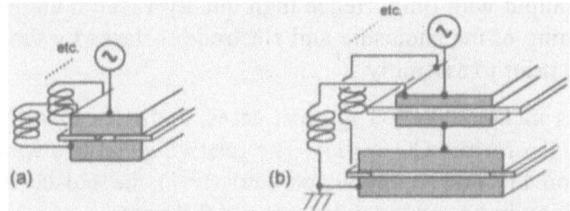

Figure 2   Distributed Parallel Resonance circuit schematics.
(a) Traditional single level (low loads) and (b) double level (very large areas)

Firstly, when the discharge lights, the creation of large sheath areas produce significant change in the capacitance of the transmission line. This in turn affects the distributed parallel resonant frequency, the appropriate selection of which is instrumental in ensuring uniformity. Hence the discharge sheath-induced frequency shift must be compensated and taken into account when selecting the parallel inductors. This frequency displacement may be very large for extended areas of discharge, which leads to large frequency offsets and possible ignition problems.

Secondly, as the area of the discharge is extended to provide very large gain volumes, the transmission line becomes highly loaded, and maintaining adequate energy storage in the parallel resonance becomes difficult; i.e. the quality factor (Q-factor) of the parallel resonance is reduced to a low value, perhaps not far from unity.

For the uniform excitation of very large area discharges e.g. an area of 770 x 95mm dissipating 9kW of rf power, a technique which has proved very useful is that of distributed double parallel resonance [5], a schematic of which is shown in figure 2(b). Here a non-dissipative transmission line is closely coupled in parallel with the electrode transmission line, which acts to store energy contributing to the parallel resonance. Hence the quality factor may be increased and discharge uniformity achieved. A highly desirable by-product of this technique is the reduction in ignition frequency shift and improved lighting conditions. This technique has been successfully implemented in the construction of a 1kW CO_2 laser [6].

2.3.   GAS COMPOSITION

The vacuum enclosure for both carbon dioxide and carbon monoxide laser experiments was designed to use 'o'-ring seals and was suitable for stable operation over relatively short time scales from several hours to one day, before refill was necessary. The prime threat to long term power stability in the carbon dioxide laser is dissociation of the CO_2 bye electron impact, though this can be minimised using metallic catalysts [7] or as in the present case by including a small quantity of hydrogen in the gas mix. For the carbon monoxide laser, decomposition of the CO can occur, causing accumulation of carbon deposits on the electrode and optical surfaces. This danger can be minimised by injecting a small quantity of oxygen to promote the reverse reaction. The major hazard to longevity is the presence of water vapour or other contaminants which drastically reduce the power output with time. Hence high quality vacuum engineering is required in addition to cleaning of the enclosure and electrode surfaces by vacuum bake-out and intensive treatment prior to assembly.

Gas mixes included Xenon in most cases, and occasionally hydrogen for the CO_2 laser and oxygen for the CO device. The relative quantities were found to have a significant effect on the power output, particularly in the pulsed mode of operation where the composition impacted the pulse rise and fall times.

3.   Similarity Laws for RF Discharges

It is well known [8] that the extrinsic properties of discharges are related through the similarity laws, so that to preserve discharge or laser excitation conditions when one parameter is changed, other linked parameters must also be adjusted. Thus in an RF discharge, the gas pressure p, electrode separation d, and generator frequency f, for CO_2 laser gas mixtures are related through the relationships [9] as outlined in table 1 and plotted in figure 3.

Table 1   Relationship between $p\,d$ and f for CO_2 discharges

pd	~	280 MHz • mm
fd	~	180 torr • mm
p/f	~	0.64 torr • MHz^{-1}

As can be seen, as the electrode gap is reduced, higher operating pressures and an elevated rf excitation frequency are necessary to preserve the optimised laser excitation conditions.

While the electrode separation and gas pressure may conveniently be varied, the excitation frequency is determined by the availability of suitable tuned power generators. For the experiments described here, a suitable power supply was available operating at a frequency of 125MHz. However, the interest in operating the cw discharge at high pressure, necessitates the consideration of devices with small values of d, even through corresponding changes in f were not possible.

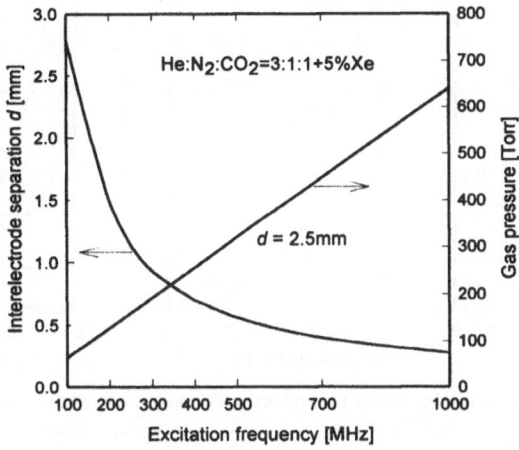

Figure 3   Interdependence of optimum excitation frequency, interelectrode separation and gas pressure for CO_2 discharges. Small interelectrode separations demand higher excitation frequency and gas pressure.

4.   Resonator issues

A consequence of reduced electrode separation is an increase in optical waveguiding and the waveguide/mirror coupling losses and an estimate of the magnitude of these losses is presented here. Furthermore, a reduced value of d gives rise to significant increase in optical power density on the resonator optics which may lead to damage and this also is considered.

4.1.   WAVEGUIDING LOSSES

Figure 4 shows the round trip waveguiding loss for a TE mode in a 390mm long narrow gap slab as a function of electrode separation for $10.6\mu m$ and $5\mu m$ radiation calculated

94

from theory [10]. The analysis assumes that the electrode surfaces are very smooth and while special efforts are taken to polish the electrodes, imperfections will cause deviation from this ideal behaviour. Experiments indicate (as expected) that in the large-aperture lateral dimension the beam profile is highly asymmetric with multiple high order modes guided by the alumina sidewalls. In the transverse dimension a low order mode is supported by the aluminium electrodes. The total waveguiding loss will be a composite of that experienced by the modes in the two dimensions, but will be dominated by that originating from the low order transverse mode.

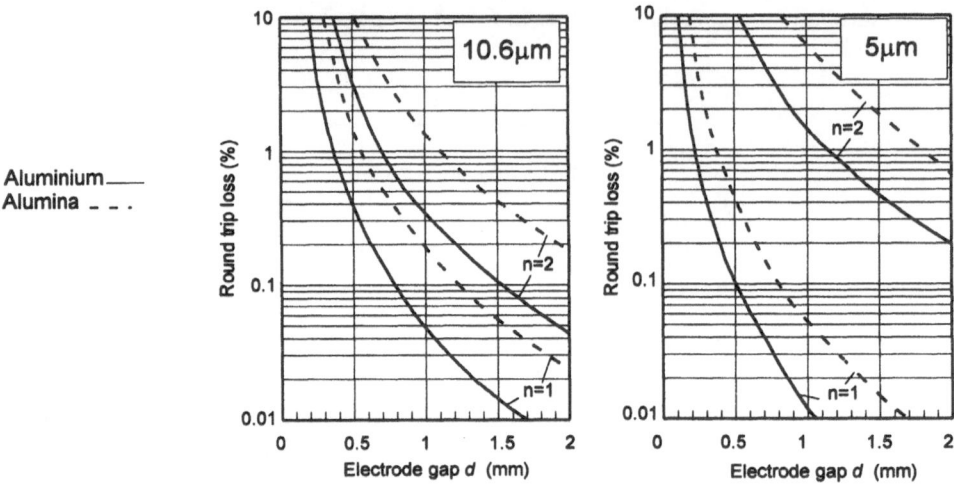

Aluminium———
Alumina _ _ .

Figure 4   Round trip waveguiding losses for low order TE modes with aluminium and alumina at 10.6µm and 5.3µm as a function of electrode separation.

The calculated results in figure 4 indicate that at 10.6µm for gaps >0.5mm the round trip losses for low order modes are negligible when compared to the 8% output coupling used in the oscillator. For dimensions less than 0.5mm, the loss for aluminium (mode order l) is still relatively small but can exceed 1% for $d < 300$µm.

The CO laser gain is less than that for the CO_2 laser and the output coupling is smaller (5%), so resonator losses will have a greater impact on the output power. However the waveguiding losses are smaller for the same interelectrode separation and for the fundamental mode the losses with aluminium only rise above 1% for separations <250µm.

These results indicate that for both wavelengths, waveguiding losses per se should not be a significant factor in the overall gain/loss situation. However, imperfections in the surfaces may scatter light which will substantially increase the losses and care must be taken in polishing the waveguide surfaces to high surface finish and flatness. These issues are particularly acute with the CO laser where losses are exaggerated by the reduced wavelength.

4.2.   COUPLING LOSSES

Most of the experiments employed a dual Case I resonator configuration, where two plane mirrors are placed close to the guide exits. However, it is impossible to put the mirror in contact with the guide aperture and some separation is essential to provide isolation of the optics and free circulation of the laser gas. Figure 5 shows the effect of varying the mirror/guide separation on the coupling losses for 5μm and 10.6μm resonators for the case of interelectrode separation of 0.5mm, calculated using a simple EH_{11} overlap integral technique.

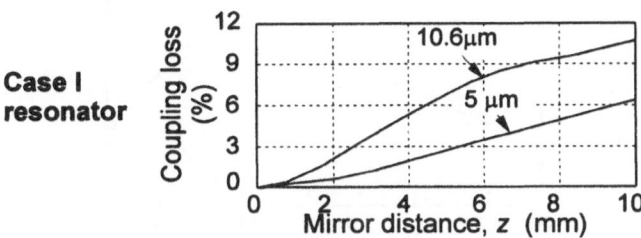

Figure 5   Coupling loss as a function of mirror/guide separation for plane mirror at 10.6μm and 5μm. Calculated using EH_{11} overlap integral.

Even for small separations (~3mm) the coupling losses rise rapidly, particularly in the case of CO_2 where the coupling loss equals the outcoupling for a mirror/guide separation of 6mm. The losses rise less acutely at 5μm but can have more of an impact on the laser output since at room temperatures, the laser gain is lower than for CO_2.

These simplified analyses of the waveguiding and coupling losses indicate that the main source of resonator losses originates from the mirror/waveguide coupling and great care must be exercised to minimise this for narrow gap lasers. Currently alternative resonator configurations are being considered to reduce the coupling losses while allowing remote location of the optics.

4.3.   DAMAGE TO THE RESONATOR OPTICS

For very small guide/mirror separations the optical power density on the resonator optics can far exceed their damage threshold leading to catastrophic damage of the dielectric coatings. Very high quality coatings are demanded and the optical power density must be compromised against the coupling losses incurred when moving the mirror from the guide aperture. Other resonator configurations may provide the key to reducing the applied power density.

5.   Experimental Results with Narrow Gap Lasers

The investigation employed several different waveguide dimensions. In all cases the output power was measured for the multimode all-waveguide resonator in order to test

96

the maximum power extraction. This may be converted into a specific output power so to normalise the power extraction to a given discharge/electrode area and the units are [kW/m^2].

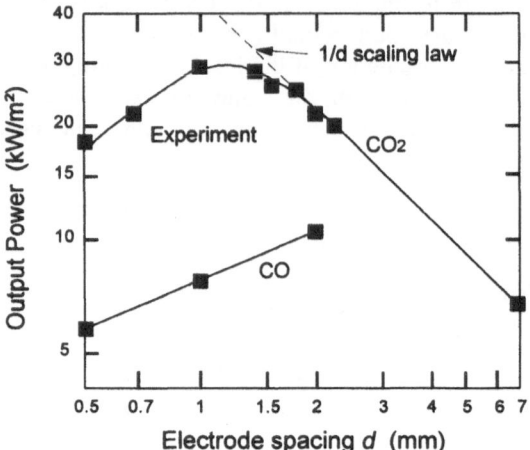

Figure 6   Specific output power dependency on electrode separation for CO and CO$_2$ lasers.

The dependence of specific output power on the electrode separation for CO and CO$_2$ lasers is shown in figure 6. The results have been derived from various laser geometries [3,11] all of which were operating in the 1-D regime where cooling of the gas is dominated by heat flow to the electrodes and not to the sidewalls.

As the electrode spacing d is reduced, several effects compete leading to the observed behaviour. Discharge cooling issues indicate that the power should scale as $1/d$ since narrow gaps provide the capability for the application of higher power densities before the axial gas temperature rises sufficiently to reduce the gain. However, for a fixed excitation frequency, the proportion of applied rf power which is deposited in the sheaths rises non-linearly and this power contributes to the average gas temperature but not significantly to the gain. Hence a turning point in laser output with electrode spacing might be expected on thermal considerations alone. An additional factor leading to reduced output power for narrow gaps is the increase in resonator losses originating from the interaction of the mode with the waveguide surfaces and from the finite divergence of the mode after the free space propagation between the guide apertures and the resonator mirrors. These are wavelength-dependent, and will also serve to reduce the extractable power at small interelectrode separations.

In the case of CO$_2$ slab lasers, as d is reduced the output power initially rises according to the $1/d$ scaling law which follows from the enhanced discharge cooling. However the turning point at ~1.2mm indicates the increasing impact of power loss in the sheaths, and optical losses originating within the resonator.

In the case of CO, the power falls monotonically with reduced electrode separation. This is surprising since the coupling and waveguiding losses are smaller at

5μm than 10μm suggesting an enhancement in output power and efficiency for reduced electrode separations similar to CO_2. Since the output power of the CO laser is very sensitive to gas temperature (see 5.2.) the impact of power loss in the sheaths will be more noticeable than for CO_2. Furthermore the lower gain CO laser will be much more sensitive to cavity losses and reduced gain. To summarise for CO, power loss in the sheaths and elevated resonator losses, combined with overall low gain serve to reduce the output power for all gaps smaller than 2mm.

5.1.   CARBON DIOXIDE NARROW GAP LASER

Two sets of waveguide dimensions have been investigated in detail; 0.5 x 20 x 385 mm, and 1.0 x 15 x 385 mm. Figure 7(a) and (b) show the variation of output power with pressure for a range of different gas compositions for the 1mm and 0.5mm electrode separations respectively. In all cases, the ratio of $0.25Xe:1CO_2$ was maintained.

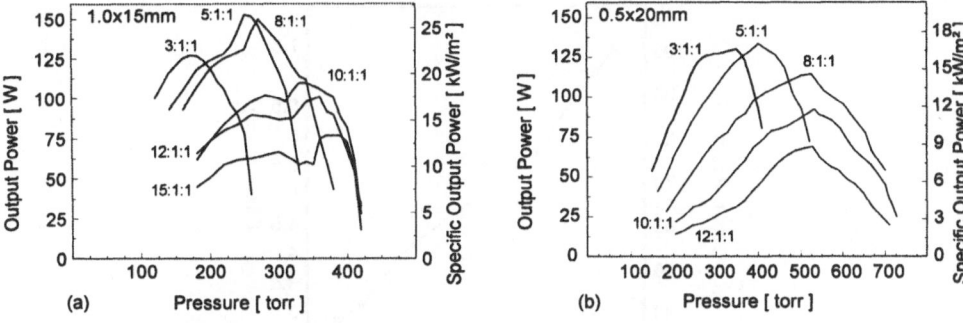

Figure 7   Variation of output power with pressure for two CO_2 lasers
(a) 1.0 x 15 x 385 mm laser and (b) 0.5 x 15 x 385 mm laser

The output power rises with pressure to a maximum which is generally slightly above the conditions for maximum efficiency. Occasionally the curves change gradient, particularly in the case of the 1.0mm laser and this may be attributed to discontinuous expansion of the discharge at higher input powers to fill the entire volume available.

The highest power output was achieved using the 5:1:1 gas mixture for both lasers, though higher pressure operation was achieved using helium rich mixtures which extended operation to close to one atmosphere in the 0.5mm case. For the 0.5mm laser the maximum extracted power output is reduced by 30% compared to the 1.0mm case due to the increased resonator losses, probably dominated by coupling losses.

Damage to the resonator optics, and in particular the silicon high reflector, was evident following operation of the 0.5mm laser. This could be minimised by using very clean mirrors placed several millimetres away from the aperture, and ensuring that the laser is well aligned before turning on.

5.2.   CARBON MONOXIDE NARROW GAP LASER

Electrode separations of 2.0mm, 1.0mm and 0.5mm have been investigated, with discharge areas of 20 x 386mm in each case. While high beam quality has been achieved in the past using an unstable resonator [12] here the all-waveguide multimode resonator was deployed to characterise the gap scaling properties.

The CO laser is designed with an additional feature, namely the capacity to cool the electrodes to temperatures below 0°C using cooled methanol. It is well know that gas temperature has a marked impact on the output power and efficiency. Figure 8 shows the variation in output power/efficiency for the 0.5mm slab waveguide CO laser over the temperature range of -25°C to +20°C. The resulting change in laser power is much more dramatic than for an equivalent CO_2 laser. Clearly, operating the laser below 0°C is preferable for high power and efficiency, but for simplicity most experiments were carried out using tap water, and an electrode temperature of 20±2°C.

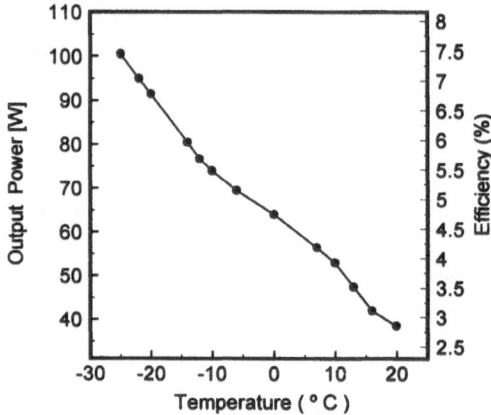

Figure 8   Variation of output power and efficiency on electrode temperature for 0.5mm CO Laser

The results for the 0.5mm laser are summarised in figure 9 where the maximum output power and corresponding efficiency is plotted as a function of pressure for a typical gas composition. As is usually the case, the maximum efficiency (3.08%) is achieved at an output power slightly below the maximum. Operation at a pressure close to one atmosphere has been demonstrated though at reduced output power and efficiency. The power and efficiency results can be scaled by operating at reduced temperatures according to the trend shown in figure 8.

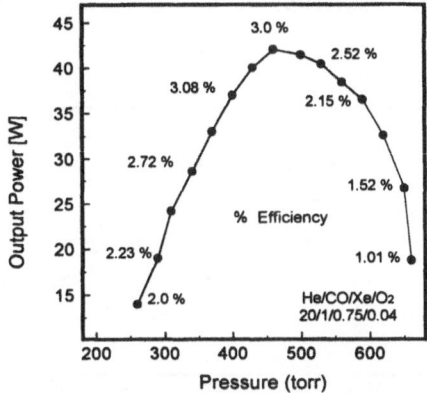

Figure 9   Pressure dependence of maximum output power for 0.5mm laser indicating the conversion efficiency at each pressure.

5.3.   PULSE MODULATED LASERS

Both the CO and CO_2 lasers may be pulsed by modulating the rf power and some applications benefit from or even rely on this mode of operation. The maximum repetition rate which may be achieved is determined by the rise and fall times of each pulse and these may be reduced by operating the laser at higher pressure where the collisional transfer rates are enhanced. The narrow gap slab waveguide lasers investigated here provide access to pressures in excess of those achievable by square bore waveguides and can enhance the pulse capability of the laser.

       The pulse-modulated narrow gap slab CO_2 and CO lasers were characterised using a chopper and a fast pyroelectric detector in order to determine the pulse decay times and maximum repetition rates while monitoring the depth of modulation. Pulse trains for the CO_2 laser are shown in figure 10 for 100% and 90% modulation. The non-square trains are a result of the finite beam size when passing through the chopper.

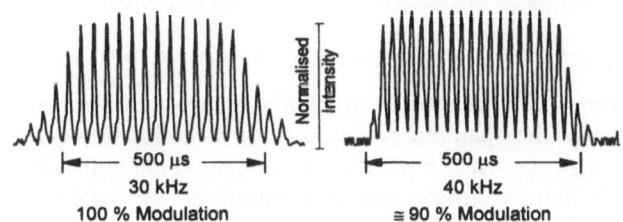

Figure 10   Pulse trains for 0.5mm laser operating at 350 torr, 260W Peak/130W Average output power, 50% duty cycle and 7% rf to optical efficiency.

Figure 11 shows the reduction in pulse fall time as a function of gas pressure for various gas compositions for the 1mm CO_2 laser. The fall time is reduced for higher pressures,

100

for example falling to 10μs from 30μs in the case of 5:1:1; the composition which provided the highest power.

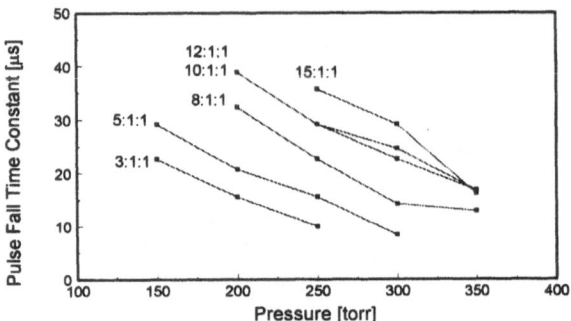

Figure 11 Variation of pulse fall time constant with pressure for 1mm electrode separation

In the case of CO lasers, relaxation times are much longer due to the cascade nature of the laser kinetics. Figure 12 shows pulse trains for 100% and 80% modulation for the 0.5mm gap slab CO laser. While not shown here, the pulse shape is wavelength dependent with some wavelengths contributing to the leading edge of the pulse, others to the decay and some to

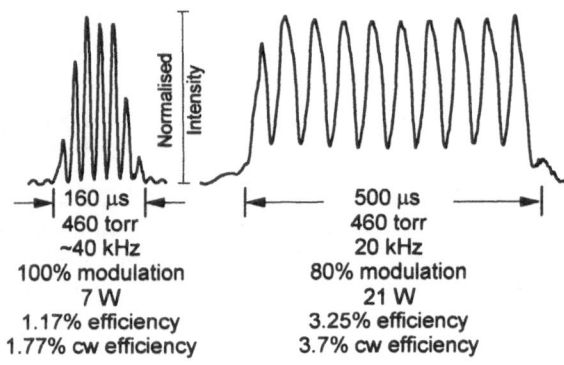

Figure 12 Pulse train for 0.5mm CO laser for 100% and 90% modulation.

The relaxation times also benefit from high pressure operation as demonstrated in figure 13 which maps the operating regime for 100% modulation depth as a function of repetition rate against electrode separation. Complete modulation above 10kHz is only achievable using 0.5mm electrode separations though at reduced efficiency.

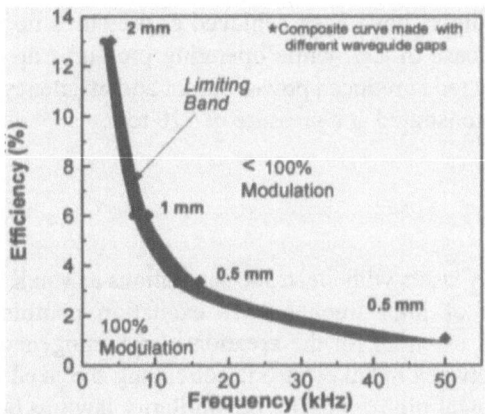

Figure 13 Pulsed operating regime for the CO laser with 2.0mm, 1.0mm and 0.5mm electrode separations.

5.4.   SUMMARY OF RESULTS

Table 2 summarises the results for three different electrode separations / CO and CO_2.

Table 2   Summary of results for CO and CO_2 lasers for three different electrode separations

Electrode Separation	2.0 mm	1.0 mm	0.5 mm	
CO_2				
Optimum Pressure	90-100	190-210	330-350	torr
Input Power Density	92	296	455	W/cc
CW Output Power	22	30	18	kW/m^2
Efficiency	12.5	10	7	%
Max Rep. Rate	10	20	40	kHz
Pd Product	~190	~200	~170	torr.mm
CO				
Optimum Pressure	100-110	240-260	450-470	torr
Input Power Density	42	130	350	W/cc
CW Output Power	10.5	7.8	2.6	kW/m^2
Efficiency	13	6	3	%
Max Rep. Rate	3	6	18	kHz
Pd Product	~210	~250	~230	torr.mm

Some of the salient points are as follows;

- Pd is approximately constant despite the common rf frequency used throughout the investigation.
- The maximum pulse modulation rate is considerably increased at higher pressures.

- Maximum power outputs have been achieved at pressures up to 350 torr for CO_2 and 470 torr in the case of CO, while operating pressures up to near atmospheric have been demonstrated at reduced power output and efficiency.
- Cw laser action demonstrated at a pressure of 710 torr.

6.   Conclusions

Area scaled CO and CO_2 lasers with electrode separations as small as 0.5mm have been demonstrated. The use of high frequency rf excitation combined with a suitable distribution technique is essential for the creation of a homogeneous stable discharge. This can be achieved between metal electrodes obviating the need for dielectric ballast and the associated technical problems. The *pd* similarity law has been loosely born out in practice even though the laws relating to the excitation frequency have not been adhered-to. For both CO and CO_2 lasers operating at maximum power, the value of *pd* is approximately 200 torr.mm.

The pulse fall times have been reduced by operating at elevated pressures and the maximum modulation rate has thus been extended. The maximum repetition rate for CO_2 has been increased from 10kHz to 40kHz by moving from 2.0mm to 0.5mm electrode separation, and similar configurations for CO have increased the maximum repetition rate of 3kHz to 18kHz.

The employment of even smaller electrode separations i.e. <0.5mm raises four issues; coupling losses, waveguiding losses, mirror damage and discharge stability. It is anticipated that the coupling losses will dominate over the waveguiding losses providing highly polished electrodes are employed. Hence improved resonator configurations are required, and for resonators utilising remote mirrors optical power loading on the resonator optics may be reduced with a diminished risk of optical damage.

Unless the rf excitation frequency is elevated according to the similarity laws, the limiting factor in using electrode separations <0.5mm originates in discharge instabilities. Gamma discharges coexist with the alpha discharge for an electrode separation of 0.5mm when the input power density is in excess of that optimised for lasing. For an electrode separation of 0.4mm, this two-mode discharge is experienced for power levels below those expected for optimum lasing conditions. The gamma discharge leads to damage of the waveguide surfaces which intolerably increases the cavity losses.

High pressure, high power slab-waveguide lasers may find new applications in materials processing, remote sensing and other applications which demand high beam quality, fast discharge-switched modulation and possibly increased tunability. The extension to even higher pressures may be possible by careful selection of the extrinsic operating parameters of the laser and the development of suitable resonators.

7.   References

1.   K M Abramski, A D Colley, H J Baker & D R Hall, Appl. Phys. Lett., **54**, 1833 (1989)

2.   H Zhao, H J Baker & D R Hall, Appl. Phys. Lett. **59**, 1281 (1991)

3.   C J Shackleton, K M Abramski, H J Baker & D R Hall, Optics Comms. **89**, 423 (1992)

4.   P E Jackson, H J Baker & D R Hall, Appl. Phys. Lett. **54**, 1950 (1989)

5.   A D Colley, H J Baker & D R Hall, Patent GB9316282.4

6.   A D Colley, H J Baker & D R Hall, Appl. Phys. Lett., **61**, 136 (1992)

7.   M B Heeman-llieva, Yu B Udalov, K Hoen & W J Witteman, Appl. Phys. Lett. **64**, 673 (1994)

8.   P P Vitruk, H J Baker & D R Hall, J. Quantum Electron, **30**, 1623 (1994)

9.   E A J Marcatili & R A Schmeltzer, Bell Syst. Tech. J. **43** 1783 (1964)

10.   A von Engel, *Ionized Gases*, Clarendon Press, Oxford (1955)

11.   X S Zhang, H J Baker & D R Hall , J. Phys. D: Appl. Phys **26**, 359 (1993)

12.   A D Colley, F Villarreal, K Abramski, H J Baker & D R Hall , Appl. Phys. Lett. **64**, 2916 (1994)

HIGH POWER MULTI-CHANNEL WAVEGUIDE CO$_2$ LASER ARRAYS

K M Abramski*, H J Baker, A D Colley, D R Hall

*Department of Physics, Heriot-Watt University
Edinburgh EH14 4AS United Kingdom
Tel: +44 131 451 3081   Fax: +44 131 451 3088*

1.   Introduction

The development of laser array technology is one of a number of laser-related concepts which in part owes its genesis to earlier work in the microwave spectral region, where phased array antennae have been in use for many years. In the optical arena, the array laser concept provides an attractive approach to laser power scaling, under conditions when technological constraints limit the power which may be extracted from an individual elemental laser device. In the conceptually simplest form of array, consisting of a set of independent laser oscillators, the output from each laser is uncorrelated with that of its neighbours, each operating at a different frequency and/or random phase. The total output power from N such identical oscillators in an 'incoherent' array is then simply N times the power of an individual laser, and spatial interference effects in the combined beam are avoided if the individual laser frequencies are well separated and time-dependent, due to high speed averaging. By contrast, when appropriate coupling is provided across the elements of the array to 'phase-lock' the individual elements so producing a collective coherent output, then the increased effective coherent aperture area is also increased by a factor N, so that the on-axis intensity (W.cm^2) scales as N^2. In addition, if some form of controllable phase modulation were to be provided to individual oscillators, then in principle, it would be possible to produce a self-scanning beam, analogous to a phased array microwave radar system. These potential advantages, plus the interest in the physics of coupled oscillation has stimulated considerable interest in developing technology capable of providing efficient coherent coupling in array lasers.

In recent years, the array concept has been extensively applied to semiconductor diode lasers as a route to laser power scaling [1]. In the diode laser, the onset of filamentation and/or device end facet damage imposes severe physical limits on the geometric scaling (length, width) of individual laser diode devices, whereas modern planar integration technology permits the fabrication of multi-element laser array

W. J. Witteman and V. N. Ochkin (eds.), Gas Lasers - Recent Developments and Future Prospects, 105–124.
© 1996 *Kluwer Academic Publishers.*

devices, either as 'end face' emitters or as 'surface' emitters (VCSLs). Arrays of carbon dioxide lasers have also been demonstrated. For example, Kozlov *et al* [2] reported the construction and operation of an array of some 78 parallel lasers in a 2-D array format which produced a total output power of 3kW. In addition, phase-locked operation of a pair of parallel dc discharge excited waveguide CO_2 was demonstrated with mutual coupling through a partially transmitting common waveguide wall [3], while Antyukhov *et al* used the free space Talbot effect to phase-lock a 2-D array of glass tube lasers [4]. However, this technology, based on longitudinal dc discharges and flowing gas is rather unwieldy, and it was not until the advent advances which occurred in the 1980s in transverse capacitive radiofrequency discharge technology, coupled with hollow waveguiding structures in the 1980s that significant progress was made in array integration, monolithic construction and phase-locked operation.

Stable phaselocking of a carbon dioxide waveguide laser array was demonstrated by Newman *et al* using RF discharge excitation and a ridge waveguide mutual coupling technique [5], and subsequently the same group were successful with a staggered waveguide structure [6]. Abramski *et al* demonstrated the extreme precision of the frequency offset [7] which may be maintained between adjacent channels of a twin waveguide laser arrangement excited by a single RF power supply in the transverse mode of excitation. Colley *et al* later determined the relative frequency shift [8] associated with amplitude modulation of the RF drive power, showing that frequency tuning as large as 1 MHz/Watt of RF power could be produced. A number of other techniques have been demonstrated for the production of phaselocked operation of waveguide CO_2 laser arrays, including simple (though inefficient) diffractive coupling [9], the use of spatial filters [10], continuous coupling in a modified 'diamond-shaped' waveguide laser array [11], and coupling produced from an optical intensity induced 'gas grating' in a 4-wave mixing experiment [12]. In addition, exploitation of the waveguide-confined Talbot effect, both in empty hollow planar structures [13], and in hollow slab structures with optical gain [14] has been used to produce coherent operation of planar arrays. This effect has also been used to generate 'array-like' optical beams from a slab laser oscillation in a single transverse, single (high order) lateral waveguide mode [15].

In this paper, we review some of the underlying features of a monolithic technology for the construction of high density (close-packed) linear and 2-D carbon dioxide waveguide laser arrays [16.17], discuss the scaling of incoherent arrays to kilowatt power levels, and briefly outline the array requirements and limitations for producing phase-locked operation.

2. CO_2 Waveguide Laser - Basic Module of the Array

The basic building block for the construction of multi-element planar 2-D carbon dioxide arrays is the square-bore waveguide laser excited by transverse radiofrequency discharges [18-20]. Several options exist for the construction of the single element waveguide laser, as illustrated by the possible device cross-sections shown in Figure 1. In (a) the metal electrodes are external to the waveguide, which consists of four ceramic surfaces, based on a square groove machined in a ceramic plate with a matching ceramic 'lid'. The 'lid' may be removed as in the structure shown in (b).

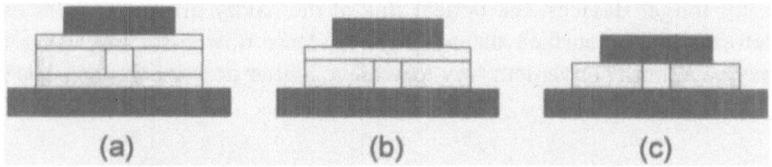

Figure 1   Schematic diagram of waveguide laser structures

which has three ceramic and one metallic waveguiding surfaces. Finally in (c) a pair of ceramic spacers is positioned between the metal electrodes, providing two ceramic and two metallic waveguiding surfaces. Although other options exist, the most convenient combination of materials for such fabrication is alumina (Al_2O_3) ceramic and aluminium alloy. Suitable waveguiding surfaces may be produced using conventional machining/grinding processes with waveguide dimensions typically in the 1.0-2.5mm range.

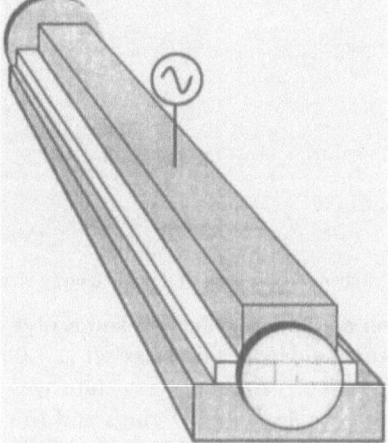

Figure 2   Schematic diagram of rf excited waveguide laser

When such a waveguide forms part of a dual Case I resonator design [21], as shown in Figure 2, with a pair of plane mirrors of suitable reflectivity positioned close to each end of the waveguide, then laser specific power output values of ~ 0.8-1.0W/cm of discharge may be achieved for waveguide gain lengths which allow the oscillators to perform well above threshold when excited by an RF generation of appropriate frequency [22,20]. Such devices operate conveniently sealed off, with gas cooling by heat diffusion to the electrodes, which require active water cooling. Under such conditions conversion efficiency of ~12% may be routinely achieved. Moreover, waveguide carbon dioxide lasers can readily be designed to operate on a single rotational-vibrational line, may exhibit highly stable outputs both in amplitude and frequency, and produce near perfect EH_{11} beams with M^2 values of ~ 1.1.

It is practical to construct single element waveguide lasers at lengths up to about 0.75m, beyond which the fabrication difficulties and costs begin to be excessive. Moreover, for longer devices, the optical flux at the cavity mirror surfaces may cause thermal deformation or surface damage. Thus, laser power scaling using the array concept may be a more convenient way to achieve higher power output ($\geq$ 100W).

3. Planar waveguide laser arrays

3.1. PLANAR ARRAY CONSTRUCTION

Using conventional ceramic machining techniques, an array of multiple parallel waveguides, of the type shown in Figure 1 (b) may be machined in a ground alumina plate and mounted between a pair of metallic electrodes, as illustrated by the schematic diagram in Figure 3. In the experiments to be described here, a square waveguide

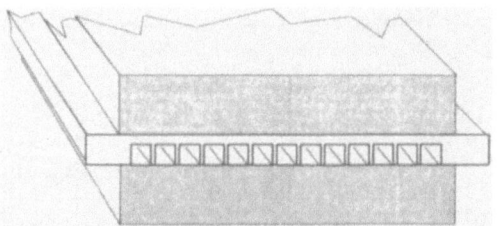

Figure 3   Schematic diagram of rf excited array waveguide laser

cross-section of 2.25mm on a side was selected, and arrays of 7 and 13 elements were fabricated, with the waveguide sidewall thickness set at 1.0mm for the 7-element array and 0.5mm for the larger device. The array structure was mounted in an aluminium vacuum enclosure, which was sealed with 'O' rings and filled to pressures in the range 65-110 torr with a gas mixture consisting of $3He:1N_2:1CO_2$ plus 5% Xe. The laser gas was cooled by diffusion to the waveguide walls, which were in turn cooled by thermal conduction to the aluminium electrodes. The latter were fabricated with internal water cooling channels. A single pair of plane mirrors, with reflectivity 99.8% and 92% respectively, simultaneously provided a Case I resonator configuration for all elements in the array. All the discharges were operated in parallel from a single 125MHz RF power generator and LC matching network, using distributed parallel resonance coils mounted inside the vacuum envelope to ensure longitudinal uniformity of the transverse interelectrode voltage.

3.2.   PLANAR ARRAY LATERAL UNIFORMITY

A major objective in designing array lasers is to construct an array which is close-packed (to achieve both device compactness and acceptable beam quality), but where the mutual proximity of adjacent elements does not result in significant reduction in the performance of individual channels in the array. In addition, a second design goal aims to minimise the element-to-element variation across the array in both power output, and (to achieve phaselocked operation) in the oscillation frequency of individual elemental lasers. These aims in turn demand the achievement of close control of the following parameters:

1) the precise waveguide channel cross section and length dimensions,
2) the matching of the RF electrical power input to the array to achieve equal power deposition in each element,
3) cavity mirror distortion
4) the waveguide wall temperature

The impact of the channel dimensional tolerance on the precise laser frequency, v_{mnq} may be determined from the well-known expression [23] for the frequency of the mnth waveguide mode, in a waveguide of cross-section axb and length, l:

$$v_{mnq} = \frac{qc}{2\ell} + \frac{c\lambda}{8}\left[\frac{m^2}{a^2} + \frac{n^2}{b^2}\right] \tag{1}$$

where q is an integer and c is the velocity of light. To find the dependence of the oscillation frequency of the EH_{11} mode on the channel dimensional tolerance in a square bore of side a, we have:

$$\frac{\Delta v}{\Delta a} = c\lambda/4a^3 \approx 70 \text{ kHz}/\mu\text{m} \quad \text{for } a = 2.25\text{mm} \tag{2}$$

Thus, to achieve phaselocked operation across multiple waveguide channels, considerable precision is required to bring the spread of natural frequencies within the ambit of the system locking range. For example, a precision of $\pm30\mu\text{m}$ in the uniformity of the dimensions of the guides is needed to achieve a frequency spread of 5MHz. It is estimated that dimensional variations due to temperature changes will introduce additional frequency shifts of the order of ~1MHz . K^{-1}.

The equality of balance of the RF power delivered to the individual channels has been estimated at about $\pm2\%$ from measurements of the channel-to-channel laser power output variation in a multi-element array. In addition, the impact on the laser oscillation frequency of variations in the RF power deposited in adjacent waveguide channels has been determined using heterodyne techniques [8], and values of the frequency shift of 1MHz/Watt have been measured.

The achievement of minimal lateral variations in the surfaces of the laser mirrors is also an essential to minimise the natural spread of the oscillator frequencies, as a pre-requisite to the achievement of stable phaselocking. Thus, it is necessary to exercise considerable care with respect to avoiding polishing errors, mounting distortion and opto-thermal effects in the mirror substrate. Tests on our laser resonator mirrors using an interferometer in the visible indicated that the distortions other than those associated with intracavity optical power (see below), were of a magnitude to produce effects which were smaller than those discussed above.

It is well known that the achievement of high specific power and efficiency in carbon dioxide lasers is strongly dependent on controlling the gas temperature [24]. Arguments based on the laser kinetics indicate that the axial temperature should be maintained below about 600K to avoid severe reduction in gain. There has been considerable effort in the past to model the temperature distribution in the discharge region [25-27], and in all cases, the profile is seen to approximate to a parabolic function. However, the assumption, which is valid in many practical cases, is that the walls of the discharge vessel are maintained at some constant and equal temperature which is close to that of the coolant. However, in the case of the close packed array cooled as described here,

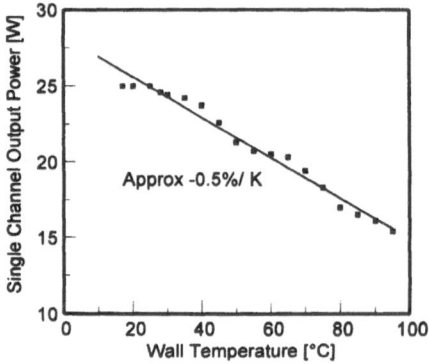

Figure 4   Power output of a single channel waveguide laser with temperature

there are several reasons why this assumption may not be valid. Thus, it is necessary to take account of several sources of thermal impedance, including that of the ceramic wall material, of the ceramic-metal electrode interface, and of the electrode-water interface. To assess the dependence of the laser power output on the wall temperature, a symmetric waveguide (2.25 x 2.25 x 386mm) was constructed with a dual Case I resonator, an 8% output coupling mirror and gas mixture and pressure similar to those used in the array experiments. The laser output power was measured as a function of the electrode temperature, which was varied by controlling the water coolant temperature. The results of laser power versus electrode temperature are shown in Figure 4, from which a power de-rating coefficient of $0.5\%.K^{-1}$ may be determined. Although this cannot be considered as an 'absolute' figure due to difference in laser output coupling, gain etc, it indicates the need to limit the temperature rise, due for example to internal thermal impedances to ≤10K, in order to achieve high performance laser devices.

3.3.   PLANAR WAVEGUIDE LASER ARRAY OUTPUT CHARACTERISTICS

The total laser power output of the two arrays has been measured as a function of the RF input power for a range of gas pressures, to determine the optimum conditions in each case. As shown in Figure 5, the 7-element array produces a maximum of 186W, corresponding to an average specific power per channel of 70.7Wm^{-1}, while for the 13-element array, the corresponding figures are 340W and 69.1 Wm^{-1}, both achieved at a

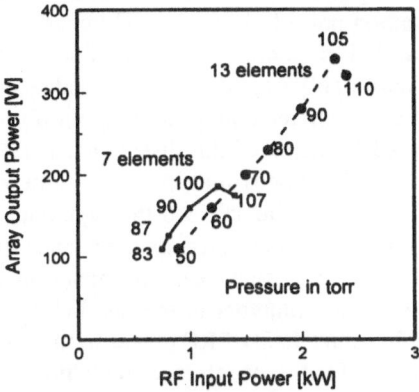

Figure 5   Power output for 7 and 13 channel arrays with input power optimised for the indicated gas pressure

gas pressure of 110 torr. As indicated above, the channel-to-channel laser power variation was ~2%. The variation across the array was investigated further by an experiment to determine the spread in oscillation frequency of the individual elements in the 7-element array. The seven outputs were allowed to spread and overlap on a room temperature HgCdTe detector (*Labimex* Model R004.1), which was connected to an RF spectrum analyser. The frequency spread of the heterodyne beat spectrum was measured to be 6MHz under optimum laser mirror alignment conditions. This rather narrow spread is consistent with the relatively high precision of the ceramic channel machining tolerances which has been achieved, as well as RF power input distribution uniformity of ~1% and wall temperature balance to within a few degrees. In section 6, thermally induced distortions of the laser mirrors produced by the intracavity optical flux is shown to be a significant contributory factor in increasing the frequency spread in the 7-element array, and is thought to be a severe problem for the achievement of phase-locking of large arrays. All elements in the array were observed to display virtually identical cavity tuning signatures, consisting of several rotational transitions near 10.6μm with the peak power obtained on the P20 line.

4.   Two Dimensional Waveguide Laser Arrays

Having demonstrated satisfactory uniformity in the power output and cooling in a 13-element monolithic planar array, it was decided to use this 'module' as the basic building block in the construction of a 2-D array, within the overall usable aperture limits set by the available 50mm diameter laser mirrors and output window. Thus, arrays of 2 x 13 and 3 x 13 elements were constructed with the aim of improving understanding of the rf distribution and cooling issues, and investigating the beam propagation properties of the 2-D laser array format. A major issue for the 2-D array design is the manner of electrical connection of the individual discharges for use with a single high power rf generator. All elements in a planar row of the matrix are electrically in parallel, with the channels sharing a common pair of electrodes. However, for a stack of two or more levels, either series or parallel rf connection is feasible. Series connection has the advantage of providing a reasonably high electrical impedance which is close to that of the 50Ω rf generator, and which aids in producing uniformity of the longitudinal voltage, due to the increased Q-factor of the distributed parallel resonance. However, the high striking and operating voltages may produce some difficulty with unwanted internal discharges, particularly at the rf feed-through and between electrodes and mirrors. In contrast, an all-parallel feed arrangement leads to a very low net overall load impedance, which gives rise to significant problems in obtaining longitudinal rf voltage uniformity using discrete inductors in the parallel resonance network. For the (convenient) case of a single centre feed of RF power to the array structure, it also leads to very high currents in the feed through and matching circuit components, with consequent rf power losses.

For the experiments described here, the series connection technique has been used, and precautions taken to minimise the associated problems. Figure 6 shows a schematic diagram of the cross-section of the three-level 2-dimensional array. The

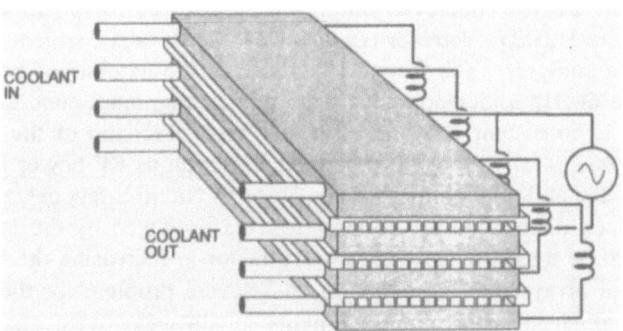

Figure 6   Schematic diagram of 2-D (3 x 13) waveguide laser array

rows of thirteen channels, machined into alumina plates, are clamped mechanically between water-cooled aluminium electrodes. No particular efforts have been made to minimise the thickness (8 mm) of the electrodes in the current design. Because of the high value of the striking voltage (~1 kV) in the series circuit, it was necessary to reduce the previously-used 376 mm electrode length to 340 mm, replacing the metal by ceramic inserts to maintain waveguiding to the mirrors. This course of action inhibits the ignition of unwanted discharges between the ends of the electrodes and the cavity mirrors, at some cost in reduced cavity fill factor. Figure 7 shows a comparison of the maximum output laser power versus rf input power for the single level, two level and three level arrays. The 26 element array produces a combined maximum output of 550W at 12% conversion efficiency, while the 39 element array correspondingly produces 750W at 11% efficiency. With allowance for the changed active length, the specific power output per channel reduces from the value of 69.1 W.m^{-1} for the 13 element, single level device to 61.8 W.m^{-1} for the two level array, and further reduces to 56.5 W.m^{-1} for three levels. In a similar manner to experiments with the 1-D array, a fast photodetector was placed in the focal plane, to perform beat frequency analysis of a sample of the combined beam from all elements of the 39 element array. The resulting beats spectrum from the multiple pairs of beams operating on the same rotational line shows a spread of 30-40 MHz, which is much larger than for the linear array. Unlike the 7-element linear array, it is not possible to obtain operation of the entire array on a single rotational line for any cavity length or mirror alignment. However, laser spectral signatures taken for individual beams by length tuning the resonator through a free spectral range, show switching between lines in the range P18-P24, in the same way as for a single channel laser of the same length.

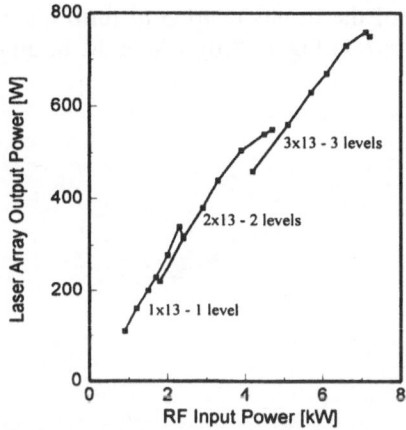

Figure 7   Power output for single (1x13), double (2x13) and triple (3x13) layer array laser

The combined output power varies by only ±0.5% when the cavity-length is tuned over a free spectral range, showing net power stability which is very much better than for a single channel laser. Most of the power variation which occurs *between* channels in the 39 element array can be accounted for by random detuning and line changes associated with fairly random positions of individual channels in the cavity length-tuning signature. The array could not be collectively peaked on the P20 line centre for maximum power output, as is the case for the 7 element linear array. Thus, the good total power stability may be explained by the averaging of the randomly positioned laser signatures of individual beams. Unfortunately, the averaging effect also reduces the power capability per channel by about 5% relative to a single channel laser peaked at P20, and this contributes to the reduction in specific power observed for the multi-level arrays. The sensitivity of the output power to mirror tilting is ~ 5% /mrad, which is comparable with the typical misalignment sensitivity of a single-element waveguide laser. To determine the beam quality of the 2-D array, the total output was focused using an off-axis spherical mirror of focal length 500 mm.

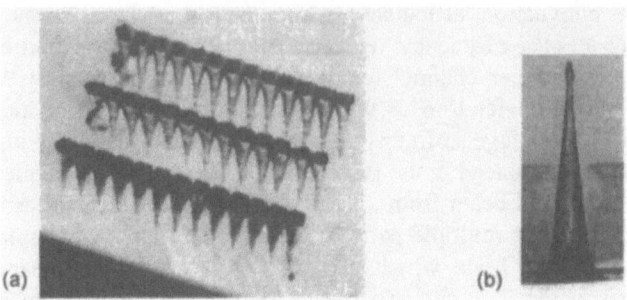

Figure 8   (a) Near field and (b) Far field burn patterns for 3 x 13 array

The near field pattern of the beam produced by the 3-level array reveals a grid of 39 equal burn patterns in a perspex block, as shown by the photograph in Figure 8(a). This result was obtained at the image plane created at a distance of 1.5 m from the laser, whereas in the focal plane all of the beams overlap to form a single lobed profile, as shown by the perspex burn pattern in Figure 8(b), where the beam diameter at the $1 / e^2$ point is ~ 4.5 mm.

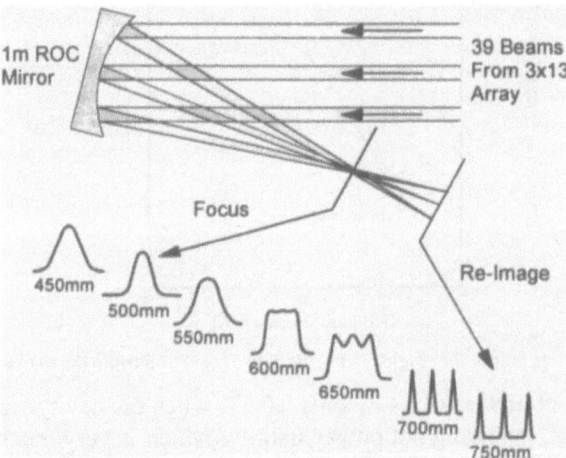

Figure 9   Beam profiles sampled at 50mm intervals through the mirror focal plane

In Figure 9, 1-D beam profiles at varying distances in a line through the focus are shown, and it can be seen that a near 'flat-top' profile is obtained in the region between the focus and re-image planes. A beam profile of this type may be particularly useful for materials processing applications, for surface treatment of large areas, or to avoid edge effects.

The results of these experimental indicate several areas which require further investigation   In particular, interest is centred on determining the causes of the significant (~30%) reduction in both specific power per channel and conversion efficiency for the largest 2-D array as compared to the values obtained with a single isolated channel, as well as for the increased spread of laser frequencies which was observed for the larger arrays.

5.   Thermal Effects in Array Structures

The observed reduction in the specific power obtained with the 2-row and 3-row array lasers was thought likely to be associated increased temperature of the channel walls, suggesting that further analysis is necessary to determine the impact of the various thermal impedances in the metal/ceramic structure referred to above.

5.1.   WALL TEMPERATURE RISE IN ARRAYS

To determine the approximate temperature distribution across a transverse section of the ceramic structure, a thermal imaging camera was used to view and make measurements of the end of the waveguide array structure under conditions where the discharge was running, but with one of the laser resonator mirrors removed. Such measurements serve to indicate the magnitude of the thermal impedance effects, although the end section of the ceramic is not fully representative of the bulk of the structure. The results showed a maximum temperature difference of approximately 40 K between the waveguide separating walls and the temperature of the water-cooled electrode, for the 1 x 13 element array, with much of the temperature rise appearing across the ceramic-metal interface.   With one waveguide wall consisting of a directly water-cooled metal electrode, the mean wall temperature is less than the peak ceramic temperature and is typically 30K, which according to the data in Figure 4 corresponding to ~15% derating of the laser power capability per channel. This is consistent with the reduction in specific output power from ~80 W.m^{-1} for a single waveguide to ~70 W.m^{-1} for a waveguide within the 13 element array.

To identify the sources of thermal losses, the array structure has been analysed in detail, using the structure and dimensions shown in Figure 10. Three thermal impedance effects are considered, those associated with the temperature gradients within the ceramic, the temperature drop across the ceramic/metal interface and the temperature drop across boundary layers within the cooling water channels. Compared to these, the thermal resistance of the aluminium electrodes is negligible. The ceramic-metal interface is considered first, since the thermal imaging camera measurements showed this to have a large temperature gradient. The microscopically rough interface surface consists of solid contact points surrounded by gas pockets, both of which can contribute to thermal conduction. Thermal contact may be improved by applying a clamping force to maximise the area of solid contact and to expel the gas layer, and by preparing a better surface finish to increase the number of contact points and reduces the gas thickness. For a relatively poor surface finish and only moderate clamping force, thermal conduction is mainly across the gas layer whose mean thickness is determined by solid contact at relatively few microscopic high points of the surface, and this worst case view is used here to estimate the interface thermal impedance. A reasonable estimate of the mean gas gap for these conditions is $\delta \approx 10\mu m$. The thermal conductivity, k, of a $3He:1N_2:1CO_2$ gas mixture is 7.10^{-2} W.m^{-1} K^{-1} at 300K [16], which reduces to $\sim 6.10^{-2}$ W.m^{-1} K^{-1} with allowance for xenon in the gas mixture. The interface thermal resistance is then estimated to be approximately $\delta / k \approx 1.67.10^{-4}$ K.m^2 W^{-1}.

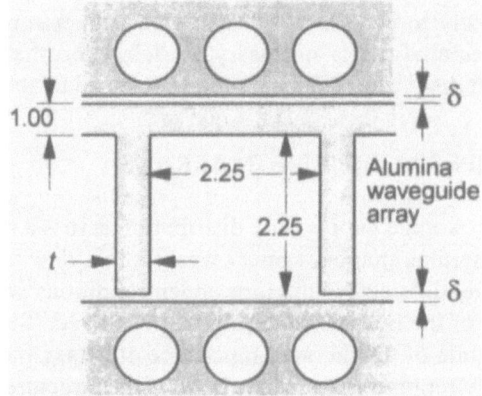

Figure 10 Waveguide channel structure and dimensions for thermal analysis

5.2. MODELLING OF WALL TEMPERATURE RISE

A commercial finite element analysis package (ANSYS) has been used to model the temperature distribution within the channel walls, assuming equal interface thermal resistance at both sides of the ceramic plate, and uniform discharge heating over the four surfaces of the waveguide. Assuming the thermal conductivity of the alumina ceramic to be 28 W.m^{-1} K^{-1}, with 530 W.m^{-1} of power dissipated in the discharge, the temperature contours relative to the metal electrode temperature within one section of the array may be determined, using $\delta / k \approx 1.67.10^{-4}$ K.m^2 W^{-1}, for the case where $t = 0.5$mm, as shown in Figure 11. To assess the power derating, the mean wall temperature rise for the four walls is calculated from the detailed contour maps. Figure 12 shows this as a function of interface thermal resistance, for values of t of 0.5mm and 1mm. For comparison, Figure 12 includes data for the single element waveguide construction with 10mm wide ceramic walls. The temperature rise in this case is much smaller, due to the efficient heat spreading within the side-walls, resulting in negligible power derating.

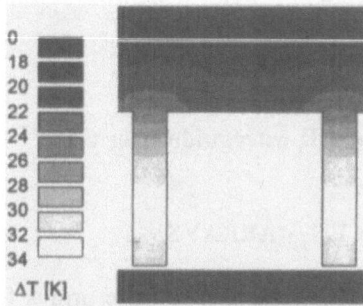

Figure 11   Temperature contours within ceramic array structure
(interface thermal impedance 1.67×10^{-4} K.m^2.W^{-1})

The principal conclusion is that the thermal impedance at the ceramic-metal interface is a major source of temperature rise and laser power derating for close-packed arrays, in agreement with the thermal imaging observations. The mean wall temperature rise predicted with reasonable estimates of the interface resistance is of the correct order to explain much of the observed derating in specific power of 1-D array devices.

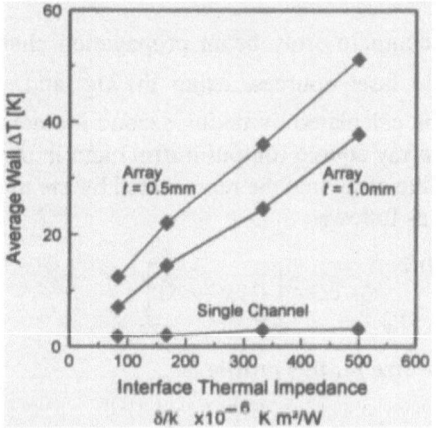

Figure 12   Mean wall temperatures of single and array structures vs interface thermal impedance

The model indicates that the impact of the small thickness, t of the dividing wall has proved to be a less important contributor to thermal impedance, consistent with the slight reduction in specific power from 70.7 W.m^{-1} with 1 mm walls to 69.1 W.m^{-1} for 0.5 mm walls.   The 0.5mm walls are close to the practical minimum for diamond grinding of the channel structure, but still provide effective "four wall cooling" of the channels with a nearly close-packed array.   Finally the effect of the impedance between the electrode material and the cooling water has been estimated to produce a temperature increase of 10-15 K for the two outer electrodes and twice this value for the inner electrodes of the 3 x 13 array, since in the present electrode design, only a moderate water flow rate is possible, preventing strongly turbulent flow. The use of a more complex water channel arrangement would produce reductions in the electrode thermal impedance, and is expected to provide a significant improvement in 2-D array output power.

6.   **Beam characteristics of 2-D waveguide laser array**

6.1.   BEAM QUALITY FOR EH_{11} ARRAYS

In order to assess the potential of 2-D incoherent laser arrays for applications such as material, it is necessary to provide a quantitative measure of the quality of the composite array beam [28]. As shown above, the array is composed of an assemblage of parallel, near perfect ($M^2 \sim 1$) EH_{11} waveguide laser beams, which can be focused to a single circular spot with a lens. However the corresponding depth of focus is much less than for a conventional laser, which in this mode of operation, restricts such beams to processing very thin materials. Alternatively, it may be used for drilling multiple parallel holes, or as a single flat-top beam for producing thermal surface modifications with good edge definition.

It is useful to compare array beam propagation characteristics with those of conventional multi-mode laser sources, using the M_x^2 and M_y^2 parameters [28,29]. These parameters may be calculated by taking second moment integrals over the beam intensity profiles at the array source (output mirror) and in the far-field (focal plane of the lens). At the source, the array may be represented by the intensity profile (for an odd number of channels, N_x as follows :

$$I(x) = \sum_{n=-(N_x-1)/2}^{(N_x-1)/2} I_0 \cos^2\left[\frac{\pi}{a}(x-nd)\right] \qquad (3)$$

The second moment spot size for this profile is:

$$W_x = \frac{N_x d}{\sqrt{3}}\left[1 - \frac{1}{N_x^2} + 0.392\frac{a^2}{N_x^2 d^2}\right]^{\frac{1}{2}} \qquad (4)$$

In the far-field of an ideal incoherent array, the time averaged intensity profile is the same as one beam, as all the beams are emitted parallel to each other. The elements each emit in a single EH_{11} waveguide mode with a far-field profile, obtained by Fourier transform, of:

$$I(\theta_x) = I_0 \cdot \frac{4\pi a^2}{\left(\pi - 4\pi a^2 \theta_x^2 / \lambda^2\right)^2} \cos^2\left[\frac{\pi a \theta_x}{\lambda}\right] \quad (5)$$

Calculation of a second moment angular size for this profile gives:

$$\Theta_x = \lambda / a \quad (6)$$

The beam quality parameter in the x direction is then given by:

$$M_x^2 = \frac{\pi}{\lambda}\Theta_x W_x = \frac{\pi N_x d}{\sqrt{3}a}\left[1 - \frac{1}{N_x^2} + 0.392\frac{a^2}{N_x^2 d^2}\right]^{\frac{1}{2}} \quad (7)$$

For the case of $N_x = 1$, equation (7) yields $M_x^2 = 1.13$, and the expression approximates to $\pi N_x d / \sqrt{3}a$ for large N_x (Note that for a close packed array with $a = d$, this becomes $M_x^2 \approx 1.81 N_x$, a result given by Siegman for coherent arrays [30]). In terms of beam quality, it is clearly beneficial for the array to be close packed, but this is only really practical in one direction since the second direction requires sufficiently large values of d for water cooling channels to be interleaved with the active regions. For the 13 x 3 channel laser, $a = 2.25$ mm and $d = 2.75$ mm in the horizontal axis, giving $M_x^2 = 32$. In the vertical direction, $d = 11.25$ mm and $M_y^2 = 25.6$. The M^2 characterisation of the array beam, combined with the second-moment-derived beam waist size, defines an equivalent multi-mode Gaussian beam which is particularly useful for the design of beam transport optics [28]. The analysis given above allows a clearer comparison to be made between the incoherent array and more conventional multi-mode lasers, particularly with respect to array format and channel spacing. The above beam quality analysis assumes ideal array behaviour, which may not always be achieved in practice. Higher order mode operation may occur in the individual waveguide channels through inappropriate choice of guide width and cavity length, or may be induced by mirror tilt relative to the guide axis. In addition, the beams may not be emitted exactly parallel due to output mirror distortion, and thermal lensing in the mirror substrate. These effects act to enlarge the far-field beam size and degrade the beam quality. The 2.25 mm alumina waveguides used in the arrays are known from previous work to produce a good EH_{11} mode, for the present cavity arrangement. The beam profile presented in Figure 8(b) confirms this view, as the measured far-field spot diameter of 4.5 mm for the focus of the *whole array* is close to the value of 4.7 mm, calculated using Equ. (6) and a focal length of 500 mm. Within a row, angular errors between the machined channels are assessed as less than 150μrad. from the machining tolerances. Between rows, angular errors of ± 0.5 mrad. are possible due to electrode parallelism and ceramic plate alignment errors. The largest of these errors is just on the edge of producing higher order mode operation and power loss through angular misalignment of the channel with the common resonator mirrors [31]. Tilt induced mode-coupling will slightly affect the direction of the external beam, but results in angular errors much less than the divergence of a single channel (9.4 mrad. full angle) and which are not sufficient to significantly spread the far-field pattern. In summary, the *fixed* errors in the present array construction are all small enough to produce no degradation of the beam quality in incoherent operation. In the following section, the effect of power induced optical errors are considered.

6.2.   THERMALLY-INDUCED DISTORTION OF THE RESONATOR OPTICS

The high intracavity circulating powers, combined with the finite losses in the resonator optics coating and substrate may lead to significant thermal distortion of both the high reflectance Si and partially reflecting ZnSe mirrors. A particular issue for large arrays is how such thermal distortion scales with array size, and the manner in which it impacts both internal mode structure and the external beam quality. The thermal gradient between the centre of the output coupling mirror and its cooled periphery causes stress-induced *bowing* of the mirrors, with consequent relative tuning of the cavity length between the individual waveguide elements, which contributes to the spread of longitudinal mode frequencies or even to operation on different lines. Extreme distortion of the resonator optics may produce increased resonator losses through misalignment and operation in a higher order waveguide mode mixture [31].

Modelling of thermal distortion effects in the resonator optics has been carried out using finite element analysis. The 50mm diameter circular resonator optics are edge-cooled by a 2.0mm overlap with the mirror mounts, and the contact thermal impedance is assumed to be constant at the periphery due to the high quality of surface finish of the mating components. Free expansion of the mirror edges is allowed. For the silicon mirror, a surface absorption of 0.2% is assumed, and the high thermal conductivity, low expansion coefficient and high modulus of elasticity lead to both very small thermal gradients and deformations of the reflecting surface. For the ZnSe mirror, the thermal loading is more complex, since the 92% reflectance coating experiences the full in-cavity power (0.13% loss), while the bulk material (4×10^{-5} mm^{-1} absorption) and AR coating (0.13%) only encounter the out-coupled radiation. For all arrays, the ZnSe mirror displays thermal distortion at least an order of magnitude larger than silicon. (The thermal and mechanical properties of ZnSe are very much inferior to silicon [32])

Results for the analysis for the ZnSe mirror in the 7 element linear array are shown in Figure 13(a). The heat load is approximated to be uniform over the area of the array beam. The centre element is displaced by 130nm relative to the end elements, corresponding to 9.3 MHz change of cavity frequency, reasonably in agreement with the 6 MHz frequency spread recorded experimentally. For the 13-element linear array, the displacements are proportionately larger, corresponding to a frequency difference of 21 MHz between centre and edge. For the 3-level array, the silicon mirror is found to have a temperature difference of 4.1 K between a corner and centre element, with a corresponding displacement of 24nm.   For the ZnSe output coupler, the central temperature is 28.2 K above the edge and the stress/strain analysis gives the axial displacement map in Figure 13(b), with a cavity length difference between central and corner waveguides of 290nm. This will induce a length tuning of ~21 MHz, and under some circumstances may cause line hopping.

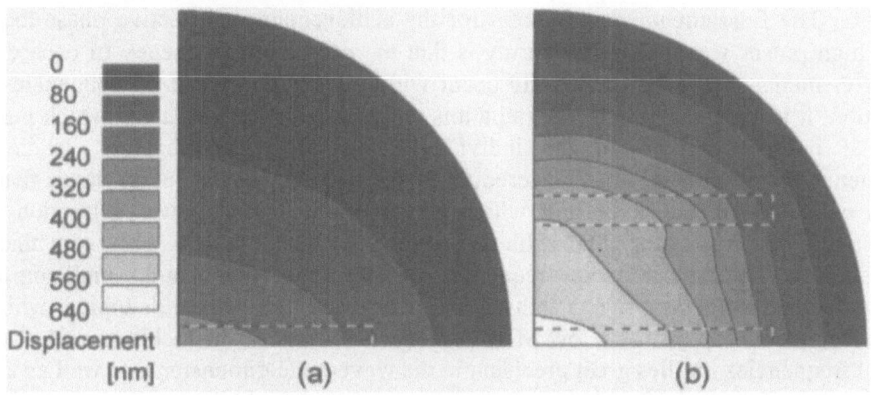

Figure 13   Axial displacement of ZnSe output mirror under thermal load of
(a) 7-element planar array   (b) 3x13 element array

With the mirror well aligned with the central waveguide, a corner element has a thermally induced misalignment of a~14 μrad. which is insufficient to lead to power loss or mode changes, and hence will not degrade the beam quality.

The temperature profile in the ZnSe, combined with the temperature dependence of its refractive index produces a thermal lens acting on the external array beam.   At the current level of in-cavity power in the array, this is weaker than the degree of output coupler lensing, which occurs in many conventional high power industrial CO_2 lasers, and is largely aberration free. Consequently, there is no significant loss of beam quality, as the slight convergence of the external array beam may be compensated in the subsequent focusing optics.   The application of finite element analysis to the resonator optics of the array laser has shown that there are no serious optical limitations on the construction of much larger 2-D *incoherent* array lasers.   The main effect will be increased thermal bowing of the output coupler which will act to spread the range of channel frequencies further, contributing to the randomisation of emission between lines in the CO_2 spectrum.   This has a major impact on the prospects for achieving stable phase-locking of large arrays, as discussed further below.

7.   Phaselocked operation of waveguide CO_2 laser arrays

The achievement of stable phase-locking of waveguide carbon dioxide laser arrays could result in major advances in high power laser technology, if the array fabrication technology and the phase-locking technique adopted were truly scalable in a practical sense. To date, no such technology has been developed, but it is instructive to examine the requirements imposed on the basic array design for the achievement of stable phase-locking.

The fundamental requirement for the achievement of effective phase-locking of a high power waveguide laser array is that the oscillation frequency of each of the array elements should be stable, and occur within a limited spread of frequencies. In addition, it is necessary that some radiation coupling technique be used, which permits mutual interaction between the individual array element resonators, leading to frequency locking of the array. Moreover to achieve high power levels, these features must co-exist with the conditions which permit efficient high power extraction from each individual oscillator. To achieve stable phase-locking, it is necessary that the spread in the 'natural' oscillation frequencies of the array elements must small compared to the locking range associated with the frequency coupling scheme adopted, which in practice means a spread of a few MHz. As discussed above, the achievement of near-equal frequencies implies great precision in the waveguide dimensions, as well as a high degree of uniformity in the electrical power deposited in, and heat energy removed from each gain element, and in the flatness achieved in the resonator optics, in the face of high levels of intra-cavity optical power.

As we have seen, these requirements are very difficult to achieve in structures such as those described in this paper. In particular, the effects of mirror distortion (mirror bowing) causes frequency shifts which are outside the locking range of existing optical coupling schemes, so that the prospects for achieving phase-locking of large arrays are not good. In previously work on phase-locking of linear arrays by the waveguide-confined Talbot effect [13], we have found that 13 and 19 element arrays phase-lock in spatial sub-groups of different frequency or spectral line. This is now understood in terms of perturbations caused by the bowing of the output mirror, and its effect on channel resonant frequency.

8.   Conclusions

The results presented demonstrate that compact, monolithic waveguide carbon dioxide laser array structures may be fabricated, which are capable of high average power levels with reasonable efficiency, using simple plane mirror resonators. The results have achieved ~70% of the maximum predicted attainable power capability for square bore waveguide channels. The study of thermal impedance derating of the specific power per channel has shown where improvements may be made in the engineering of the array structure, to achieve a specific power nearer to that of a single waveguide laser and to improve the conversion efficiency. In the future, significantly longer and wider ceramic array sections may also be used which are still within the capability of the present fabrication technique, and more levels employed to provide multi-kilowatt output power. For example an array of 60 channels, each emitting ~40 W seems possible, without significant changes in layout or fabrication techniques.

It is generally desirable that all beams in an *incoherent* array operate at widely differing frequencies to avoid interference effects in the integrated output beam, caused for example by phase-locking induced by back scatter from a workpiece in materials processing. In the present 2-D array, randomisation of the channel frequencies occurs for the range of reasons discussed earlier, but has not been deliberately controlled. In future, frequency randomisation could be ensured by design variations of the channel dimensions and by controlled static mirror distortion.

Mirror distortion problems are expected to limit the possibility for phase-locking of the large arrays, but not to be a limit on the beam quality for incoherent operation. The beam quality of the present array is close to that of an ideal array of EH_{11} waveguide emitters, with no measurable degradation due to multi-mode operation, mirror thermal distortion or constructional errors. The future development of optical techniques for close-packing and reformatting of the array beam will be aided by the good beam quality of individual array elements, and their high position accuracy and parallelism. Following reformatting, the array laser will provide an intermediate beam quality, compact and sealed-off laser source which will be particularly useful in material processing for surface treatment applications.

* **Present address**	Technical University of Wroclaw,
	27 Wyspianski Str., 50-370 WROCLAW, Poland

9.   References

[1]   M Sakamoto, J Endriz, D R Scifres *Electron. Lett.* **28**, 197, (1992)

[2]   G Kozlov, V Kuznetsov, V A Masyukov *Sov. Tech. Phys. Lett.* **4**, 53 (1978)

[3]   D G Youmans Appl Phys Lett **44**, 365 (1984)

[4]   V V Antyukhov, A V Globa, O R Kachurin, F V Lebedev, V V Likhanskii, A P Napartovich, V D Pis'mennyi *JETP Lett.* **44**, 78, (1986)

[5]   L A Newman, R A Hart, J T Kennedy, A J Cantor, A J Demaria, W B Bridges *Appl. Phys. Lett.* **48**, 1701 (1986)

[6]   R A Hart, L A Newman, A Cantor, J Kennedy Appl Phys Lett **5**, 1057 (1987)

[7]   K M Abramski, A D Colley, H J Baker, D R Hall IEEE J Quantum Electron **QE-26**, 711 (1990)

[8]   A D Colley, K M Abramski, H J Baker, D R Hall   IEEE J Quantum Electron **QE-27**, 1939 (1991)

[9]   A D Colley PhD Thesis Heriot-Watt University (1990)

[10]   C Lescroart, R Muller, G L Bourdet Paper CWJ15, *CLEO Conference*, Baltimore (1993)

[11]   K Abramski, A D Colley, H J Baker, D R Hall Appl. Phys Lett **60**, 530 (1992)

[12]   K Abramski, A D Colley, H J Baker, D R Hall Opt Comm **90**, 61 (1992)

[13]   H J Baker , A Hornby, D R Hall App Phys Lett **63**, 2591 (1993)

124

[14]   A M Hornby, H J Baker D R Hall Opt Comm **108**, 97 (1994)

[15]   K Abramski, H J Baker, A D Colley, D R Hall Appl Phys Lett **60**, 2469 (1992)

[16]   A D Colley, K M Abramski, H J Baker, D R Hall Paper CWJ6, *CLEO   Conference*, Baltimore (1993)

[17]   K Abramski, H J Baker, A D Colley, D R Hall Paper Submitted to IEEE J   Quantum   Electron (1995)

[18]   J Lachambre, J McFarlane, G Otis, P Lavigne, Appl Phys Lett **32**, 652 (1978)

[19]   D He and D R Hall Appl Phys Lett **43**, 726 (1983)

[20]   M B Heeman-Illieva, Y B Udalov, K Hoen, W JWitteman Appl Phys Lett **64**, 673 (1994)

[21]   John J Degnan, Denis R Hall IEEE J Quant Electron **QE-8**, 901 (1973)

[22]   D He and D R Hall, IEEE J Quantum Electron **QE-20**, 509 (1984)

[23]   D R Hall and C A Hill Chapt.3 (pp 165-258) in "Handbook of Molecular Lasers", ed. P K Cheo, Marcel Dekker, New York, 1987

[24]   W.J. Witteman "The CO_2 Laser" Chapt. 4, Springer Series in Optical Science, Springer Verlag, Berlin, 1987,

[25]   A J Laderman, S R Byron *J. Appl. Phys.* **42**, 3138 (1971)

[26]   R C Sharp *J. Appl. Phys.* **64**, 545 (1988)

[27]   X S Zhang, H J Baker, D R Hall   *J. Phys. D: Appl. Phys.* **26**, 359 (1993)

[28]   M W Sasnett in *The physics and technology of laser resonators*, pp132-142, Ed. D R Hall and P E Jackson, Adam Hilger, Bristol, (1989)

[29]   A E Siegman *Proc. SPIE* **1224**, 2 (1990)

[30]   A E Siegman, *Opt.Lett.* **18**, 675 (1993)

[31]   C A Hill and A D Colley, IEEE J. Quantum Electron. **QE-26**, 323 (1990)

[32]   D J Dyson in *The physics and technology of laser resonators* Chapt. 13. ed. D R Hall and P E Jackson , Adam Hilger, Bristol (1989)

RF EXCITED SLAB WAVEGUIDE ARRAY CO_2 LASER WITH MUTUAL INJECTION PHASE COUPLING

J.G. XIN S.Q. ZHOU Z.Y. WANG P. WANG
X.Y. PENG G.Z. FANG
Engineering Optics Department
Beijing Institute of Technology
P.O.Box 327, Beijing, china

1. Abstract

In this paper, a new technique of slab waveguide array lasers is presented, which combines the advantages of square waveguide array lasers and area scaling techniques to provide a volume scaling laser device and to show the prospect of producing more compact and sealed off gas lasers.

2. Introduction

Since D.Youmans first[1] reported a leaky mode waveguide array CO_2 laser in 1984, various techniques of square waveguide array lasers have been reported[2-4]. For a square waveguide array laser, each square waveguide produces a fundamental mode and the waveguides of the array are phase coupled to produce a coherent beam output. The square waveguide array lasers have shown the prospect of achieving high power and compact sealed off gas lasers.

In 1989, Denis Hall[5] presented a RF excited diffusively cooled area scaling technique, a slab waveguide laser coupled with an off-aixis telescope optical resonator, with this technique, he has reported a kilowatt CO_2 laser in 1992[6].

In this paper, we shall present a technique of RF excited diffusively cooled slab waveguide array lasers which combines the advantages of square waveguide array and area scaling techniques to provide volume scaling laser devices and has the potential to achieve multikilowatts diffusively cooled compact and sealed-off CO_2 lasers.

In our experiments, two types of RF excited diffusively cooled slab waveguide array lasers with mutual injection coupling were employed. In the experiments of the first type of the slab waveguide array laser, an 1×2 slab waveguide array with a case I optical waveguide resonator[7] and mutual injection phase coupling is demonstrated, with which a phase coupled power output of 44 watts and a phase uncoupled power output of 66 watts were obtained from an only 200 milimeters gain length device, a two dimensional interference pattern has been observed at the far field of the output beam which demonstrated the phase coupling of the slab waveguide array. In the experiments

W. J. Witteman and V. N. Ochkin (eds.), Gas Lasers - Recent Developments and Future Prospects, 125–133.
© *1996 Kluwer Academic Publishers.*

126

of the second type of the slab waveguide array laser, an 1×2 slab waveguide array with two off-axis telescope optical resonators and mutual injection coupling is demonstrated, with which a phase coupled power output of 860 watts and a phase uncoupled power output of 1020 watts were obtained from an only 500 milimeters long electrode device.

3. Experiments

In this section, the experiment setup and results of two types of 1×2 slab waveguide array lasers with mutual injection phase coupling are presented, the first is a slab array with a case I waveguide optical resonator, the second is a slab array with two off-axis telescope resonators.

3.1. SLAB ARRAY WITH A CASE I WAVEGUIDE OPTICAL RESONATOR

The schematic diagram of the slab array laser with a case I waveguide optical resonator and mutual injection coupling is shown in Fig.1. The structure consists of three polished

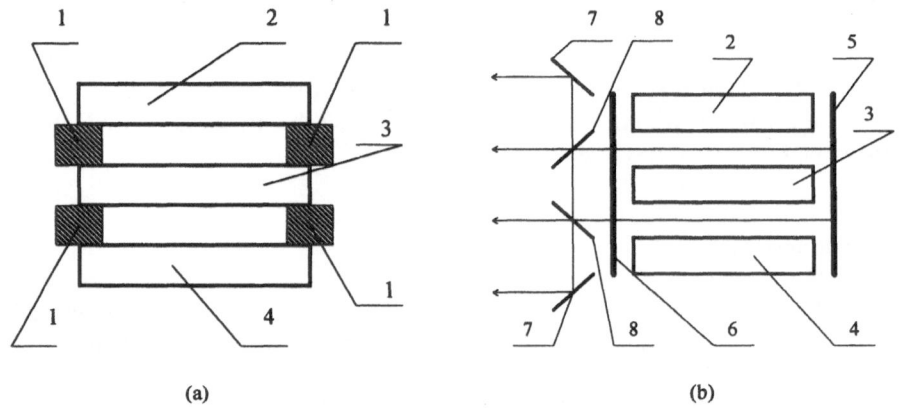

(a)   (b)

Fig.1 Schematic diagram of slab waveguide array laser head

(a) transverse cross section of the array   (b) optical resonator setup of the array
1. ceramic side wall   2. top electrode   3. central electrode
4. bottom electrode   5. total reflector   6. output coupler
7. 45 degree total reflector   8. 45 degree beam split mirror

aluminium electrodes and four alumina plates, which form two slab waveguides (discharge regions). Each discharge region has the cross section of 14.5 milimeters wide and 1 milimeter high, the length of the electrodes is 200 milimeters. The two discharge regions are separated by 9 milimeters. The top and bottom electrodes are grounded, and the central electrode is a RF feed electrode. The three electrodes were all cooled with tap water.

In the experiments, a technique of one dimensional limit parameter waveguide array[4] was used to obtain the spatially suppressed output beam intensity distribution at far field in a horizontal direction from each slab waveguide, and a technique of mutual injection phase coupling was employed to couple the field of the two slab waveguides. The phase coupling is determined by observing the interference field in the overlapping area of the beams from the two slab waveguides, because only when the two optical beams are in the same frequency and phase coupled, can an interference pattern be observed. The output beams from the two slab waveguides were focused with a lens of 1200 milimeters focal length, and the intensity distributions at the focus point were measured by using a linear laser beam intensity distribution scanner. The experimental setup is shown in Fig.2.

In experiments, a resonator, as shown in Fig.1, consisting of a ZnSe flat mirror with a reflectivity of 94.5% as an output coupler and a Si flat mirror with a reflectivity of 99.89% as a total reflector, which were separately positioned a few milimeters apart from the waveguide ends. Two 45 degree Ge flat mirrors with a reflectivity of 20%, as coupling mirrors, were used to direct the split output beams of each slab waveguide into the neighbouring slab waveguides for mutual injection coupling. Two 45 degree copper mirrors with a reflectivity of 98.5% were employed for the split output laser beam collection.

In our experiments, a RF power supply at a frequency of 99 megahertz was used to excite the gas medium in the two slab waveguides synchronically, a

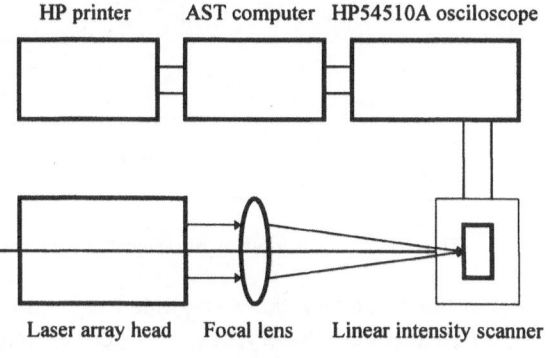

HP printer   AST computer   HP54510A osciloscope

Laser array head   Focal lens   Linear intensity scanner

Fig.2

matching circuit of LC was employed to deliver the RF power effectively into the discharge plasma and to keep the reflected RF power under 5% of the input RF power, and a gas mixture of CO_2:N_2:He in a ratio of 1:1:3 with an addition of 5% Xe was used. With the above laser head structure, a phase uncoupled power output of 60 watts and a phase coupled power output of 44 watts were obtained at a gas pressure of 14×10^4 Pa and a RF power input of 600 watts.

In the experiments, when the coupling mirrors (see Fig.1) were removed, the two slab waveguides were not coupled with mutual injection, the intensity distributions of the ouput fields of the slab waveguide array at the focal point are measured across the field center in the horizontal and vertical directions respectively, which are shown in Fig.3. It can be clearly seen that a spatially suppressed intensity distribution in a horizontal direction can be obtained (see Fig.3a), which is the overlapped field of the output beams of the two limit parameter waveguide arrays[4], and a near waveguide fundamental mode intensity distribution in a vertical direction was obtained (see Fig.3b), which is the intensity distribution of the overlapped field of two fundamental waveguide modes.

When the coupling mirrors were not removed , the two slab waveguides were coupled by the mutual injection, a pattern of interference fringes can be observed at the focus point of the focal lens as shown in Fig.4, it clearly indicates that the two beam from the two slab waveguides are phase coupled. The intensity distribution of the output beams in the horizontal direction at the focus point remained the same as shown in Fig.3a. In the experiments, the separation of the two lab waveguides (2d) is 9 mm, the distance between the waveguide end and the focal lens (Z) is 760 mm, the focal length of the lens is 1200 mm and distance between the lens and the detecting plane (L) is 1200 mm.

For the understanding of the pattern of interference fringes at the focus point as

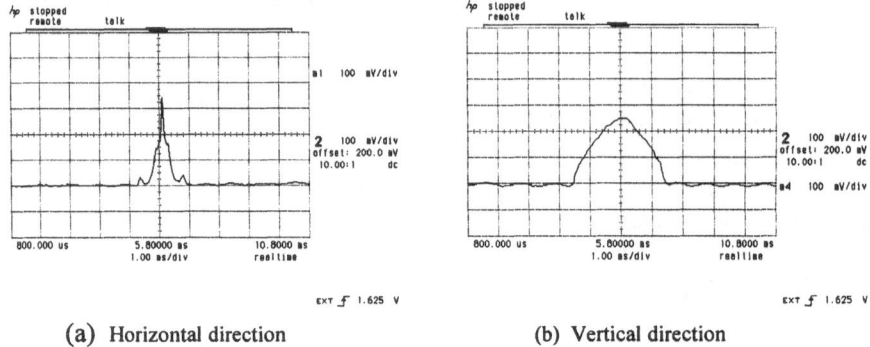

(a) Horizontal direction   (b) Vertical direction

Fig.3 Intensity distribution of slab array output without mutual injection coupling

shown in Fig.4, we take the waveguide EH_{11} modes as Gaussian fundamental modes in terms of Henderson's theory[8] and cosider an interference of two Gaussian beams at

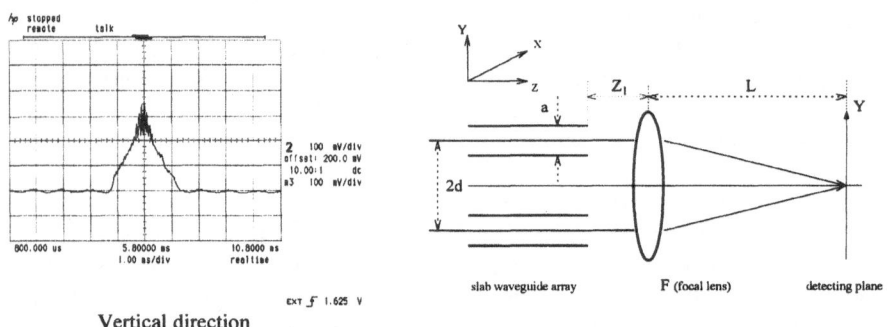

Vertical direction

Fig.4 Intensity distribution of slab array
with mutual injection coupling

Fig.5 illustration of beam overlapping
at detecting plane (focus point)

the focus point as illustrated in Fig.5, because in practice the side two output beams directed by the two 45 degree total reflectors are very week compared with the two central output beams (see Fig.1) and can be neglected. Therefore, the interference field at the focus point can expressed as:

$$I(y) = |E_1(y)|^2 + |E_2(y)|^2 + 2E_1E_2 COS(\phi_1 - \phi_2) \qquad (1)$$

$$E(y)_m = E_{0m}\frac{w_0}{w(t)}\exp[-(\frac{yL}{w(t)\sqrt{d^2+L^2}})^2] \qquad (m=1,2) \qquad (2)$$

where w_0 is the beam waist radius at the waveguide outlet ends, $w(t)$ is the beam radius at the detecting plane, E_0 is the magnitude of the optical wave at the waveguide outlet ends, $E(y)$ is the manitude of the optical wave at the detecting plane, and the other parameters are defined in Fig.5.

$$\phi_1 = kt + \frac{yd}{L} - \tan^{-1}(\frac{t\lambda}{\pi w_0^2}) + (\frac{yL}{\sqrt{d^2+L^2}})^2(\frac{k}{2R(t)}) \qquad (3)$$

$$\phi_2 = kt - \frac{yd}{L} - \tan^{-1}(\frac{t\lambda}{\pi w_0^2}) + (\frac{yL}{\sqrt{d^2+L^2}})^2(\frac{k}{2R(t)}) \qquad (4)$$

$$w^2(t) = \frac{\lambda}{\pi}\{\frac{(1-\frac{Z}{f})^2(\frac{\pi w_0^2}{\lambda})^2 + [Z+\sqrt{d^2+L^2}(1-\frac{Z}{f})]^2}{(\frac{\pi w_0^2}{\lambda})}\} \qquad (5)$$

$$R(t) = \frac{(\frac{\pi w_0^2}{\lambda})^2(1-\frac{Z}{f})^2 + [Z+\sqrt{d^2+L^2}(1-\frac{Z}{f})]^2}{\frac{1}{f}(\frac{Z}{f}-1)(\frac{\pi w_0^2}{\lambda})^2 + (1-\frac{\sqrt{d^2+L^2}}{f})[Z+\sqrt{d^2+L^2}(1-\frac{Z}{f})]} \qquad (6)$$

$$t = \sqrt{d^2+L^2} + Z \qquad (7)$$

From the above equations, it can be seen that the interference patterns at the focus point of the focal lens are related with the parameters of L, Z, f and d. In Fig.6 and Fig.7, we

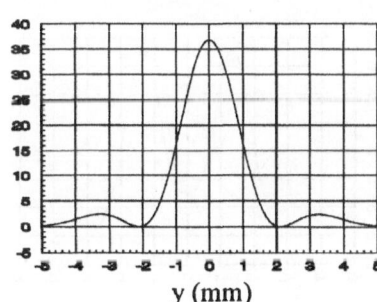

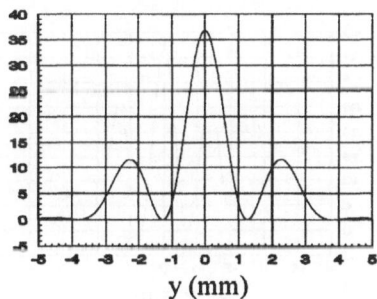

Fig.6a
d=1.5 mm, L=1200 mm, f=1200 mm

Fig.6b
d=2.5 mm, L=1200 mm, f=1200 mm

have presented the calculated intensity distribution of interference patterns in y direction

130

for the different separation of the slab waveguides (see Fig.6) and focal lengths (see Fig.7), the parameters are designated in Fig.3. The dimension of the slab waveguide channels is 1mm high and 14.5mm wide, the distance between the waveguide ends and the focal lens is 760 mm, the other parameters are given in Fig.6a-d and Fig.7a-d.

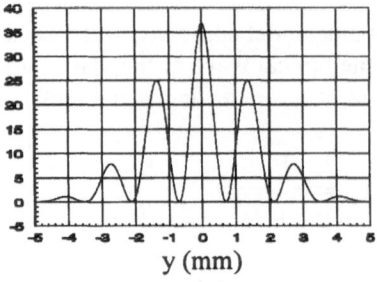

Fig.6c
d=4.5 mm, L=1200 mm, f=1200 mm

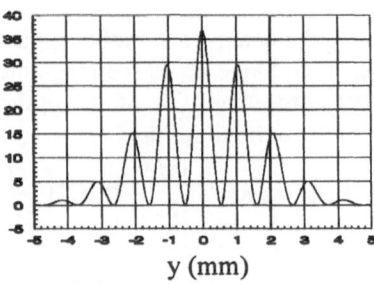

Fig.6d
d=6.0 mm, L=1200 mm, f=1200 mm

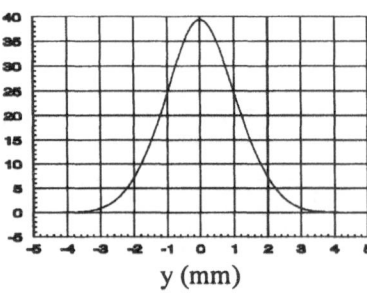

Fig.7c
d=4.5 mm, L=1200 mm, f=1500 mm

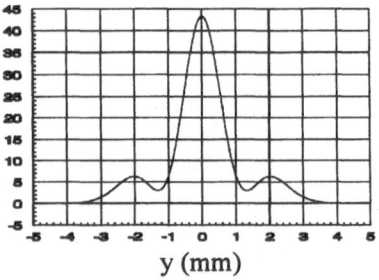

Fig.7b
d=4.5 mm, L=1200 mm, f=1400 mm

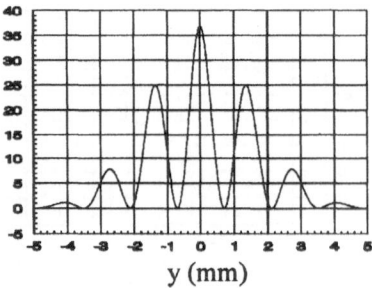

Fig.7a
d=4.5 mm, L=1200 mm, f=1200 mm

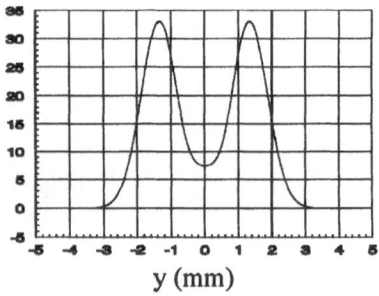

Fig.7d
d=4.5 mm, L=1200 mm, f=1700 mm

From Fig.6, it can be seen that the numbers of the interference fringes at the focus point of the focal lens will increase with the increase of the separation of the slab waveguides for a coupled waveguide array when the other parameters are fixed, a beam of a single lobe can be obtained by reducing the separation of the slab waveguides. And from Fig.7, it can be seen that the intensity distribution of a single lobe can be obtained even off the focal point for a big separation (2d) of the slab waveguides, and a pattern of interference fringes can be seen at the focal point. The different selection of the parameters defined in Fig.3 will present the different patterns of interference fringes in the vertical direction (Y direction) at the focus point.

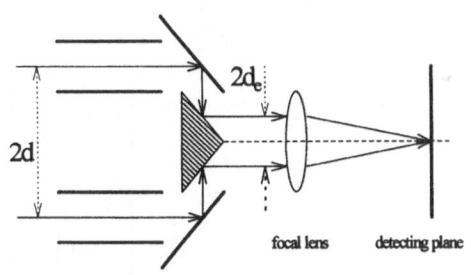

Fig.8 the experimental setup for reducing the separation of the ouput beams of the slab waveguide array

But in practice, it is desirable to obtain a beam of a single lobe with a small beam radius, therefore an optical setup is needed to reduce the separation of the output beams of the slab waveguide array as shown in Fig.8. By using this technique, a beam of a single lobe with a small beam radius can be obtained from the slab waveguide array in terms of the above theoretical analysis.

3.2. SLAB ARRAY WITH TWO UNSTABLE RESONATORS

As we know that an output beam from an unstable optical resonator is a superradiation, which has the poor coherence, because there is no feedback field in an unstable resonator. In our experiments, we add two beam splitter into the optical resonator to direct the part of each slab waveguide output beam into the neighbouring waveguides to provide a feedback for each slab waveguide and to couple the two slab waveguides in the aim to obtain a better coherent beam. A schematic diagram of a slab array with two unstable resonators and mutual injection coupling is shown in Fig.9.

The two slab waveguide channels are constructed with three aluminium electrodes and four alumina plates. The top and bottom aluminium electrodes are grounded, RF power is feeded through the central aluminium electrode. All electrodes are cooled with tap water. The electrodes are separated by the 2 mm thich alumina plates to define a waveguide transverse cross section of 2 mm high and 99 mm wide for each discharge channel, the separation of the two slab waveguide channels is 26 mm. The length of the electrodes is 500 mm. Each discharge channel is matched with a telescope resonator. Two sets of the optical resonators are employed. Each optical resonator is a hybrid laser resonator which consists of a positive branch confocal resonator in the plane of the slab with an 1.2 magnification, and a waveguide resonator of flat, producing a field of E_{11} mode in the vertical direction. Each set of the optical resonators uses two gold coated cylindrical copper mirrors separated by 530 mm, with 4650 mm convex and 5710 mm concave radius of curvature respectively. Each optical resonator has a coupling outlet of

18 mm to provide a 18% geometrical output coupling which has not been optimized. Two 45 degree beam splitter with a transmitivity of 95% are used to separately split each output beam from the slab waveguides and to reinject the split part of output beams respectively into the neighbouring waveguides for field couplings.

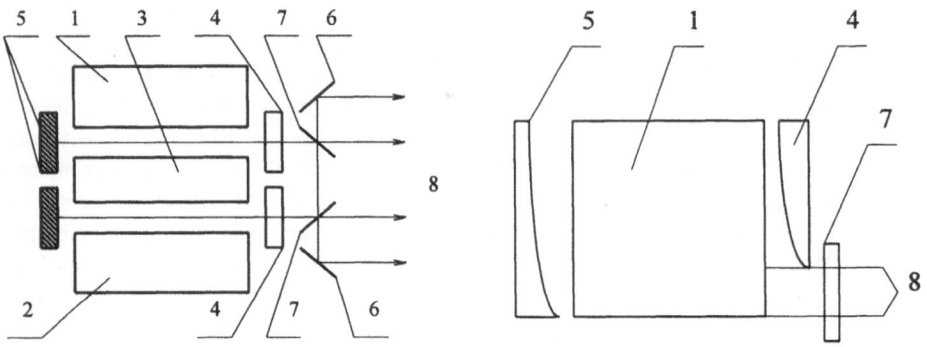

Fig.9 the Schematic diagram of the slab waveguide array with two unstable resonators

1. top electrode
4. convex reflectors
7. coupling mirrors

2. bottom electrode
5. concave reflectors
8. output beams

3. central electrode
6. collecting mirrors

In the experimental setup, the whole electrode set are mounted within a stainless steel vacuum chamber which is vacuum sealed with "O" rings, three invar bars with a diameter of 15 mm were used to support and separate the resonator mirrors from the discharge structure. A ZnSe optical window of 50 mm in diameter is employed for the beam output. A RF power supply, capable of a maximum output power of 10 kilowatts at a frequency of 93 megahertz, is connected to the central electrode of the laser head by a single 50 ohm coaxial cable. A matching circuit of LC was used between the laser head and the RF power supply to efficiently couple the RF power into the discharge plasma to simultaniously excite the gas medium in the two slab waveguide channels, the reflected RF power was kept under 2% of the RF input power through all RF input power ranges. The used gas is a mixture of CO_2:N_2:He:Xe in a ratio of 1:1:3:5%.

An analysis[9] has shown that for a parallel flat electrode structure, the gas plasma uniformity along the electrodes produced by a RF gas discharge is not accordant with the RF voltage longitudinal uniformity, it is accordant with the longitudinal uniformity of RF power deposition in the plasma. Therefore, in our experiments, a new technique of longitudinal inductor matching with periodical varying inductance values has been used to obtain the RF gas discharge uniformity.

In the experiments, when the feedback coupling mirrors were removed, a laser power output of 1020 watts was obtained at a gas pressure of 9.21×10^3 Pa and a RF power iuput of 9 kilowatts, when the feedback coupling mirrors were added on, a phase.

coupled laser power output of 860 watts was obtained.

4. Conclusion

In the above sections, we have presented a technique of slab waveguide array lasers with mutual injection coupling. In the experiments, we have demonstrated two types of slab waveguide array lasers with mutual injection. In the first type of the slab array laser, an optical resonator with flat mirrors was used, a phase coupled field distribution was obtained, an uncoupled power output of 66 watts and a coupled power output of 44 watts were achieved from a 200 mm long electrode device. In the second type of the slab array laser, two unstable optical resonators, with two 45 degree feedback coupling mirrors, were employed, a coupled power output of 860 watts and an uncoupled power output of 1020 watts were obtained from a 500 mm long electrode device.

5. References

1. Youmans, D. G. (1984) Phase locking of adjacent channel leaky waveguide CO_2 lasers, *Applied Physics Letters,* 44(4), 365–367.
2. Newman, L. A., Hart, R. A., Kennedy, J. T., Canter, A. J. and Demaria, A. J. (1986), High power coupled CO_2 waveguide laser array, *Applied Physics Letters,* 48(25), 1701–1703.
3. Hornby, A. M., Baker, H. J., Colley, A. D. and Hall D. R., (1993) Phase locking of linear array of CO_2 waveguide lasers by the waveguide-confined Talbot effect, *Applied Physics Letters,* 63(19), 2591–2593.
4. Xin, J. G., Zang, R. J. and Wei, G, H., (1994) Radio-frequency excited one dimensional limit parameter waveguide array CO_2 laser, *Optical Engineering,* 33(4), 1142–1145.
5. Jackson, P. E., Baker, H. J. and Hall, D. R. (1989) CO_2 large-area discharge laser using an unstable-waveguide hybrid resonator, *Applied Physics Letters,* 54(20), 1950–1952.
6. Colley, A. D., Baker, H. J. and Hall, D. R. (1992) Planar waveguide, 1kw CW, carbon dioxide laser excited by a single transverse RF discharge, *Applied Physics Letters,* 61(2), 136–138.
7. Degnan, J. J. and Hall, D. R. (1973) Finite-aperture waveguide laser resonators, *IEEE Journal of Quantum Electronics,* vol. QE–9(9), 901–910.
8. Henderson, D. M. (1976) Waveguide lasers with intracavity electrooptic modulator: misalignment loss, *Applied Optics,* 15(4), 1066–1070.
9. Xin, J. G. and Peng, X. Y. (1995) Analysis of discharge uniformity in RF excited gas lasers, *Proceedings of international conference on electrooptics and lasers '95,* Hang Zhou, China, B67, 265–268.

GAS DISCHARGE LASERS WITH COMBINED PUMPING

N.A.YATSENKO
Institute for Problems in Mechanics,
Russian Academy of Sciences
101, Prosp. Vernadskogo, Moscow,
117526, Russia

1. Introduction

The combined electric gas discharge, supported by superposition of high-frequency (RF) and direct current (DC) or low-frequency (AC) electric fields [1-6], or combination of electric and magnetic fields [7-9] for gas lasers excitation can essentially improve their characteristics. Nevertheless such discharges are practically not studied yet despite of the good results of their application in laser engineering.

In this paper main attention is paid to the combined discharge in alternating and static electric fields. However the discharges with additional steady-state magnetic fields are also considered, because of their importance for the excitation of active medium in slab gas lasers [10].

It is known, that there are some specific requirements to electric discharge, used as an active medium of a powerful gas laser, which one can fulfill using the combined discharge. The most important of them is the following: in order to get high power laser with high efficiency and beam quality it is necessary to excite uniformly a large gas volume with the value of E_{pl}/N close to optimum for the excitation of the upper laser level (E_{pl} - electric field in plasma, N - neutral gas particles density).

However it is well known [11], that the optimum value $(E_{pl}/N)_{opt}$ to build up the inversion (for example, in a CO_2 laser) [11] is much lower than the value E_{pl}/N, which is necessary to support local ionization balance in self sustained discharge plasma. Furthermore it is very difficult to obtain a stable powerful self-sustained discharge at moderate pressure with volumetric plasma column, because of the discharge instabilities in plasma [11-13], result in contraction of the volumetric discharge. The attempts to overcome these difficulties have resulted in the development of systems with the double discharge or discharge in combined fields [11-13]. In this case the function of one discharge (usually it is a pulse-periodic discharge) is to produce and to maintain necessary electrons concentration in the whole volume of laser resonator. The second discharge is generally non-self-sustained, because of the low steady-state value E_{pl}/N, and it makes the pumping of the upper laser level more efficient.

W. J. Witteman and V. N. Ochkin (eds.), Gas Lasers - Recent Developments and Future Prospects, 135–154.
© 1996 *Kluwer Academic Publishers.*

Despite of a basic simplicity, the systems with double discharge have rather complex construction. It is, first of all, due to the development of various types of instabilities, just as in case the self-sustained discharge. To eliminate of these instabilities it is necessary to have sectioned electrodes with a specific profile. The lifetime of such electrodes is small, as a rule, and the uniformity of the excitation of gas mixture is worsening gradually. There is another way of electron productive in gas laser active medium. It is electron beam ionization. But the lasers with ionization of active medium by electron beam are more complex, and require heavy and expensive systems for protection against x-ray radiation.

One can avoid some of the above mentioned difficulties, using radio-frequency capacitive discharge (RFCD) for laser mixture ionization, and additional DC field for upper laser level pumping. The successful works [1-6] in this field were mentioned above. But despite of the good practical results, the authors [1-6] didn't obtain convincing experimental data about combined discharge physics and the reasons of contraction threshold growth in the DC discharge in powerful RF-field.

Moreover, the problem of formation of active medium in the gas laser with a narrow slit gap (slab gas lasers) becomes important. The best results in this field were obtained by using transverse RF discharge with well chosen parameters [10, 18-20]. However RF sources which are needed to sustain high-power RF discharges are rather complex and expensive, and it is necessary to search for alternative methods of slab gas lasers pumping. The combined discharge is the perspective way of the solution of this problem. It's a must to mention the work of J. Macken [8], who was the first to propose and successfully realize an original method of slab gas lasers pumping by glow DC discharge with magnetic stabilization of flat plasma column (Macken Discharge).

2. Experimental conditions

The experimental set-up, shown in Fig.1,2, was used to study the characteristics of the combined DC + RF discharge as a source of excitation for gas lasers active medium. The experimental setup for studying the magnetically stabilized slab glow discharge will be described in section 6 (see Fig. 8).

As the RF and combined discharge's parameters depend strongly on a method of RF field energy coupling into the discharge plasma [18], the design of discharge chambers provided us with a possibility of flexible change of the configuration: flat, profile and coaxial, or made of several copper rods, 1-2 mm in diameter, which were placed on the outer part of a dielectric channel (quartz tube) along its axis equally spaced from each other. In particular, long (up to 2 ms) pulse and cw RF and combined discharges in tubes of various diameters from 5 mm up to 80 mm, as well as in planar channels of rectangular section (width 10-200 mm, length up to 1 m) were experimentally investigated (Fig.1).

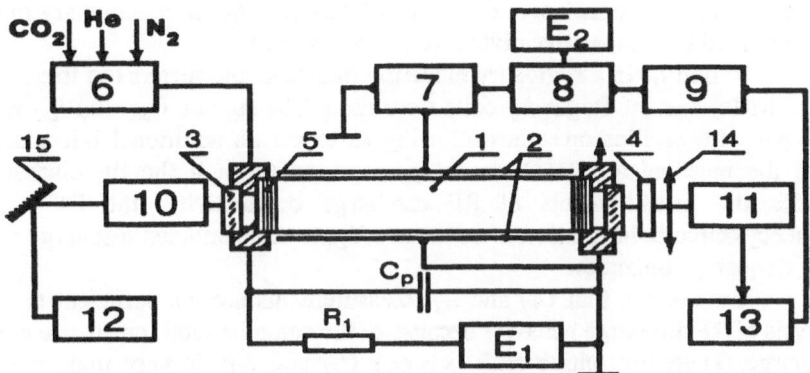

Figure 1. Experimental setup. E_1, E_2 - DC-power supplies; 1- discharge chamber; 2 - RF electrodes; 3 - outcoupling window 4 - infrared filter; 5 - DC electrodes; 6 - gas mixer; 7 - RF generator; 8 - modulator; 9 - pulse generator; 10 - attenuator; 11 - photodetector; 12 - probe laser; 13 - oscilloscope.

For the excitation of the RF and combined discharges the RF-generators with a frequency from 1 up to 100 MHz and the maximum power 6 kW were used. The special attention was paid to matching of a plasma impedance with an output of RF-generator. The power of DC (AC) source was limited to 10 kW.

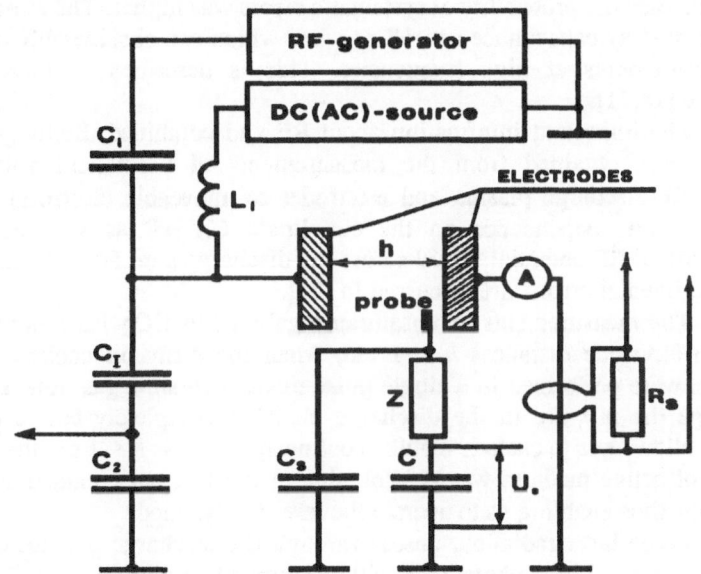

Figure 2. Schematic diagram of electrical measurements

The basic setup for the essential electric measurements is shown in Fig.2. RF voltage applied to the electrodes U_{rf} and RF current through the discharge I_{rf} were measured. The characteristics of the discharges were determined from the measurements of current I_{dc} (I_{ac}) and DC voltage U_{dc} (low-frequency voltage U_{ac}), which was applied to the same electrodes as RF-voltage through an inductance, blocking RF current. U_{rf} in a stationary mode was measured by a capacitive divider (C_1, C_2 in Fig.2).

To find I_{rf} in a stationary mode the thermo-ampermeters (on frequency $f < 15$ MHz) and the Rogovsky coils were used. The signals U_{ac} and I_{ac} were analyzed with oscilloscope, thus allowing us to get an additional information about the mechanism of AC-current propagation through the RF discharge. Besides the measurements of RF discharge conductivity, the DC (low-frequency) circuits have allowed us to investigate the combined discharge with high-frequency ionization.

It is known, that U_{rf} and I_{rf} measurements are not sufficient for the analysis of RF discharge behavior because of the complex total impedance of the discharge. Therefore, phase shift between U_{rf} and I_{rf} is very important to determine the RF-voltage drop in the plasma U_{pl}.. In this work methods of oscillographic registration of a phase shift between current and voltage signals, as well as the method of capacitive compensation of the discharge [18], were used.

An error in AC-voltage and current measurements was less than 5 % for the effective values, and 10 % for the instant ones. An additional error in measurements of effective values is due to the difference of the shape (especially of AC-current) from the sine wave in some modes of the discharge. Control of the signal shape with the oscilloscope made it possible to account for this circumstance.

An error of the RF-voltage and current measurements less than 10 %, but in this case the probability of systematic errors was higher. The reason is the presence of stray capacitance as well as wires which can considerably influence the measurements at high frequencies. This is described in more details elsewhere [18,21].

The important information about RF and combined discharge spatial structure was obtained from the measurements of permanent voltage U_0 between RF-discharge plasma and electrodes by moveable electrostatic probe (Fig.2) and its dependence on the coordinate $U_0(x)$, as well as on the distribution of RF potential $U_{rf}(x)$ across the discharge gap. Possible difficulties and experimental errors are discussed in [21].

The measurements of unsaturated gain k_0 in CO_2 laser mixtures for the inter-electrode distances $h > 1$ cm, when the diffusion cooling was not effective, were performed in a single pulse mode with slow gas flow, sufficient to change the mixture in the discharge chamber completely before the next pulse. It allowed us to exclude a bulky cooling system . At $h < 1$ cm the efficient cooling of active medium was possible due to the thermal conductivity of the electrodes, thus enabling us to operate the laser in CW mode.

Probe laser radiation, passed through the discharge and focused by a lens, was detected by a photodiode with a temporal resolution of 10^{-6} for 10.6 μm. To eliminate the influence of spontaneous discharge radiation with other wavelengths, an infrared filter with a bandwidth $10,6 \pm 0,5$ μm was placed in front of the phododetector (Fig.1). Typical oscillograms of the gain $k_0(t)$ measured in the RF discharge are shown in Fig.3: 1 - for frequency 13.6 MHz , and 2 - for the combined discharge.

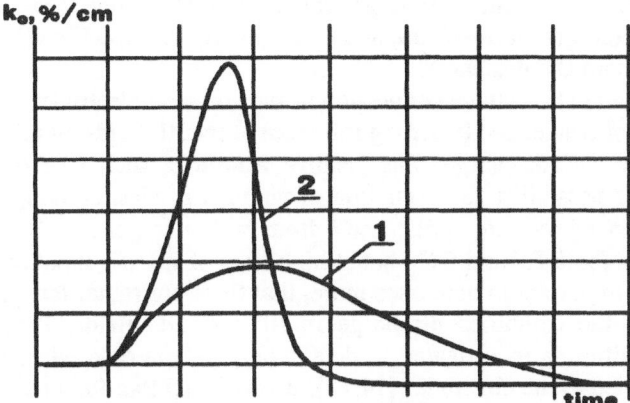

Figure 3. Gain oscillograms in RF and combined discharge in mixture $CO_2:N_2$ =1:1 at p
= 5 Torr. Scale: horizontal - 0,5 ms per division, vertical - 0,4 %/ cm per division.

To study a DC field influence on the gain values, copper ring electrodes were mounted inside the laser chamber. The capacitor C_p charged to the necessary value U_{po} served as a DC source. A DC voltage U_{dc} was applied to the electrodes. The change of the voltage on this capacitor during its discharge through RFCD plasma was registered by an oscilloscope. If the capacitance C_p and the temporal dependence of the voltage on it $U_p(t)$, measured with the oscilloscope, are known, it is possible to calculate the electric energy input in the discharge plasma for any moment, without measuring the current. We can neglect the recharge of C_p during the time of the electric pulse τ if we choose the value R_1 (Fig.1) large enough to satisfy the following condition: $R_1 >> C_p/\tau$.

The exclusion of the current value from measurements in the C_p circuit has allowed to increase the accuracy of definition of energy input in the discharge plasma, because the accuracy of U_p measurement is substantially higher than the accuracy of current measurement. It is due to the fact, that the characteristic time resolution was 10^{-5} - 10^{-6} s in our experiments. But in such conditions one can get considerably larger signal from a voltage divider, than from a current shunt. Since the pickup in both cases is identical, a signal to noise ratio in $U_p(t)$ measurements will be higher than the corresponding signal to noise ratio in $I_p(t)$ measurements.

3. Specific features of RF + DC combined discharge excitation

There are two essentially different methods of RF and DC sources connection to the electrodes for the excitation of the combined discharge in RF + DC fields: 1)DC-electrodes have a high impedance with respect to RF electrodes (Fig.1); 2)RF and DC electrodes are joint, or have small impedance between them (Fig.2). Let's consider that the impedance is high, when it is much higher than the impedance between RF electrodes, and small - if it is much lower.

The first method of connection is used, as a rule, for stabilization of lengthy DC discharges by RF field [1-3]. In this case the RF field essentially influences plasma in a positive glow. Due to a better homogeneity of plasma column parameters along discharge chamber section, one can manage to increase the contraction threshold, and consequently, the specific electric power input in the main DC discharge.

RF and DC voltage connection to the common electrodes differs from the first type of connection by strong influence of the RF fields on near electrode sheaths of the DC discharge. The positive feature of such type of combined discharge is a possibility for excitation of volumetric plasma along large area electrodes, located on a small distance h from each other, at small (hundreds of volts) voltage (both RF and DC), but at high level of specific input power [6].

It is important to note once more, that there are rather few publications dealing with the combined discharge in RF and DC fields. Process of RF discharge ignition is investigated in details only for the case when a DC field was also applied to the electrodes [14,15]. It was found that for small DC fields, it is necessary to increase the amplitude of RF voltage to breakdown the gas. This fact is explained by additional removal of electrons from the inter-electrode gap, as a result of drift in the additional field E_{dc}. For example, if E_{dc} is uniform and it is directed along the X axis, the condition for the breakdown will be as follows

$$d^2 n_e /dx^2 + (\mu_e\, E_{dc} /D_e)dn_e /dx + (v_i /D_e)\, n_e = 0 \qquad (1)$$

Here μ_e, D_e, n_e are the mobility, the diffusion coefficient and the concentration of electrons, v_i - the frequency of ionizing collisions.

Equation (1) differs from a usual equation for the breakdown in RF-field in diffusion mode [16] by a term $(\mu_e/D_e)E_{dc}(dn_e/dx)$, which accounts for the electron drift in the additional DC field.

It is possible to consider a new condition of the breakdown in a combined field (according to (1)):

$$v_i /D_e = 1/L^2 \qquad (2)$$

where $1/L^2 = 1/L_0^2 + (b_e E_{dc} /2D_e)^2$; it is easy to see, that it is equivalent to ordinary condition of high-frequency breakdown in a chamber of small size.

When DC and RF voltages become comparable the contribution of a DC field to ionization prevails, according to [14-15]. Therefore a decrease in RF breakdown voltage is observed.

4. The features of the current-voltage characteristics and the spatial structure of combined discharge in DC+RF fields

Let us consider typical voltage current characteristic (CVC) in combined discharge with high-frequency ionization, when RF and DC (or AC low-frequency) voltages are applied to the same electrodes. The experimental data were measured for flat water cooled electrodes (Fig.2). DC, or low-frequency (50 Hz) voltage was applied to electrodes, exciting the RF discharge, through blocking inductor L_i.

The typical voltage current characteristics of the self-sustained AC and combined discharges in air at $p=7,5$ Torr are shown on Fig.4. The effective values of current and voltage are presented. Curves 1, 2 describe the self-sustained AC-discharge at $U_{rf}= 0$, whereas 3 - 6 illustrate measurements in the combined discharge: 3, 4 - at RF power input P_1 = 100 W and inter electrode distances 20 and 10 mm respectively; 5,6 - at the RF power input P_2 = 800 W and the same inter-electrode distances.

The positive slope of voltage current characteristics is an important feature of the combined discharge (contrary to the self-sustained DC discharge). It specifies the stability of such type of discharge to perturbations. Actually, the reduction of ballast resistance to zero value in a low-frequency circuit did not result in the development of instabilities. The RF field imposed on the self sustained AC discharge causes the reduction of AC voltage across the electrodes and the growth of AC current I_{ac} (transition from "A" to "B" and back in Fig.4).

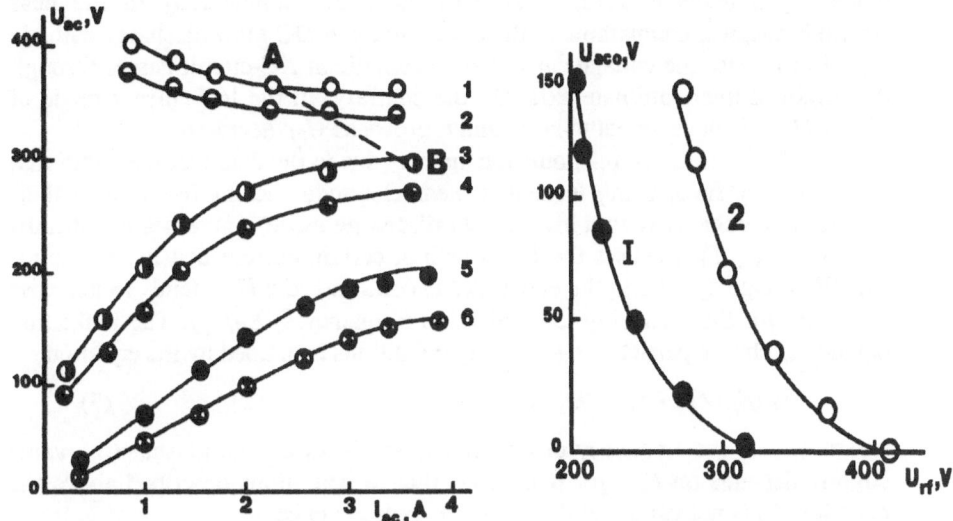

Figure 4. Current-voltage characteristics   Figure 5. U_{aco} as a function of U_{rf}

The conditions in the combined discharge, when the value U_{ac} is less than the voltage of stationary discharge, are interesting. In this case all depends on the value of RF power input into the discharge. It is noticeable, that when this value is high (P_2 = 800 W, curve 5, 6 on Fig.4), the Ohm law for low-frequency circuit at low U_{ac} (< 120 V) can be used with a good accuracy, i.e. the conductivity of the discharge gap is totally defined by the RF-field.

The interpretation of these voltage current characteristics is not difficult if one takes the advantage of modern knowledge about specific features of the spatial structure and forms of the RF capacitive discharge [17,18].

There is a difference between the curves 3, 4 and 5, 6, because of the existence of two different modes of RFCD [17,18]. It is clear, that when the contribution of RF power to the combined discharge is small (P_1 = 100 W), the corresponding voltage current characteristics of the low-frequency discharge begin at I_{ac} = 0 with a certain not equal to zero low-frequency voltage U_{aco}. One can notice, that when the RF voltage applied to the electrodes increases, the value of U_{aco} decreases gradually to zero at some U_{rf}. The corresponding experimental dependence U_{aco} as a function of U_{rf} is presented in Fig.5. It is clear that when $U_{aco} = 0$ the value of U_{rfo} corresponds to U_{cr} - the RF voltage applied to the electrodes, when the breakdown of near-electrode space charge sheaths occurs, and RF discharge transforms into a high-current mode [17]. The experimental curves in Fig. 3 measured for the combined discharge in air (curve 1) and in He (curve 2), at the inter electrode distance $h=1$ cm and pressure $p=7,5$ Torr have confirmed that in all cases the values U_{rfo} and U_{cr} were the same for the above mentioned gases, p and h values within the accuracy of 5 - 7%.

So, we can explain the difference in the behavior of voltage current characteristic of low-frequency circuit, when the currents are small, by the existence of two modes of the RF capacitive discharge. In the high-current mode, U_{aco} tends to zero, because of the active conductivity in the near electrode sheaths, comparable to the conductivity of DC glow discharge cathode area. In this case the voltage current characteristic of AC-circuit passes through the origin of the coordinate axis. On the contrary, in the low-current mode of RFCD U_{aco} is never equal to zero, and it grows as U_{rf} decrease.

The considered phenomenon appears due to the fact, that the additional low-frequency field, being located in near electrode sheaths (because of their low active conductivity in a low-current discharge mode), increases the sheaths thickness d_{sl}. That causes the U_s growth at certain current density. But since the RF voltage applied to the electrodes is constant, the U_{pl} tends to decrease (i.e. E_{pl} / p) thus reducing the ionization frequency $v_i(E_{pl} /p)$. The ionization balance of charge particles in stationary conditions is defined by the equation

$$v_i \ (E_{pl} / p) - v_r = 0, \tag{3}$$

where v_r - is the frequency of electron losses in discharge plasma (the value slightly depends on E_{pl} /p). It is clear, that in conditions described above the equation (3) is not valid and the discharge cease to exist.

When $U_{ac} > U_{aco}$, the breakdown of the capacitive near-electrode sheaths occurs in combined field, causing the more efficient electron production. As a result the high-current RFCD spatial structure (negative glow and the Faraday dark space, i.e. zones with low electric fields) is established. Therefore U_{rf}, which is necessary for support of the high-current RFCD, decreases; so the situation differs from the case when $U_{ac} < U_{aco}$.

Let us consider in detail the dependence of U_{ac} on I_{ac} when the external parameters of the high-current RFCD are varied. A set of CVC in low-frequency circuit at the fixed $h = 2$ cm and air pressure $p = 15$ Torr is shown in Fig. 6. RF current through the discharge plasma I_{rf} was the parameter.

It is interesting to note, that when I_{rf} increases, the decrease of U_{ac} /I_{ac} takes place. This means that the total active resistance R_a decreases when I_{rf} growth. The typical behavior of the R_a dependence on the RF current in the combined discharge is shown in Fig. 7.

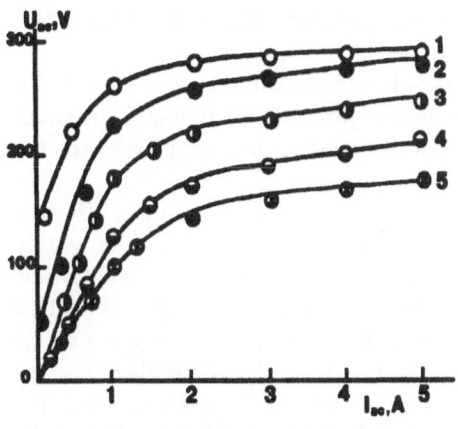

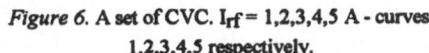

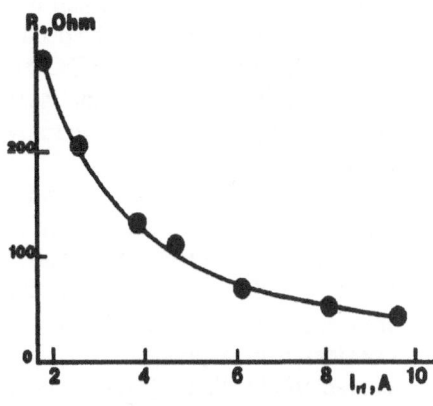

Figure 6. A set of CVC. I_{rf} = 1,2,3,4,5 A - curves
1,2,3,4,5 respectively.

Figure 7. Active resistance
R_a as a function of I_{rf}

An interesting effect, related to the RF-current growth, is the increase of the longitudinal length of the spatial structure equivalent to Faraday dark space L_F. It is especially clear at low (about several Torr) pressure. This phenomenon is less pronounced at higher pressures. The experimental results, illustrating the forcing out of a plasma column from electrodes are tabulated below:

I_{rf}, A	1	1.2	2	3	4	6	8
L_F, mm	7	7	10	12	16	20	26

The experiments were carried out in the high-current RFCD in He at p = 15 Torr for the inter-electrode gap h = 60 mm at the excitation frequency of 13,6 Mhz. It is noticeable that the DC voltage imposed on the high-current discharge results in the same effect: the positive plasma column forces out from negative electrode.

If a low-frequency current (50 Hz) is applied in RFCD it looks as if the plasma column expands in the direction perpendicular to the electrodes, i.e. Faraday dark space disappears. However, this is only the result of an optical illusion.

There is an experimental fact, which we could not explain in this paper. When the current density of direct current exceeds the value $j_{ocr}(I_{rf})$ a bright structures on the anode of the combined discharge, similar to a torch, appear. Their light in the direction perpendicular to the cathode becomes more weak, and they gradually disappear at some distance from the anode. There was no such effect in the regular RF-discharge. When I_{rf} increased, j_{ocr} grew also. This phenomenon is likely to be associated with the boundary conditions on the anode.

The experimental data considered here show that the problem of the analysis of DC in RF discharge can be separated in two independent ones.

1) To estimate the value of direct current flowing through a low-current RFCD when a DC voltage, low enough to break down the capacitive near-electrode sheaths is imposed on it.

2) To find a correlation between the direct current and the value of DC voltage in the high-current RFCD in the case when RFCD exists in low-current

mode till the time an additional field is applied (it is the value of the field, when the breakdown of the sheaths occurs).

Let us consider the DC current flow in the conditions, mentioned in 1). It's clear, that this problem is similar to the problem of non-self-sustained electric current in gas slightly ionized by an independent source (here by the RF field).

But there is an essential distinction. It is connected with the different physical conditions in transition layer of the space charge between the weakly ionized gas and the chamber wall. Let us consider the example. The fast particle beam provides a uniform ionization of the gas volume with the ionization frequency v_i. If v_i has such a value, that the weakly ionized gas can be considered as a plasma, i.e.. $L >> r_D$ (here r_D is the Debye radius, L - the characteristic size of the plasma), then an intermediate layer with the characteristic size of order r_D appears between the plasma and the wall, and the value of potential barrier for the electronic component will be set to $U \sim kT_e/e$, where k is the Boltzmann constant, T_e is the electron temperature in the plasma.

Different behavior is observed in the near electrode sheaths in the case of volume ionization of the gas by the RF field. The plasma possesses a constant potential U_O positive with respect to the electrodes, more than by an order of magnitude larger than the characteristic value of kT_e/e [17-18]. It is located in the near electrode space charge sheaths (which thickness is much larger than r_D ($d_{s1} >> r_D$)). In other words, even when the external DC fields are absent, the phenomenon can be interpreted as follows: one manages to switch on symmetrically two DC sources with the voltage U_0 between the plasma and the electrodes (the positive terminations of the DC sources are connected to the plasma, and the negative to the electrodes).

Therefore, in order to make use of the results of research in non-self-sustained discharge with an independent ionization it is necessary to take into account high constant voltage U_0, existing in the near-electrode sheaths. We should not consider this problem in details, because: a) if we take into account U_0, it is similar to the well known process - the DC flow through the gas with independent source of ionization, and b) this case is not interesting for the high power gas laser applications because of the small conduction currents in the near electrode sheaths.

Let us consider the second variant of the problem. The experimental results show, that the basic difference between the combined discharge with high-current RFCD and those of the first case is the absence of DC current limitations imposed by the regions of space charge in the near electrode sheaths (the characteristic concentration of ions n_+ in it is not more than the electron concentration in plasma column $n_+ < n_{epl}$). This makes it possible to apply an additional conduction current of considerably larger value through the plasma column of such combined discharge, and consequently to make larger power input into discharge plasma from the DC-source, than in the case considered above.

To our opinion the explanation of this fact is as follows: The spatial structure of the RFCD changes after the breakdown of the sheaths, maxima of n_e, and consequently maxima of n_+ on the sheaths boundary with plasma appear [17-18]. I.e. after the sheaths breakdown $n_+ >> n_{epl}$ (contrary to the previous case), and since the active conductivity of near electrode sheaths is determined by n_+, it increases. Moreover the near electrode sheaths thickness d_S sharply decreases, when the RF discharge undergoes transformation into a high-current mode in the conditions, typical for CO_2 lasers [17]. These two features together cause a strong reduction in the near electrode sheaths resistance and an increase of power input in the plasma of combined discharge, mentioned above, as far as

$$R_S = (1/qn_+\mu)d_S/S_S \qquad (4)$$

where q, μ - charge and mobility of ions, respectively, S_S - the sheaths area .

It is possible to conclude also, that the effect of forcing out of the plasma column from the electrodes, observable experimentally, (symmetrically from both electrodes in high-current RFCD, and from negative DC electrode - in the combined discharge) has the same nature in both DC and RF discharges, and it is determined by an increasing role of the near-electrode areas for the processes of electron production. Let us consider this important question in detail.

As it is known from the DC glow discharge physics [13], the Faraday dark space is determined by specific character of ionization processes in the cathode area. This results in the appearance of the electron concentration maximum in negative glow, production of fast electron beams, etc. The Faraday dark space length L_f in a normal mode of glow discharge is determined basically by a mechanism of electron losses, which defines a profile of n_e concentration lowering in the direction towards anode, and consequently a shape of electric field growth, since far from the cathode, when it is possible to neglect the diffusion, the following equation is valid:

$$j = en_e\mu_e E = const \qquad (5)$$

where j is a discharge current density. From (5) we can see, that for large n_e corresponding E are small. This fact explains the absence of the glow, since the electron energies are insufficient for the excitation. Certainly if one should consider this question more strictly (especially at low gas pressure and in an abnormal mode), it is impossible to ignore the influence of fast electron beams, generated in the cathode layer, on the balance of charged particles in the Faraday dark space.

If we neglect a role of electron beams, the following equation could be used to define the structure of the electron distribution along the discharge gap in the Faraday dark space

$$D_a (d^2 n_e /dx^2) - k_r n_e^2 = 0 \qquad (6)$$

It is commonly accepted, that the electrons penetrate into the Faraday dark space as a result of ambipolar diffusion from the end of negative glow, and they disappear in dissociative recombination process (D_a is the ambipolar diffusion coefficient, n_e is the electron concentration, the axis x coincides with the direction of current). We can neglect the ionization of neutral particles by the electrons, heated up by the electric field, being present in the considered zone of the discharge, since the role of this process becomes essential only near the plasma column. Then the electronic concentration profile is described by the expression

$$n_e(x) = n_{e0} \left[x(k_r n_{e0}/6D_a)^{1/2} + 1 \right]^{-2} \tag{7}$$

where n_{e0} is the initial electron concentration near the end of negative glow. Taking into account that the electron concentration in plasma column $n_{epl} \ll n_{e0}$, and the using (7), we derive an expression for the Faraday dark space length L_F (accounting for the negative glow)

$$L_F = (6D_a / k_r n_{epl})^{1/2} \tag{8}$$

From (8) we can see, that the electron concentration growth in a plasma column should reduce Faraday dark space size, since the plasma decay to the large value n_{epl} occurs at smaller distances. It is in a good agreement with the experiments.

The experiments were carried on for different gases, however they had the most pronounced effect in He, where the Faraday dark space is rather large. Therefore the results will be described only for He, although they remain valid for other gases. It turns out, that in the RF-discharge, excited in He at pressure, for example, 10 Torr on the frequency $f = 13{,}6$ Mhz, the length of Faraday dark space is several times shorter than in the discharge, supported by DC field, provided both discharges are in the normal current density mode. The explanation is as follows. The value of normal current density in the RF discharge at pressures and frequency being under discussion is more than an order of magnitude higher than the corresponding value for the DC glow discharge (as it was already pointed out above), because of the displacement current in the sheaths. And due to the weak dependence (logarithmic relationship) of the electric field in a plasma column electron concentration, n_{epl} in the high-current RFCD is approximately an order of magnitude higher than n_{epl} in the DC discharge.

To make it sure that L_F decreases because of the n_{epl} increase, the following experiment with a glow DC discharge was conducted. The cathode - a disk of 50 mm in diameter - had multiple holes with diameter, essentially exceeding double thickness of the cathode potential drop zone. The area of the cathode working surface was increased, and the diameter of a plasma column remained as before. The number and the sizes of the holes were chosen from the consideration that the cathode area should increase ten times. It enabled us to pass a ten times higher current through a plasma column while remaining in a normal density mode. As a result a Faraday dark space length was substantially shorter just in the high-current RFCD.

Let us now consider an influence of the electron beam, formed in the near electrode sheaths. It is known [13] that there are two groups of electrons with different energy in a zone of cathode potential drop, where the beam is formed,. The first large group consists of electrons, responsible for the ionization within a layer. But they can not make an electron beam, since they have a small free path, relative to the ionization $l_i \sim d_s / \ln (1 / \gamma)$. Assuming $\gamma = 10^{-1}$, we obtain for He $L_F \sim 0.1$ cm at $p=10$ Torr, (since $p d_s = 1.5$ cm Torr [16]), that is much less than experimental value L_F ($L_F > 1$ cm). The electrons from the second group can have a longer free path. They cross cathode sheath without collisions and acquire energy, close to $e U_s$. Accepting for simplicity of the analysis, that the electrons escape from the cathode only due to interaction of positive ions with its surface, and taking into account, that in the cathode layer of DC discharge, the current is flowing practically only due to ions, we shall receive the following approximate expression for the value of initial electron beam intensity I_e in the end of negative glow

$$I_e = \frac{\gamma j}{e} exp\{-\int_0^{d_i} \frac{d x}{\lambda_{e\,i}[U(x)]}\} \qquad (9)$$

where λ_{ei} is the electron free path with respect to the scattering due to ionization of neutral particles (this process is considered to be the main), $U(x)$ - an instant value of electric field potential with respect to cathode, which determiner electron energy at a given place, and j is a density of the discharge current (in the case of RFCD j - a conduction current component in a cathode layer).

Although the formula (9) is only approximate one, it gives the certain information about the intensity of electron beam and about the electron concentration, which the beam can produce in the Faraday dark space area. For DC discharge in He, when the recombination electron losses take place, as a result of interaction with He molecular ions, the value of electron concentration is defined as

$$n_e = \{ I_e / [\lambda_{ei} (U_c) k_r]\}^{1/2} \qquad (10)$$

From (10) we can see that $n_e = 10^{11}$ cm^{-3} for $p = 10$ Torr, $U_c = 160$ V, (i.e. $\lambda_{ei} \sim 0.1$ cm) and $\gamma \sim 10^{-1}$. Therefore the beam could influence the Faraday dark space, however, λ_{ei}, is rather small, its influence has only the minor importance in comparison with the diffusion processes, due to the rapid beam attenuation as a result of electrons scattering.

However there are certain cases when the role of an electron beam becomes significant. In the case considered (discharge in He at intermediate pressure) electron beam can be more important, than the diffusion processes, if the discharge mode is abnormal, i.e. for the high values $U_s >> U_c$. Since practically in all interesting cases the maximum of ionization section is located at the energies, smaller, than eU_s, for $U_s=U_c$, it's clear, that when an electron beam energy increases, the ionization cross-section decreases, hence, λ_{ei} increases finally. When an electron beam determines the processes in a discharge we can find L_F from a condition

$$(I_e / \lambda_{ei})exp(-L_F/\lambda_{ei}) = k_r n_{epl}^2 \qquad (11)$$

(the electron diffusion influences only on the structure of initial area). Assuming $[I_e/(\lambda_{ei} k_r)]^{1/2} \sim 10^{11}$ cm^{-3}, $n_{epl} \sim 10^{10}$ cm^{-3}, $\lambda_{ei} = 1$ cm (see above), one can estimate $L_F \sim 5\lambda_{ei} \sim 5$ cm, that qualitatively explains the experimental results (see table).

It is noticeable, that in low-current RFCD the analogue of the Faraday dark space is absent [17], and the border of the plasma column is separated from the electrode by thin compared to the Faraday dark space dark sheath. This is not strange as far as in a low-current RFCD the secondary emission electrons are not essential for the charge particle balance. Therefore it is possible to consider, that the electron's distribution in inter-electrode gap of a low-current discharge has the form, similar to a bell, as at low pressures [18]. As for high-current RFCD, the n_e (x) distribution will be the same, as in the DC glow discharge with the only difference - it is symmetric with respect to the electrodes [17].

5. Processes responsible for the high stability of RF + DC combined discharge

The experiments show [2,3,6], that the RF-field imposed on a plasma column of DC glow discharge increases its stability with respect to contraction, i.e. it permits to increase a maximum electric power input into the volumetric plasma w_{max} , at which the ionization instability does not appear and the plasma remains uniform. In this part of paper some physical interpretations of the specified phenomenon are critically analyzed, and the techniques of contraction elimination in the combined discharge sustained RF and DC electric fields are discussed.

Several explanations have been suggested for the stabilizing influence of RF fields on the combined discharge volumetric plasma. According to [22], high stability of weakly ionized plasma in radio frequency field (when the frequency of RF field is much larger than the characteristic frequency of the setting up of electron concentration) was explained by the fact, that the discharge plasma remains in recombination phase during a large part of RF field period. That means that the ionization of a gas takes place only during a short time (in comparison with the RF field period), when the value of E_{rf} is close to amplitude value. This explanation is so attractive, that, despite of a justified criticism in [23], (where it has been shown, that the impact of this effect on the stabilization process can not be more than 20 %), this mechanism is still used for the explanation of high stability of RF discharge to contraction [24].

Let us notice, that the mechanism of stabilization, proposed in [22], should take place at any orientation of vectors E_{rf} and E_{dc},, whereas in experiment [2,3] the increase of stability of the volumetric combined discharge was observed only at the perpendicular orientation of E_{rf} and E_{dc}. Due to this experimental fact the idea of the positive influence of oscillation effect on volumetric plasma stability [2], and effect of total vector of electric field $E = E_{dc} + E_{rf}$ rotation in a volume plasma [25] became attractive.

To understand better the mechanism of high stability of DC+RF combine discharge with respect to contraction, the following experiments were performed. DC (AC) voltage and RF one were delivered to joint electrodes. By this mean the direction of vectors E_{rf} and E_{dc} was the same. Nevertheless in this case we could provide an electric power input into plasma w_{pl} up to 10 W/cm^3 and more, which is substantially larger, than the results, observed [25], where the maximum specific power input w_{max} = 4 W/cm^3 was fixed in a rotated electric field. Thus, one should mention especially the simplicity of realization of the combined discharge, compared to the method of production of volumetric plasma in rotating RF electric fields [25]. There are only two electrodes without dielectric cover, and two phase-shifted RF sources are not required.

Let us analyze the experimental conditions of [2,3], where the negative result of the stabilization of glow discharge with parallel E_{dc} and E_{rf} was obtained, and compare them with the conditions in this work. We can see, that the difference sizes from the inter-electrode distance.

Special experiments were conducted, which have confirmed the conclusion about the influence of the inter-electrode distance on the RF and combined discharge stability. The shorter is the inter-electrode gap, the better is the RF or combined discharge stability. Really, the inter-electrode gap h in the experiments [3] was 52 cm, whereas in the experiments described above it did not exceed 2 cm. It is interesting to note, that the similar experimental fact was observed also in [24], but it remained unnoticed. The increased RFCD stability was explained according to [22].

An attempt to explain high stability of the combined DC+RF discharge on the basis of high-current RFCD features, presented in [17], was made. The plasma column of such discharge was separated from both electrodes by the areas with high conductivity (low electric fields). It allowed to consider the electron concentration fluctuation δn_e approximately equal to zero at the ends of a plasma column, and, at this approximation, to investigate the stability of differential equations system, describing combined discharge. In detail the specified differential equations system and the procedure of its stability investigation were described in [26].

Two cases were considered. 1) RF electrodes were close so the plasma column could not appear ($2d_{s2} < h < 2d_{s2} + 2d_t$), where d_{s2} is the thickness of the near electrode sheaths in high-current discharge mode, d_t is an intermediate zone between space charge sheath and plasma column (the Faraday dark space analogy). 2) In the inter electrode interval a plasma column with the length d_{pl}, i.e. $h > 2d_{s2} + d_t$. appeared.

It was found that in the first case the combined discharge is stable, if

$$w_{max} = \frac{NC_pT}{2}q\frac{2k_rn_e - Nk_i}{2k_rn_e - Nk_i + Nk_i\underline{k_i}} \quad (12)$$

where q is the factor that determines heat removal speed for both convection and diffusion cooling, C_p is the specific heat, T, $3/2\Theta_e$ - the temperature of the gas and average electron energy respectively, k_i, k_r, k_u are the constants of ionization speed, dissociative recombination and energy transfer from the electrons to the neutral particles in a gas, $\underline{k_j} = d\,lnk_j\,/dln\Theta_e$. .

For the second case it was found, that the threshold power input w_{max}, in a specially uniform plasma column of the combined discharge, can be determined from the equation

$$w_{max} = NC_pT\kappa\frac{\pi^2}{d_{pl}^2}\frac{E_{rf}^2}{E_{dc}^2 + E_{rf}^2}\{1 + [\frac{1}{2\underline{k_i}}(1 + \frac{q_kd_{pl}}{\pi^2\kappa}\frac{E_{dc}^2 + E_{rf}^2}{E_{rf}^2})]^{1/2}\}^2 \quad (13)$$

Here E_{dc} is the DC electric field strength, $q_k = 1/\tau_r$, where τ_r is a time of heat removal due to convection cooling only, $\kappa_a = \lambda_a/NC_p$, λ_a is a heat conductivity, d_{pl} is the length of a plasma column.

150

Comparison of the first experimental results with the ones calculated from (12-13) has shown the similar trends of the w_{max} increase as h is reduced. But the results were similar trends. In [26] it was explained by the fact, that the experimental data were obtained in the conditions not adequate to calculated ones.

However other experiments also came into the conflict with the calculated results, given above.

1) In low-current RFCD as the RF frequency from 13.6 up to 81 Mhz, w_{max} increased up to 100 W/cm^3 at $h = 0.5$ cm, that was more than 50 times larger, than w_{max} according to (13). But another fact is more important. In low-current RFCD the areas with high conductivity at the ends of plasma column [10,17,18] are absent, i.e. the main assumption (the basis of (12,13)) is not fulfilled, but the effect of stabilization (even stronger) is nevertheless present.

2) The dielectric sectioning of the negative glow area of the DC discharge with a continuous cathode permitted to form a volume plasma column at small h. When sectioning was removed, the contraction of a plasma column was observed in the same conditions and at the same discharge current.

Indicated arguments confirm, that the mechanism of influence of inter-electrode distance on stability of the RFCD volume form are not completely correct.

Nevertheless it is possible to consider that the following facts are firmly established:

1) one can manage to excite the volumetric discharge with a large specific power input in a plasma column irrespective of RFCD mode with the reduction of inter-electrode gap.

2) in the combined discharge with joint electrodes, the electrode, which causes contraction is the anode in a DC circuit. It can be shown in the following way. When DC current I_{dc} increases in the combined discharge at I_{rf} = $const$, one or several bright spots appear on the anode (see above), and the number of spots increases with the I_{dc} growth. It is important to note, that the I_{rf} increase resulted in disappearance of the specified near anode inhomogeneities. But they occurred again as I_{dc} increase further and at last they could result in the contraction of volumetric combined discharge plasma.

It is possible to conclude, that to determine the mechanism of influence of inter-electrode distance on the RFCD and combined discharge stability it is necessary (apart from near electrode sheaths influence) to take into account non-uniformity of the stationary discharge in direction perpendicular to current, i.e. two-dimension process.

6. Combined discharge in a steady state magnetic field as an active medium for slab gas laser pumping.

One of efficient means of increase the power output of gas lasers is a slab resonator geometry, filled in by active medium. In this case, a minimal size of active medium determines its specific characteristics, and a total laser output power depends on two other sizes. The main difficulties in practical realization of a slab laser exist because it is difficult to fill a narrow slab gap with a plasma with necessary parameters [10].

As it was mentioned in the introduction, the specified problem is solved more efficiently if an RF capacitive discharge is used [10,18-20]. But RF pumping has disadvantages too. Therefore the transition to alternative methods of slab gas lasers pumping, which would allow, at least to reduce the value of RF power, consumed by laser, is an important. The use of combined discharge can be a perspective method to solve this problem.

New type of the combined discharge (Macken Discharge) for the pumping of slab gas lasers was suggested by J.Macken [8].

The flat uniform positive plasma sheaths was produced in [8] by the sectioning of electrodes and the application of a constant magnetic field ($B<0.1$ T) on a glow DC discharge. The orientation of magnetic field was chosen in [8] so, that the Lorentz force, vector was directed along the longer side of chamber. Essential defect of the described method of slab gas laser pumping is the necessity of sectioning of the electrodes, that makes laser design more complicated.

In this work possible way of application of continuous DC electrodes in Macken Discharge was investigated. For this purpose an additional source of gas ionization [9] was added to the discharge excitation circuit. In the simplest case it may be a RF capacitive discharge, located on periphery of the main discharge. However other physical methods of gas ionization can also be used.

The experimental set-up for investigations of influence of a steady state magnetic field on the structure of positive plasma column of DC discharge is shown on fig. 8. The discharge chamber consisted of two water cooled steel plates (used also as magnet poles), the flat working surfaces of which had the dimension $h \times b = 855 \times 197$ mm^2. To prevent shunting of direct current, the surfaces plate was covered by dielectric from the discharge side. A gap between the plates was $d = 9$ mm.

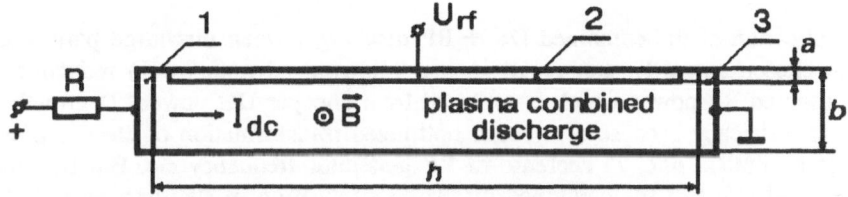

Figure 8. Experimental set-up for studying of magnetically stabilized slab glow discharge. 1, 3 - DC electrodes; 2 - additional RF - electrode.

The magnetic field induction, created by constant magnets, did not exceed 0.1 T. DC electrodes 1,3 (Fig. 8), were made of plane-parallel copper plates, and placed at the ends of the chamber. The local ionization area 2 with the characteristic size a ($a \ll b$, see, Fig. 8) was created by an additional RF discharge. For this purpose an additional electrode corrected to RF source was added. A copper tube, covered by a dielectric, served as an electrode. Steel plates functioned as the second electrode.

Plasma appeared in the area 2 when RF voltage, sufficient for the breakdown of a gas and maintenance of self-sustained RF discharge, was applied to the additional electrode. By appropriate choice of discharge mode and frequency [19] RF discharge consumed small (< 10 %) power, in comparison with main non self sustained DC discharge.

Non self sustained DC discharge appeared at the voltage Udc. At small I_{dc} it was also located in the area 2; its physical characteristics did not differ from those in DC + RF combined discharge, considered above. When I_{dc} increased discharge width l_{dc} was also increased from a up to filling up complete discharge chamber with a plasma, i.e. when $l_{dc}=b$. It is interesting to note, that DC voltage across the electrodes U_{dc} remained and was determined by the RF input power P_{rf} into the additional discharge. At large P_{rf} the smaller U_{dc} was necessary to maintain stable main discharge. However the case, when $l_{dc}=b$ was not always realized. When the induction of a magnetic field was $B < B_{cr}$ (B_{cr} - is defined by a gas composition and pressure, and values d and P_{cr}), the contraction of the combined discharge occurred, before it filled the whole area, i.e. in this case the width of volumetric plasma column l_{cr} did not exceed b. An increase of a magnetic field induction caused l_{cr} growth provided all other conditions remained unchanged.

So already for a relatively low values magnetic field induction $B < 0.1$ T, one can manage to fill in the whole volume of discharge chamber, described above, by a volumetric plasma of the combined discharge,. A pressure of the laser mixture CO_2:N_2:He = 1:1:8 was 15 Torr in that case, and the specific power input reached 4 W / cm^3, that was enough to make a slab CO_2 laser with an average output power. Furthermore, we can easily apply a considered method of gas laser excitation in a pulse-periodic mode, which is preferable for a lot of applications.

7. Conclusion

Application of the combined DC + RF discharge, when discharge parameters are chosen correctly, can result in the following advantages: 1) reduction of consumed RF power which is replaced by a cheaper DC power, 2) growth of laser efficiency because of a better optimized transformation of electric power into an optical one, 3) decrease of RF generator frequency etc. But the most essential effect of the combined discharge application in slab gas lasers is the homogeneous filling of a slab with active medium. It is impossible to solve this problem while using only independent glow DC discharge, that was shown in [1,10], and the RF pumping is too expensive. The letter is especially important for the development of high power slab lasers.

8. Acknowledgement

The author is extremely grateful to Mr. John Macken (Optical Engineering Inc., CA, USA) for his initiation of this work and permanent support.

The research presented in this publication was made possible in part by Grant # N7E000 from International Science Foundation and a Grant from the State Education Committee of Russia.

9. References

1. Crocker, A. and Wills, M.S. (1969) Carbon-dioxide laser with high-power per unit length, *Electron. Letters* **5**, No.4, 63-64.
2. Eckbreth, A.C. and Davis, J.W. (1972) RF augmentation in CO_2 closed-cycle dc electric-discharge convection lasers, *Appl. Phys. Letters* **21**, 25-27.
3. Brown, C.O. and Davis, J.W. (1972) Closed-cycle performance of a high-power electric-discharge laser, *Appl. Phys. Letters* **21**, 480-481
4. Nickols, D.B. and Brandenberg, W.M. (1972) Radio-frequency preionization in a supersonic transverse electrical discharge laser, *IEEE J. Quantum Electronics* **QE-8**, 718-719.
5. Fusayma, T. and Sekiguchi, T. (1975) Effects of radio-frequency preionization on a pulsed CO_2 laser, *Japan J. Appl. Physics* **14**, 735-736.
6. Kozlov, G.I. and Yatsenko, N.A. (1978) Combined discharge with RF ionization. *Sov. Tech. Phys. Lett.* **4**, 171-172.
7. Golubev, V.S., Krivenko, Yu.N., Leonov, P.G. and Flerov, V.N. (1988) Coaxial laser with magnetically stabilization of discharge, *Pis'ma Zh. Tekh. Fiz.* **14**, 1522-15226.
8. Macken, J. (1992) DC Slab CO_2 Lasers, *Proc. of Laser Advanced Materials Processing (LAMP '92)* **1**, 67-72,.
9. Yatsenko, N.A. and Masyukov, I.V. (1993) Glow discharge in transverse magnetic field for pumping thin-channel gas lasers, *Tech. Phys. Lett.* **19**, 480-482.
10. Yatsenko, N.A. (1992) Slot gas lasers, *Bull. Russ. Acad. Sci., Phys.* **56**, 1901-1907.
11. Witteman, W. J. (1987) *The CO_2 Laser*, Springer-Verlag, Berlin, Heidelberg, New-York, London, Paris, Tokyo.
12. Golubev, V.S. and Pashkin S.V. (1990) *Glow Discharge in Elevated Pressure Molecular Gases*, Nauka, Moscow (in Russian).
13. Raizer Yu.P. (1991) *Gas Discharge Physics*, Springer-Verlag, Berlin, New York.
14. Levitskii, S.M. (1957) Investigation of ignition potential high frequency discharge in gas in transition field of frequencies and pressures, *Zh. Tekh. Fiz.* **27**, 970-977.
15. Sen, S.N. and Bhattacharjee, B. (1966) Radio-frequency breakdown in a superimposed dc field, *Can. J. Phys.* **44**, 3270-3272.
16. Brown, S.C. (1959) *Basic Data of Plasma Physics*, Technology Press Willey, New York.
17. Yatsenko, N.A. (1981) Relationship between the high constant plasma potential and the conditions in an intermediate-pressure RF capacitive discharge, *Sov. Phys. Tech. Phys.* **26**, 678-683.
18. Raizer, Yu.P., Shneider, M.N. and Yatsenko, N.A. (1995) *Radio-Frequency Capacitive Discharges*, CRC Press, Boca Raton, Ann Arbor, Tokyo, London.
19. Yatsenko, N.A.(1993) Choice of the combustion regime and the frequency of radio-frequency capacitive discharge for the pump of the gas lasers. *Bulletin of the Russian Academy of Sciences. Physics* **57**, 2156-2163.
20. Colley, A.D., Baker, H.J. and Hall, D.R. (1992) Planer waveguide, 1 kW cw, carbon dioxide laser excited by a single transverse rf discharge, *Appl Phys. Lett.* **61**, 136-138.
21. Yatsenko, N.A. (1994) Methods of rf capacitive discharges space structure investigation, in M.F. Zhukov and A.A. Ovsyannikov (eds.), *Low Temperature Plasma Diagnostic*, Nauka, Novosibirsk, pp. 328 - 373.
22. Rakhimova, T.V. and Rakhimov, A.T. (1975) Stabilization of a gas discharge by an RF electric field, *Sov. J. Plasma Phys.* **1**, 468-470.
23. Raizer, Yu.P. and Shapiro, G.I. (1978) About thermo-ionization instability of glow discharge in alternative fields and stabilizing effect of frequent pulses. *Fiz. Plasmy* **4**, 850-857.

154

24. Hugel, H.E. (1986) RF-excitation of high power CO_2 lasers, *SPIE* **650**, 2-9.

25. Gilinsky, A.P., Kuteev, B.V., Smirnov, A.S. and Shevchenko, Yu.I. (1978) Investigation of the discharge with rotating electric field in molecular gases, *Zh. Tekh. Fiz.* **48**, 2260-2264.

26. Myshenkov, V.I. and Yatsenko, N.A. (1982) Stability of composite discharge sustained by static and RF electric fields. *Sov. J. Plasma* **8**, 397-400.

RECENT ADVANCES IN THE THEORETICAL STUDIES OF LOW TEMPERATURE PLASMAS FOR GAS LASERS

M. CAPITELLI, S. DE BENEDICTIS, C. GORSE AND S. LONGO
Centro di Studio per la Chimica dei Plasmi del CNR
Department of Chemistry- University of Bari(Italy)

Abstract

We present a review of the efforts made by the research group in Bari for understanding plasma kinetics in molecular plasmas of laser interest.
We present two theoretical models. The first is based on the simultaneous solution of vibrational master equations including the dissociation process coupled to the Boltzmann equation for the electron energy distribution function (eedf). The second model adds a master equation for the population of electronically excited states.
Results from the first model are validated by comparing them with experimental results: in particular a satisfactory agreement is found in the vibrational distributions of N_2 and CO for RF plasmas.
The second model is on the contrary used to show the importance of superelastic electronic collisions in affecting eedf in laser mixtures as well as in pure nitrogen discharges.

1. Introduction

Molecular plasmas have been extensively used for generating lasers covering a large spectral range from infrared to vacuum ultraviolet wavelengths. At the same time molecular plasma kinetics has been developed to understand and optimize the output of the relevant lasers.
Apparently all has been done in the domain of gas lasers sustained by electrical discharge even though we believe that a more sophisticated kinetic description of the laser medium could end in the development of new lasers.
In this paper we want to review the efforts made in the last decade to build up self consistent models able to describe the laser medium under consideration. The present review is mainly based on the work performed by our group even

155

W. J. Witteman and V. N. Ochkin (eds.), Gas Lasers - Recent Developments and Future Prospects, 155–167.
© *1996 Kluwer Academic Publishers.*

though excellent work has been done by other research groups in particular by russian ones.

Fundamentally we will distinguish cases in which vibrational kinetics of ground electronic state of molecules takes an essential role (i.e., infrared lasers) from those in which the kinetics of electronic states is dominant (visible, uv, vuv lasers). This distinction is only historical because, as we will show, the two kinetic have strong linking being in any case coupled to the kinetics of free electrons (Boltzmann equation).

2. Theoretical models

2.1 COUPLING OF VIBRATIONAL KINETICS AND FREE ELECTRON KINETICS

Figure 1 shows a schematic representation of the coupling between the system of vibrational master equations and the Boltzmann equation for the electron energy distribution function (eedf). The master equation includes the contribution to the population (depopulation) of the vth vibrational quantum level due to electron-molecule and molecule-molecule energy exchange processes. In particular e-V, e-D, e-I, e-E and e-da terms describe the pumping of vibrational quanta through electron-molecule processes (e-V) and the dissipation of them through dissociation (e-D), ionization (e-I), electronic excitation (e-E) and dissociative attachment (e-da) by electron impact while V-V and V-T terms describe the redistribution and destruction of vibrational quanta by vibration-vibration and vibration-translation energy exchange processes. V-T terms should include the dissipation of vibrational quanta in the collision of vibrationally excited molecules with molecules and open shell atoms, the last being the real killer of vibrationally excited molecules.

The general scheme of figure 1 also includes the deactivation on the walls as well as the pumping of vibrational quanta during atomic recombination process either in gas phase as well on the surface, the knowledge of the relevant rate coefficients being at the moment very scanty.

The kinetics of free electrons is described by the Boltzmann equation that is written in the form of a kinetic equation, the different terms appearing in it being outlined in figure 1. Note that e-M and e-e terms indicate elastic collisions of electrons with molecules and atoms (e-M) and with electrons (e-e).

Keeping in mind that the system of vibrational master equations includes also an equation for a pseudo-level (v' + 1) located above the last bound level v' of the molecule we can understand that our system describes not only the vibrational distribution but also the dissociation process.

COUPLING   Vibrational Kinetics ↔ Electron Kinetics

MASTER EQUATIONS:

$$\frac{d\,N_v}{dt} = \left(\frac{d\,N_v}{dt}\right)_{e\text{-}V} + \left(\frac{d\,N_v}{dt}\right)_{V\text{-}V} + \left(\frac{d\,N_v}{dt}\right)_{V\text{-}T} + \left(\frac{d\,N_v}{dt}\right)_{e\text{-}D} +$$

$$\left(\frac{d\,N_v}{dt}\right)_{e\text{-}I} + \left(\frac{d\,N_v}{dt}\right)_{e\text{-}E} + \left(\frac{d\,N_v}{dt}\right)_{e\text{-}da} + \left(\frac{d\,N_v}{dt}\right)_{Ric} + \left(\frac{d\,N_v}{dt}\right)_{wall}$$

N_v [molecules] [atoms] ↓

↑ $K_{e\text{-}V}$ $K_{e\text{-}D}$ $K_{e\text{-}I}$ $K_{e\text{-}E}$ $K_{e\text{-}da}$

BOLTZMANN EQUATION:

$$\frac{\partial\,n(\varepsilon,t)}{\partial t} = -\frac{\partial\,Jf}{\partial\varepsilon} - \left(\frac{\partial\,Jel}{\partial\varepsilon}\right)_{e\text{-}M} - \left(\frac{\partial\,Jel}{\partial\varepsilon}\right)_{e\text{-}e} + In + Sup + Rot$$

N_v: population density of vibrational level v
$n(\varepsilon,t)$:   number density of electrons with energy between ε and $\varepsilon + d\varepsilon$
$\frac{dJf}{d\varepsilon}, \frac{dJel}{d\varepsilon}$: flux of electrons along the energy axis due to electrical field and elastic collisions

In: Inelastic collisions   (e-V, e-D, e-E, e-da, ...)
Sup: Superelastic collisions $e + A_2\,(v) \rightarrow e + A_2\,(w)$ with $v>w$
Rot: Rotational collisions

Figure 1. General modeling (zero dimensional approach)

The coupling between the system of vibrational master equations and the Boltzmann one occurs through the superelastic vibrational collisions, the inelastic terms involving a given vibrational level and the concentration of atoms that changes the initial composition of primary "cold" mixture.

The solution of the system of equations is started in the so called "cold gas approximation", i.e., at t=0 all the molecules are in the ground state reaching different quasi stationary conditions as a function of time. Explicit equations and details of used cross sections can be found in refs. /1-2/.

The scheme illustrated in figure 1 has been used by our group to describe N_2, He-CO and N_2-CO systems.

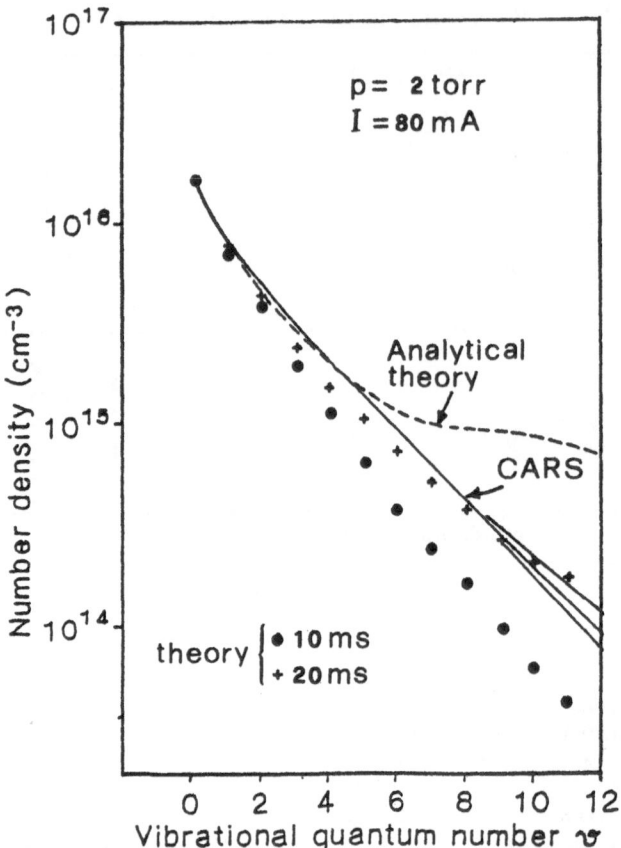

Figure 2. N$_2$ vibrational distributions: comparison between theory and experiment.

In particular figure 2 shows a comparison between the experimental vibrational distribution of N$_2$ measured by Massabieaux et al /3/ by using CARS spectroscopy with the corresponding theoretical values obtained by using the model of figure 1 (see ref. 1 for details). Note that the three experimental curves result from different hypothesis on the concentration of level v=12.

The experiment refers to a DC flowing discharge with a residence time of approximately 10 ms while the theoretical values correspond to an evolution of 10 and 20 ms, the last being in good agreement with the experimental values.

In the same figure we have also reported the theoretical vibrational distribution calculated according to the analytical theory of Gordiets and

Zhdanok /4/ based on the experimental θ_1 (the so called 0-1 vibrational temperature) and T (translational temperature) values. We note that the analytical theory overestimates the distribution due to the fact that the experiment has not reached a stationary situation.

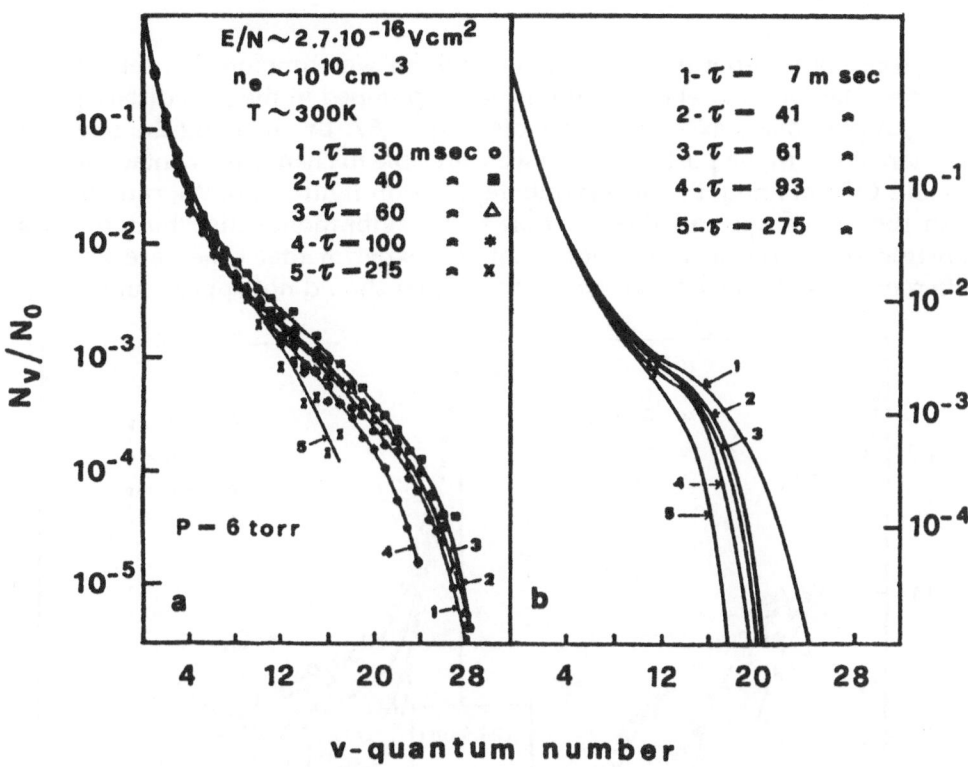

Figure 3 Vibrational population distributions at various residence times
($E/N = 2.7\ 10^{-16}\ Vcm^2$ and $n_e = 10^{10}\ cm^{-3}$),
a: experimental results, b: theoretical results.

Let us now compare theoretical and experimental vibrational distributions of CO in RF flowing discharges. Figure 3a reports the experimental vibrational distributions obtained by infrared emission spectroscopy at different residence times while figure 3b reports the corresponding theoretical values (see ref. /5/ for details). We see that the experimental values present a slight increase in the vibrational distribution in the early times of the evolution starting to decrease from 40 ms on. The theoretical distributions present a monotone decrease, experimental and theoretical behaviors being consistent from 40 ms

160

on. It should be noted that the theoretical values underestimate the populations compared with experimental values probably because of an overestimation of the rate coefficients of the following process

$$CO(v) + CO(v) \rightarrow CO_2 + C$$

Last example is reported in figure 4a-b where we compare theoretical and experimental vibrational distributions of CO pumped in the post discharge by nitrogen previously excited in a RF discharge /6/. By changing the injection position of CO in the post discharge we were able to change the contact time of N_2 and CO obtaining the nice results reported in figure 4a-b. We can observe both the activation and deactivation of CO vibrational distributions as a function of residence time. Note that the experimental times are slightly different from the theoretical ones, a result that should not appear surprising

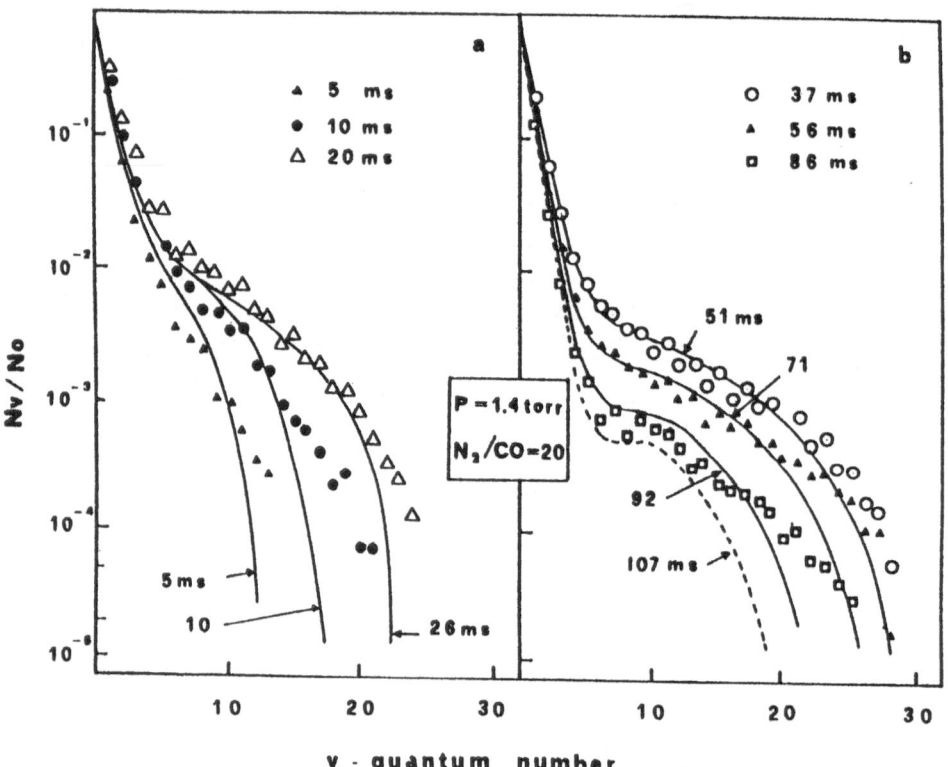

Figure 4 CO vibrational distributions at several contact times. Points: from LN$_2$ cooled experiments. Curves : from theory.

due to the enormous number of processes inserted in the code (see ref. /6/ for details).

The three examples above reported can be considered as indicative of the success made by the plasma kinetics based on the scheme of figure 1 in reproducing the experimental behavior of the vibrational distributions of N_2 and CO in molecular plasmas used for the production of infrared lasers.

2.2 COUPLING OF VIBRATIONAL KINETICS, BOLTZMANN EQUATION AND KINETICS OF ELECTRONIC STATES

The results reported in the previous section have been obtained by completely neglecting the electronic states, which indeed were considered as inelastic losses in the Boltzmann equation. Of course electronic states do exist in a discharge having a noticeable effect on the whole kinetics.

First of all we should consider the coupling of electronic states and free electrons through the superelastic electronic collisions

$$e + M_2^* \ ----> \ e + M_2$$

In these collisions, electrons recover part of the energy they have lost in the corresponding inelastic process.

Superelastic electronic collisions are as more important as lower is the reduced electric field E/N. This point can be appreciated by looking at the results of figure 5a-b /7/ where the electron energy distribution functions of the laser mixture $He-CO_2-CO-N_2$ have been reported for different E/N values and different (parametric) concentrations of the metastable states of $N_2(A^3\Sigma_u^+)$ and $He(^3S)$ states. Note that in this figure T_1 represents the vibrational temperature of 100 and 010 modes of CO_2 while T_2 represents the vibrational temperature of 001 mode of CO_2 as well as those of CO and N_2. We see that the presence of metastable states strongly affects eedf: in particular they shape eedf, the effects being much more evident in the presence of He (^3S). The strong dependence of eedf on the concentration of metastable states propagates in the corresponding rate coefficients as can be appreciated in figure 6 where the rate coefficient for the dissociation of CO_2 by electron impact has been reported as a function of E/N.

The results reported in figures 5a-b and 6 as well as the possible linking of electronic and vibrational states through the so called E-V collisions /8/ indicate the necessity of including a kinetics of electronic states in plasma kinetics. The general scheme reported in figure 1 should be therefore changed in the scheme of figure 7, where we have added a system of electronic master equations describing electronically excited states. Once more the kinetics of an A state includes its pumping trough free electrons as well as bimolecular

162

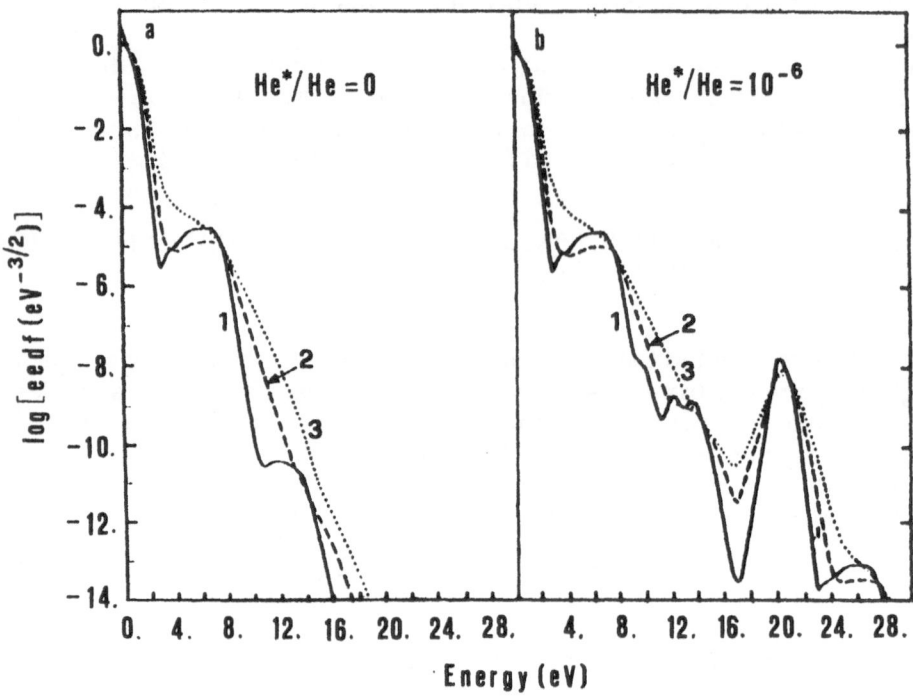

$\log[\text{eedf}(\text{eV}^{-3/2})]$

He*/He = 0

He*/He = 10^{-6}

Energy (eV)

Figure 5 eedf in the laser mixture with $T_1 = 500$ K, $T_2 = 1500$K,
$N_2^*/N_2 = 10^{-4}$, at different values of reduced electric field:
1) E/N = 5 Td, 2) E/N = 10 Td, 3) E/N = 15 Td.

collisions (A-A), quenching (A-X), vibration to electronic conversion (v-A), recombination, spontaneous emission and so on.

This general scheme has been used by our group to understand the kinetics of N_2 discharges. We have developed a sophisticated model that can be used parametrizing the pressure, the E/N value and the electron density /9/. A recent development has also inserted an ionization kinetics that allows us to eliminate the electron density as a parameter /10/.

Typical results of this code have been widely discussed in refs. /9-10/. As an example figure 8 reports the temporal behavior of eedf calculated with the code including the ionization kinetics. We can note the role of superelastic vibrational collisions and superelastic electronic ones in shaping eedf. The strong decay of eedf at longer times is due to the fact that, for the considered conditions, electrons are completely lost at these times being unable to sustain large populations of vibrationally and electronically excited states so that eedf is coming back to the cold gas approximation.

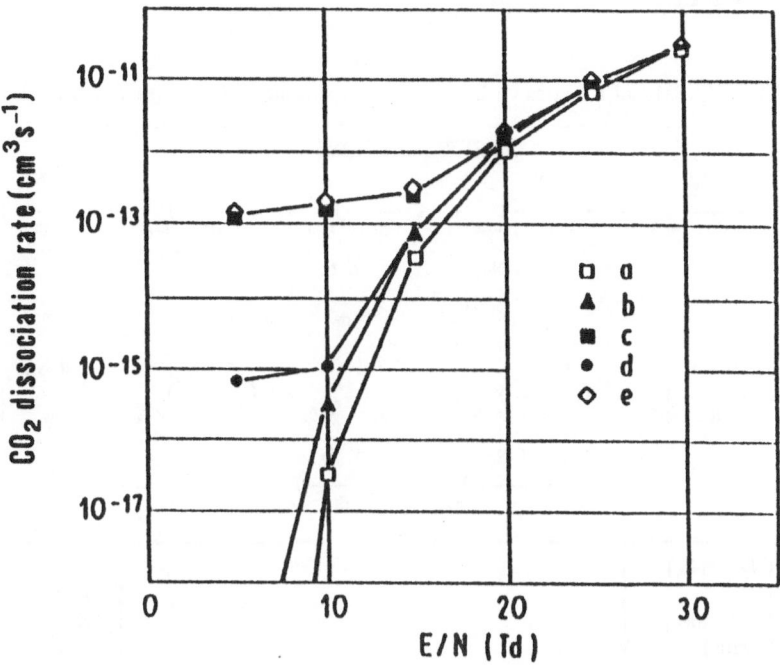

Figure 6 CO_2 dissociation rate coefficient as a function of the reduced electric field in the following conditions: a) $T_1 = 0$ K, $T_2 = 0$ K, $He^*/He = 0$, $N_2^*/N_2 = 0$,   b) $T_1 = 500$ K, $T_2 = 1500$ K, $He^*/He = 0$, $N_2^*/N_2 = 0$,   c) $T_1 = 500$ K, $T_2 = 1500$ K, $He^*/He = 0$, $N_2^*/N_2 = 10^{-4}$,   d) $T_1 = 0$ K, $T_2 = 0$ K, $He^*/He = 10^{-6}$, $N_2^*/N_2 = 0$,   e) $T_1 = 0$ K, $T_2 = 0$ K, $He^*/He = 10^{-6}$, $N_2^*/N_2 = 10^{-4}$.

The results of excimer laser kinetics sustained by an RCL circuit have been reviewed by Longo /11/.

It should be noted that we have studied the coupling of excimer laser kinetics and eedf in the RF bulk plasma, i.e., by solving the Boltzmann equation in a sinusoidal field $E = E_0 \cos\omega t$ /12/.

RF capacitively coupled discharges are now widely used for laser generation: the theoretical approach is completely different due to the formation of space charge near electrodes. To this end a PIC-MCC (particle in cell with Monte Carlo collisions) has been developed in our laboratory and coupled with the

164

vibrational kinetics /13-14/. This model is able to furnish non-local eedf as a function of the distance between the electrodes (parallel plate configuration), non-local vibrational distributions as well as the distribution of electrical field in the reactor.

COUPLING   Vibrational Kinetics ↔ Electronic Exited States Kinetics ↔ Electron Kinetics

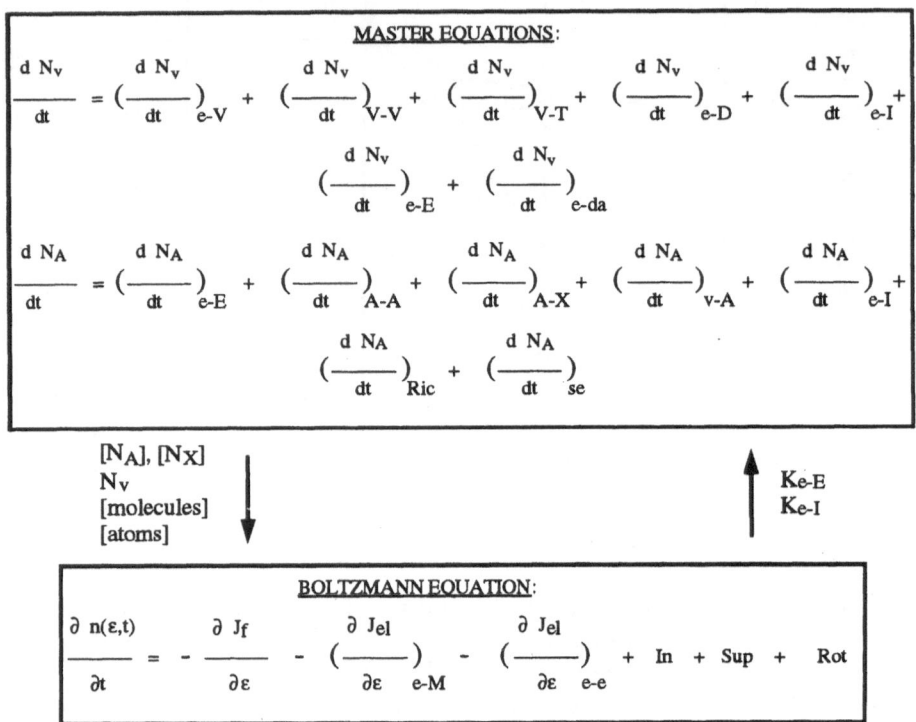

$[N_A]$ $[N_X]$: population density of electronic excited state A_2^*, (X_2^*)

$n(\varepsilon,t)$:   number density of electrons with energy between ε and $\varepsilon + d\varepsilon$

$\dfrac{dJ_f}{d\varepsilon}$, $\dfrac{dJ_{el}}{d\varepsilon}$:   flux of electrons along the energy axis due to electrical field and elastic collisions

In: Inelastic collisions   (e-V, e-D, e-E, e-da, ...)

Sup: Superelastic collisions $e + A_2(v) \rightarrow e + A_2(w)$ with v>w
                                     $e + A_2^* \rightarrow e + A_2(v)$

Rot: Rotational collisions

Figure 7. New modeling scheme (zero dimensional approach)

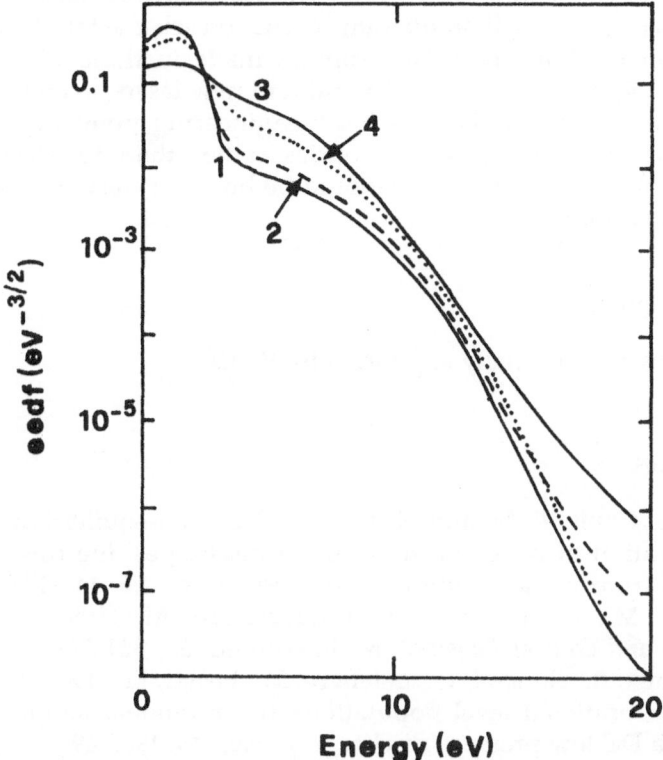

Figure 8 Temporal evolution of the N_2 eedf ($ne_{(t=0s)} = 10^{11} cm^{-3}$, E/N = 80 Td,
p = 2 torr, R = 1 cm), (the different curves correspond to
1: t = 4.5 10^{-8}s, 2: t = 4.3 10^{-4}s, 3: t = 1.6 10^{-3}s, 4: t = 1.0 10^{-2}s).

3. Conclusions

In this paper we have briefly review the efforts made by our group to describe
the kinetics of molecular plasmas. A very sophisticated plasma kinetics has
been developed in the past years. This approach could be used to rationalize
(and to optimize) infrared (He-CO; He-CO$_2$-CO-N$_2$) and uv (N$_2$) lasers, the
development of which were based on crude models.
Moreover this approach could also be used to better understand laser
instabilities /11/.
As a further example one could suggest that the vuv H$_2$ laser could be
improved at the light of the large experience accumulated by plasma kinetics

in describing negative ion sources /15/. The vibrational excitation of H_2 in this kind of discharges is based on the same excitation scheme for H_2 laser.

In conclusion we hope that the progress made in these years by plasma kinetics can help the development of old and new lasers, even though we are aware of the fact that today the laser engineering point of view is more linked to classical and quantum optics rather than to plasma kinetics. Divergence of the different fields should be however very dangerous for the future of gas lasers. ·

4. Acknowledgments

This work has been partially supported by ENEA.

5. References

1. Cacciatore, Capitelli, M. and Gorse, C. (1982) Non-equilibrium dissociation and ionization of nitrogen in electrical discharges: the role of electronic collisions from vibrationally excited, *Chem. Phys.* **66**, 141-152
2. Capitelli, M., Celiberto, R. and Cacciatore, M. (1994) *Adv. Atomic, Molecular and Optical Physics,*, M. Inokuti ed. **33**, 321-372.
3. Massabieaux, B., Gousset, G., Lefebvre, M., Pealat, M. (1987) Determination of $N_2(X)$ vibrational level populations and rotational temperatures using CARS in a DC low pressure discharge, *J. Phys.* **48**, 1939-49
4. Gordiets, B. and Zhdanok, S. (1986) Analytical theory of vibrational kinetics of anharmonic oscillators, *Top. Curr. Phys.* **39**, 47-84
5. De Benedictis, S., Capitelli, M., Cramarossa, F., d'Agostino, R., Gorse, C. and Brechignac, P. (1983) Vibrational kinetics in liquid nitrogen cooled 5% He-CO radio frequency discharges, *Optics Comm.* **47**, 107-110
6. De Benedictis, S., Capitelli, M., Cramarossa, F. and Gorse, C. (1987) Non equilibrium vibrational kinetics of CO pumped by vibrationally excited nitrogen molecules: a comparison between theory and experiment, *Chem. Phys.* **111**, 361-370
7. Colonna, G., Capitelli, M., De Benedictis, S., Gorse, C. and Paniccia, F. (1991) Electron energy distribution functions in CO_2 laser mixture: the effects of second kind collisions from metastable electronic states, *Contrib. Plasma Phys.* **31**, 575-579
8. Wallaart, H.L., Piar, B., Perrin, M. Y. and Martin, J. P. (1995) Transfer of vibrational energy to electronic excited states and vibration enhanced carbon production in optically excited V-V pumped CO, *Chem. Phys.* **196**, 149-170

9. Gorse, C., Cacciatore, M., Capitelli, M., De Benedictis, S. and Dilecce, G. (1988) Electron energy distribution functions under N_2 discharge and post-discharge conditions: a self-consistent approach, *Chem. Phys.* **119**, 63-70

10. Gorse, C. (1993) Non equilibrium plasma modeling, *Proceedings III - XXI International Conference on Phenomena in Ionized Gases*-G. Ecker, U. Arendt and J. Boseler eds, 141-148

11. Longo, S. (1996) Excimer laser kinetics, *this book*

12. Capitelli, M. and Gorse, C. (1990) *Non-equilibrium processes in partially ionized gases*, Capitelli, M. and Bardsley, J. N. eds, Plenum Press, 45-61

13. Longo, S. and Capitelli, M. (1994) Coupling of space-dependent electron dynamics and vibrational kinetics in radio frequency discharges in nitrogen, *Phys. Rev. E* **49**, 2302-2306

14. Longo, S., Gorse, C. and Capitelli, M. (1994) Coupling vibrational kinetics of nitrogen molecules and electron dynamics in a PIC-MCC model of a parallel plate RF discharge, *12th ESCAMPIG*, 119-120

15. Gorse, C., Celiberto, R., Cacciatore, M., Lagana', A. and Capitelli, M. (1992) From dynamics to modeling of plasma complex systems: negative ion (H^-) sources, *Chem. Phys.* **161**, 211-227

XeCl* LASER KINETICS

S. LONGO
Centro di Studio per la Chimica dei Plasmi del CNR and
Dipartimento di Chimica dell'Università
Via Orabona 4, 70126 Bari (Italy)

Abstract

We discuss some modern issues in the XeCl* laser kinetics, which appear as open problems arising from detailed analysis of the results of complex computer models. In particular we consider the problem of vibrational kinetics of HCl molecules in the discharge plasma and the mechanism of discharge instability. The enphasis is on fundamental problems, mainly on the limitation of the macroscopic approach to study such far-from-equilibrium systems

1. Introduction

The aim of this contribution is to discuss some recent results in the theoretical study of the non-equilibrium chemical kinetics of the XeCl* laser. This laser has a number of important applications related to its high power and the fact that it produces UV radiation (λ=308 nm) useful for photo ablation processes and dye laser pumping. In recent times it has been proposed to use the highly focusable radiation of these devices to produce very high temperature plasmas interacting with various materials, to be used as novel sources of soft X-rays [1]. Despite this remarkable properties, XeCl* lasers suffer from some limitations, that could be solved only with a better understanding of the non-equilibrium chemical kinetics of the discharge plasma. This understanding is the goal of up-to-date theoretical research, mainly founded on extensive computer models.

This work deals with the results obtained in this direction in the last few years. To these results our group has given a valuable contribution. In the following we will focus specially on the electric-discharge pumped XeCl* laser based on a Ne/Xe/HCl mixture, because of its importance from the points of view of both fundamentals and applications. Two open problems will be considered in the next sections: the vibrational kinetics of HCl molecules in

W. J. Witteman and V. N. Ochkin (eds.), Gas Lasers - Recent Developments and Future Prospects, 169–183.
© 1996 *Kluwer Academic Publishers.*

the laser mixture and the chemical induced instability of the discharge. In the last part of the paper we propose a view of the discharge instability in XeCl* as a stochastic process characterized by non-Poisson distributions of concentration of chemicals, and point at the weakness of the traditional rate equation approach in this light.

2. Vibrational Kinetics of HCl molecules

In the XeCl* laser mixture the chlorine is provided usually by including HCl in the laser mixture composition: it is the most practical way, since pure chlorine cannot be used because of the high absorption cross section of Cl_2 molecules to the laser radiation. Alternative chlorine donor molecules such as BCl_3 were also considered, but at the present stage of development they cannot be considered technologically relevant. For these reasons a good understanding of the HCl molecules kinetics in the XeCl* laser medium is needed.

The knowledge of the non-equilibrium vibrational distributions of HCl is important to calculate the rate of formation of Cl^- which together with Xe^+ is believed to be the main precursor of XeCl* excimer molecules. At the same time, the vibrational kinetics of HCl has an important role in the development of discharge instabilities.

It is quite surprising that the literature has a strong tendency to underestimate the importance of this question, surely because the introduction of a really complete vibrational kinetics of HCl molecules increases the computational effort.

The main point of complexity is that the population of different vibrational levels of HCl molecules does not follow the Boltzmann law: this is not surprising since the laser operates far from equilibrium. It follows that the rate coefficients of different elementary processes in the plasma involving HCl molecules cannot be characterized by a single parameter such as the vibrational temperature T_V, and one has on the contrary to use a state-to-state approach by solving the vibrational Master Equation, that is the equation for the time evolution of the population of different vibrational levels [2]. More strictly speaking, the equation to solve is the rate equation, becuse the Master Equation would involve multiple-particle distribution functions, which are reduced to products of single-particle ones by invoking the molecular chaos: we get in this way the nonlinear rate equation from the linear Master Equation.

In any case, the transition probabilities entering this equation are *linear functionals* of the eedf, this last being obtained by solving the Boltzmann equation, and not simply (non-linear) *functions* of macroscopic parameters.

This approach has been followed in the literature only partially, by taking into account from one up to three vibrational levels, and very simplified descriptions of the kinetics itself. All these descriptions have been considered somewhat equivalent, even if it has been shown by several authors that changing the vibrational kinetics scheme in a given XeCl* laser model has a strong effect on macroscopic quantities of interest, especially the time evolution of the discharge current and the value of the self-sustaining voltage in the quasi-steady state.

In the model developed by our group in successive steps in the last few years [3,4,5], the non equilibrium vibrational kinetics of HCl is taken into account by solving the vibrational Master Equations coupled to the Boltzmann equation for the electron energy distribution function, taking into account a quite comprehensive list of processes involving 7 vibrational levels of HCl molecules.

The following processes have been inserted in our model:

    a) introduction of vibrational quanta by e-V processes

$$e + HCl(v) \quad \text{<->} \quad HCl^- \quad \text{<->} \quad e + HCl(w)$$

this process raises a very important conceptual problem, because the scaling laws connecting the rate coefficients for excitation processes

$$e + HCl(1) \text{ --> } e + HCl(v), \text{ and}$$

$$e + HCl(n) \text{ --> } e + HCl(v+n)$$

to each other are unknown. It was shown by Gorse at al. [4] that different guesses for the form of these scaling laws lead to strongly different results for the vibrational energy distribution function.

    b) redistribution of vibrational quanta by V-V processes

$$HCl(v) + HCl(w) \quad \text{<->} \quad HCl(v') + HCl(w')$$

    c) destruction of vibrational quanta by V-T (vibration-translation) processes

$HCl(v) + HCl$	->	$HCl(v') + HCl$	v>v'
$HCl(v) + H$	->	$HCl(v') + H$	v>v'
$HCl(v) + Cl$	->	$HCl(v') + Cl$	v>v'

    d) direct dissociation by electron impact, dissociative attachment, recombination as well as detachment (M is a buffer gas atom)

$e + HCl(v)$	->	$e + H + Cl$
$e + HCl(v)$	->	$H + Cl^-$
$H + Cl^- + M$	->	$HCl + M$
$H + Cl^-$	->	$HCl + e$

172

Extensive computations performed using this model, whose results are reported in the literature, show that:

1) even if the vibrational distribution function of HCl molecules is close to a Boltzmann one in the steady self-sustained stage of the discharge, it is strongly non-Boltzmann in the initial stage, because of the different characteristic times necessary to the populations of different vibrational levels to reach the steady state

2) at the end of the steady state associated with the discharge voltage plateau the vibrational deactivation of molecules is mainly due to VT processes, specially the H + HCl(v) molecule collisions: this observation invalidates the long-time results for the most of the models discussed in the literature

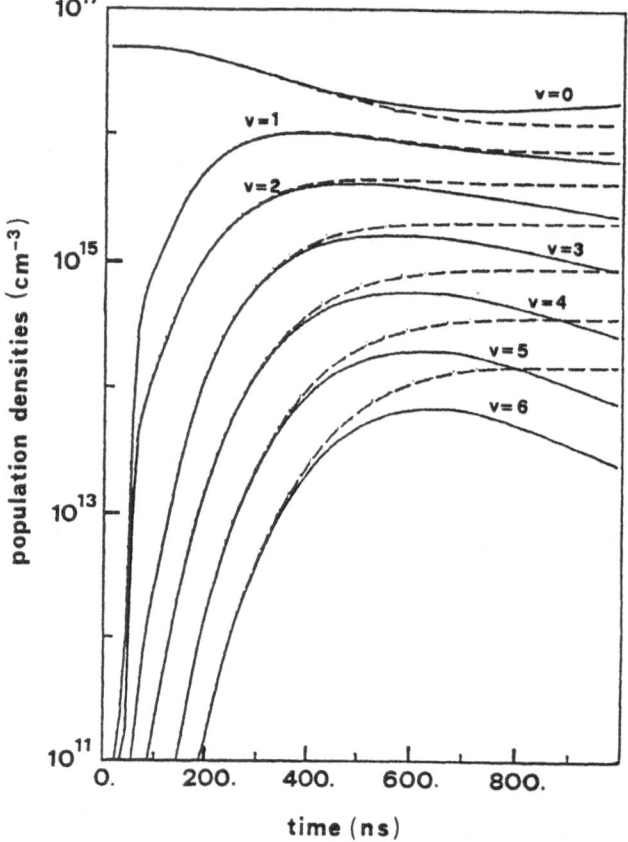

Figure 1 concentrations of ground state and vibrationally excited HCl molecules calculated using a self-consistent model [5], reported as a function of time. The vth curve from the above refers to molecules in the $(v-1)^{th}$ vibrational level. Dashed curves refer to a model neglecting VV and VT processes

3) inclusion of the process of direct dissociation of HCl molecules by electron impact has a non-negligible effect on the results. This effect is due probably to the vibrational deactivation of HCl molecules in collisions with H atoms which are in turn produced by dissociation, but can also be the effect of the reduction of the total amount of HCl molecules.

These results can be better appreciated by inspecting figure 1, which shows the temporal evolution of the concentration of the HCl molecules in different vibrational levels, calculated for the 10-liter XeCl* laser HERCULES dveloped by the group of Letardi *et al.* [1]

For the operating conditions considered and other details the reader is invited to refer to the paper by Longo *at al.* [5]. What is important to observe here is that the full curves refer to a comprehensive chemical kinetics, while the dashed curves are obtained neglecting VT and VV processes: the difference in the results of the two approaches is evident in the last stage of the discharge, where the vibrational temperature for the full model starts to decrease, while it keeps a constant value in the second model. The most important process responsible for the difference between the two results is the VT due to HCl+H collisions.

This last effect allows giving a tentative answer to the problem: 'It is necessary or not to include a highly detailed vibrational kinetics of HCl in XeCl* models?'. The answer is that if the mole fraction of HCl is low, which is the case for normal XeCl* mixture, problably the attachment rate can be calculated not too inexactly by using three vibrational levels: this is because the cross section for attachment does not increase too much with the vibrational quantum number v if v>2. The situation is completely different with dissociation kinetics: for all cases in which dissociation of HCl by electron impact is important, such as high-current devices and developed instabilities, one must be aware that we are presently not able to calculate the rate for this process. Infact, dissociation rate depends strongly on the shape of the vibrational distribution function especially in the region of high v, which in turn cannot be accounted without knowing the exact scaling laws of cross section, the detailed effect of VT and VV processes, and many other informations still unknown.

It follows from the above conclusions that our understanding of the vibrational kinetics of HCl is too poor to claim for quantitative results specially for high-current or large aperture devices.

3. Current positions on the problem of HCl kinetics

We can now spend some words on the impact on the literature of the results reported in the last years about the importance of the vibrational kinetics of HCl in XeCl* laser discharges. We underline that our group was the first one to raise *explicitly* this question, and precisely in the two papers by Gorse *et al.* [4], in collaboration with the french group directed by J.Bretagne et Longo *et al.* [5], in collaboration with the russian group directed by A.P.Napartovich. Before the work of Gorse *et al.* the only attempt to include a detailed vibrational kinetics of HCl in a XeCl* laser model was in the work of Kannari *et al.* [6], which however was devoted to an e-beam pumped laser, and (this is the main difference with respect to our work) no discussion on the real relevance of the vibrational kinetics of HCl was included. The situation was completely different, as it is easily understood, for papers devoted to the modeling of pure HCl laser: in these papers, largely due to the group directed by M.Capitelli at the Bari University, it was considered a very refined vibrational kinetics of HCl molecules, by solving coupled Master / Boltzmann equations and considering tenth of vibrational levels, whose populations were affected by electron impact processes, VT, VV, and so on. These pioneer papers provided the insights for the successive work on the XeCl* lasers. It is the case to acknowledge at this point the important successive contribution of Estocq, Delouya and Bretagne [7], the first and last authors being co-authors of our 1991 work: this paper is in some respects parallel to our work of 1992 [5], but it is more extensive, it refers to a different device and, principally, it includes the discussion of other phenomena and open problems, such as the importance of the metastable channel in the generation of XeCl* molecules and the memory effect (though not called in such a way in the work) in the parametric time evolution of the electron drift velocity as a function of the reduced electric field, due in turn to the generation in the discharge plasma of atomic and molecular metastable states. In this last paper was also observed a further time that at *low mole fractions* of HCl the effect of highly vibrationally excited molecules (v>3) on the *attachment* rate (certainly not on the dissocation one) coefficient is also low, even if four levels are in any way essential (and this absolutely is not a trivial point, as it can be appreciated by a birdeye look at the literature). On the contrary, some authors have considered critically these results in successive papers, but regrettably not without some misunderstandings. For example, in a paper by Luck *et al.* [8] it is observed that the effect of highly vibrationally excited HCl molecules on model results is low: this is absolutely not surprising, however, since in their model there is no VT process to deactivate the molecules, and therefore the production of H atoms by direct dissociation of HCl could not affect the kinetics! In another paper by Riva *et al.* [9] there was an attempt to demonstrate that detailed vibrational kinetics of HCl molecules is not essential by comparing model and experimental results for some quantities characteristic of the electrical pumping circuit for a the XeCl* laser discharge: this paper is very interesting and the calculations reported are rigorous, but unfortunately it has the important flaw to consider conditions in which our kinetic model also gives as a result that highy excited levels of HCl molecules are inessential (in fact the authors used essentially for the vibrational kinetics the scheme used by our group and the one of J.Bretagne). All these works considered our work in critical light. This criticism, however, fails in a very important point: it does not touch the essential arguments of our two works of 1992 and 1993. The point is that we do not know sufficiently well the vibrational kinetic of HCl molecules in order to provide quantitative results: we have observed for example at the present stage the scaling laws for the vibrational excitation cross sections (see above) are unknown: so which is the origin of the cross sections used for cascade excitation by all these authors which are all able to reproduce their experimental results? In fact they use, in this respect, completely different schemes. With these question open we pass to the next section of this work, devoted to another (far more recognized) fundamental problem: the (in)stability of XeCl* laser discharge.

4. XeCl* discharge instabilities

In the last few years has been increasingly better established that the limitation to the pulse duration in XeCl* discharge-pumped lasers is due to the development on the discharge bulk of micro streamers-like instabilities due in turn to chemical induced plasma instabilities. This view of the XeCl* discharge is very characteristic, because the usual processes claimed to generate micro instabilities in discharge plasmas are of more physical nature (e.g. ion-acoustic instabilities, electrode effects, thermal instabilities). In the case of XeCl* discharge, the discharge instability seems presently to be due to a so-called Halogen Depletion Instability (HDI). This instability mechanism is assumed to be the result of the following circumstances:

a) electrons in the discharge are mainly destroyed due to dissociative attachment to HCl

b) there is no mechanism active in the XeCl* laser kinetics to restore in the laser operation times the HCl molecules destroyed by the dissociative attachment

c) the external circuit drives the gap voltage to the self-sustaining quasi-stationary value in order to realize the *global* balance of electron density gain and loss in the whole reactor

As a result of statements (a), (b), and (c) it follows that the XeCl* laser discharge is unstable, even with flat electrode and uniform electric field, with respect to any perturbation of the electron density in a plane normal to the applied electric field. it is very interesting to observe the local nature of this instability, in the sense that no instability process seems to take place by looking only at space-averaged quantites, such as the discharge current and voltage. The pioneering work in this field was the one by Coutts and Webb [10], in which the effect of halogen depletion on the local stability of the XeCl* plasma was demonstrated by using a macroscopic approach based on balance equations for HCl concentration. This paper was really clarifying, but at the same time it openend the new question to test the proposed mechanism in the framework of a chemical kinetics scheme more realistic than the highly simplified one used by Coutts and Webb. For example, these authors considered only one vibrationally excited level of HCl, and they ignored many alternative mechanism of HCl destruction in the discharge, such as direct dissociation by impact with electrons or excited Xe atoms, enhanced by vibrational excitationand.

Many theoretical works in the literature have shown that the HDI effect actually produces a large-scale discharge instability in one or two-dimensional models of XeCl* laser. The most important change in the general point of view in the last few years concern the scale of the instability phenomenon: the HDI can in fact be viewed either as a volume contraction of the discharge or as a trigger for micro-instabilities. In the first case, it is assumed that the instability is a macroscopic phenomenon with smooth space variation of th related quantities, triggered by large-scale disuniformities in the electron density due in turn to limitations of the pre-ionizing device: this large scale non-uniformity have been evaluated through the space variation of the X-ray dose due to the pre ionizing device. In the second case, the HDI is an intermediate stage in the development of micro streamers in the discharge gap: these micro streamers could be triggered by microscopic defects of the electrode surface leading to small-scale isolated regions characterized by strongly higher than average electric fields, and would continue their evolution through physical plasma instabilities. Holographic investigations [11] of a large-volume XeCl* laser discharge showed both volume contraction of the discharge and micro streamers propagating form the cathode and gradually filling the plasma. The effect of these phenomena on the laser pulse is mainly of optical nature: far before the plasma stability is seriously compromised, laser pulse will terminate due to the spoiling of optical quality of the medium, due in turn to absorption and scattering of light by the non homogenous plasma regions. Approximately two years ago our group performed a comparative study [12] in order to establish the relevance of the two kinds of instability to determine the length of the laser pulse. The study was performed by using an 1D model based on the Parallel Resistor Network (PRN) concept [13], which consists in dividing the discharge volume into plasma elements which are not connected in no way but through their contribution to the total resistance of the plasma, as 'seen' by the pumping circuit. The model calculations have shown that the concept of volume instability, with realistic chemical parameters and initial electron density perturbations, is quantitatively unable to explain the difference between theoretical and experimental pulse lengths for large-volume devices, leaving as only explanation of the discrepancy the development of a micro streamer instability into the discharge plasma. In the same work it was also shown that the necessity of invoking an effective discharge instability rules out from the kinetics any fast recombination proce ss between positive ions and electrons, such as the $NeXe^+ + e$, because these processes would provide a strong stabilizing mechanism for the discharge, specially in the case of volume instability. In any case, it should be understood that since HDI is due to dissociative attachment process, and since the dissociative attachment rate coefficient cannot be calculated quantitatively without a state-to-state approach based on the vibrational Master Equation, it can be stated that at present stage only very few reliable calculations exists of the HDI development in the XeCl* discharge plasma, even for the simpler volume instability. As a demonstration of this problem, we can quote the results obtained in recent calculations by Dem'yanov A.V. et. al [14] by using a very complete self consistent 1D model, which show that the HCl in the laser mixture can, to some extent, *stabilize* the discharge in same conditions, because a local excess of electrons leads to stronger vibrational excitation of HCl molecules, therefore increasing the dissociative attachment rate coefficient and consequently the rate of electron loss. The name Halogen Attachment Stability (HAS) has been proposed for this phenomenon, in analogy to HDI.

5. Beyond rate equation models for XeCl* lasers

Very recently we have considered the possibility that the mechanism of XeCl* laser discharge instabilities could be a stochastic one. This idea arised by working on particle models of plasma kinetics based on the Monte Carlo method with some insights from the works of Van Kampen [15] and I.Prigogine (see for example the recent popular review by Nicolis and Prigogine [16]) on the stochastic description of the chemical reactions out of equilibrium.

The basic idea is to look for the fluctuations of the concentrations of chemicals as a function of space, that is to study the distribution function $f(N/\Omega)$ of the number of particles N in a given volume Ω, onsidered as a random variable: the mean value of this variable divided by Ω, that is $<N>/\Omega$ is by definition the particle number density n that enters into the macroscopic rate equations. The point is that looking at $f(N/\Omega)$ instead of n gives us access to a lot of informations which are hidden in the macroscopic approach, in particular the fluctuations of N/Ω with respect to n.

The approach to the problem we are going to propose is the following: attachment controlled discharges are characterized by larger fluctuations of the electron $<N>\Omega^{-1}$ with respect to recombination controlled or simply diffusion dominated, and this explains the triggering of microinstabilities which are subsequently amplified by HDI. This is a very complex research program, which cannot be completed in a short time, it is just possible here to provide the recipe which could be followed:

1) get a 1D particle-in-cell with Monte Carlo collisions model (PIC/MCC), with periodic conditions ,

2) add an electric field in a direction perpendicular to the mathematical mesh axis, in order to sustain the discharge,

3) sample particle number densities in different mesh point and time and construct $f(N/\Omega)$.

To anticipate some possible conclusions of this program, we have perfomed a preliminary calculation by using a Cellular Automaton (CA) model of an attachment-controlled kinetics. The CA considered is a 1D version of the more involved 2D one of Dab and Boon [17], which was used to study the effect of fluctuations in different systems. As it is known, the CA are mathematical models of the world based on intrinsically discrete space and time: in CA the 'quantum' of space is the CA cell, which has a limited number of available 'states'. The CA cell changes from one state to another according to the status of the same cell and the closest ones. In the Dab and Boon CA the cell states (here and in the following we limit ourselves to the discussion of the 1D automaton which is of our concern, while the oridinal Dab and Boon CA is two-dimensional) are chosen to represent up to two particles in it, moving with opposite velocities: the states are therefore: 1) (empty), 2) (->), 3) (<-), 4) (<-/->). We assume that these particles represents electrons. the diffusion process is introduced by a random rotation of 180° of the whole cell, that is according to the table 1->1, 2->3, 3->2, 4->4 with probability p_1: the higher p_1, the lower the diffusion coefficient D. Our aim is to study the properties of a first-order controlled kinetics, therefore we introduce the two processes:

'ionization':	'attachment':
1 -> 2 or 3 (random choice)	1 -> 1
2 -> 4	2 -> 1
3 -> 4	3 -> 1
4 -> 4	4 -> 2 or 3 (random choice)

It is clear that the entry 4->4 in the 'ionisation' table is nonphysical. It is the manifestation of the so-called 'exclusion principle' in CA, which follows from the necessity to limit the number of states available to the cell. To avoid conceptual problems in the interpretation of the results, we took care to keep the CA gas dilute, that is, more than 90% of the cells are in the status 1 (empty) at any time. To simulate the control on the electron concentration of the pumping circuit we have introduced the probabilities of 'ionization' p_2 and attachment p_3 according to:

$$p_3 = p_2 \text{ cost } N^{-1}$$

where N is the total number of electrons in the CA.
According to this formula, the number of particles in the CA is controlled by the balance of ionization and attachmnent up to the stability point one, which would be N=cost if the exclusion principle were completely ruled out, but is infact a little lower. Let us now examine the results: the characteristics of the CA considered are the following:

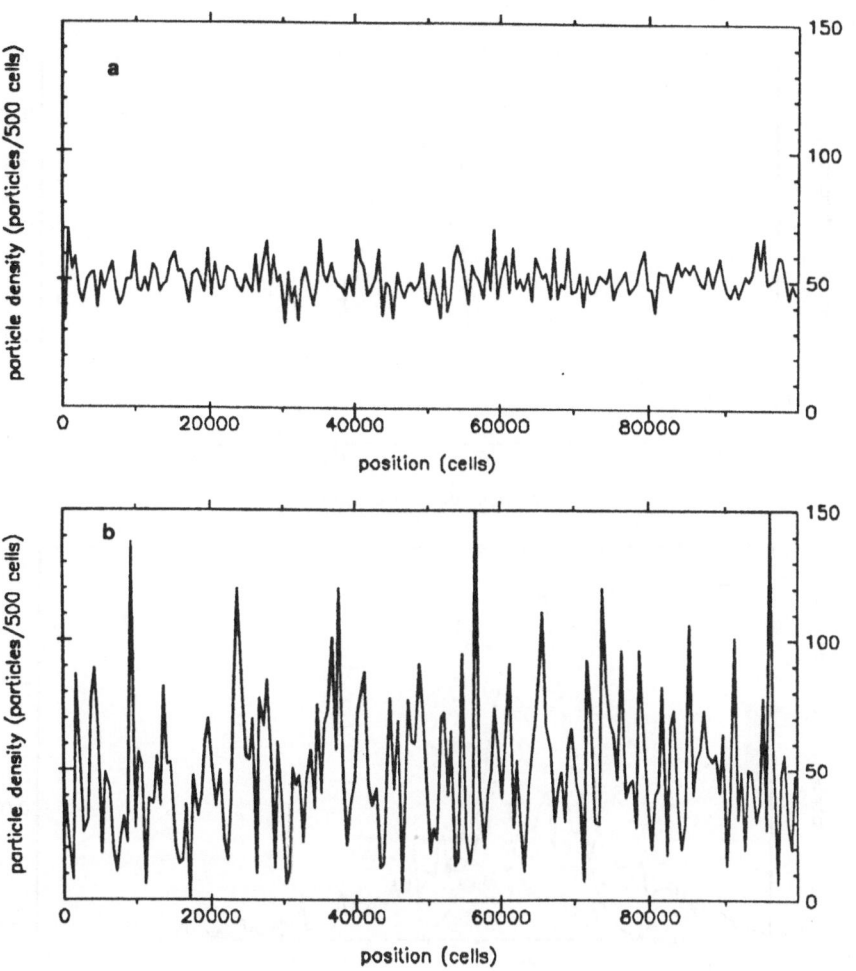

Figure 2 particle concentration in a simplified reaction-diffusion model of attachment-controlled kinetics based on a one-dimensional Cellular Automaton. (a): no kinetic process, only diffusion (b): result obtained including kinetic processes

180

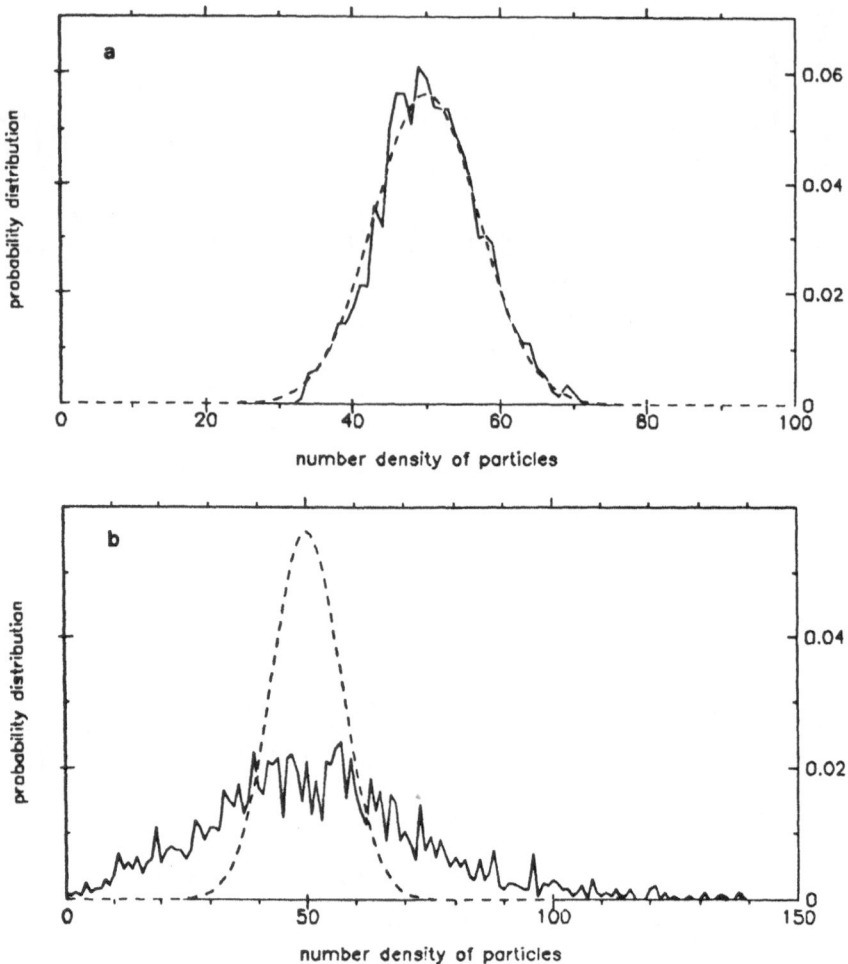

Figure 3 particle distribution functions in a simplified reaction-diffusion
model of attachment-controlled kinetics based on a one-dimensional Cellular
Automaton. The dashed line is the corresponding Poisson distribution (a): no
kinetic process, only diffusion (b): result obtained including kinetic processes

number of cells = 10^5

$p_1 = 0.2$

$p_2 = 0.02$

$p_3 = 0.025 \ 10^4 \ N^{-1}$

In figure 2-3 it is possible to see the comparison of the result after 10^3 time steps. In particular figure 2 shows the values of N/Ω with $\Omega=500$ cell widths, as a function of position. The upper part of the picture shows the result when only diffusion processes are active, while the lower part shows the same results with the kinetic processes inserted.

It is clear that the amount of fluctuation is higher in the presence of fast I-order chemical processes, with unefficient mixing. This point can be better appreciated by inspecting figure 3, which shows the probability distribution of N/Ω as sampled from a collection of results of the kind presented in figure 2. It is clear that, while the diffusion-controlled system follows a Poisson statistics (the dashed curve is a Poisson distribution with $\mu = 50$), this is not the case for the kinetic-controlled one. By the way, as it was shown by Nicolis and Prigogine [16], the non-Poisson fluctuation spectrum rules out the use of macroscopic equations containing only concentrations of chemicals in the form $n=<N>\Omega^{-1}$, i.e. the rate equations.

In conlusion: *if our view of the attachment-controlled discharge instability is correct, this phenomenon cannot be studied by using the rate equations* . We leave also this problem, hopely interesting enough to deserve further studies for example by PIC/MCC methods, completely open, and shift to the conclusion of this work.

6. Conclusions

In this paper some recent issues in the XeCl* laser kinetics have been examined in a critical perspective as open problems, deserving further investigations. The problem when developing computer models for technologically interesting devices is in fact that any fundamental problem spontaneously arising during the study and not previously scheduled in research programs is usually 'solved' is some ways just in order not to spend too much time on. The consequence is that the different authors in the literature concerning XeCl* laser modeling seem to accept without any problem the fact that their simplified kinetics is completely different from the kinetics of any other author, and nevertheless all of them are able to reproduce the respectively available experimental results. These approach to the problem excludes, by the way, any possibility to really discover weak points in the foundations of generally accepted modeling technique. The analysis of the discharge instability presented above, for example, is a (very rough) attempt to solve a problem of the XeCl* laser modeling by completely changing the modeling technique, and not simply the values for some rate coefficients or cross sections. Paradoxically, the present situation, in which the interest in modeling of these lasers is decreasing after the frantic production of last years could be the best one to understand many critical issues in the XeCl* laser kinetics, while waiting for the new interest which certainly will arise when operating these lasers to pump X-ray sources. Only a time-consuming but rigorous approach based on complex self-consistent models with detailed state-to-state vibroelectronic kinetics of molecules and electron one based on the solution of Boltzmann equation and possibly particle models of the XeCl* laser plasma could clarify the many uncertain points still present in this field.

Acknowledgment

This work was partially supported by ENEA

References

1. Letardi T. (1996) Large Aperture Discharge Pumped Excimer Lasers as Drivers for Soft X-Ray Sources, *this volume*

2. Capitelli M., De Benedictis S., Gorse C. and Longo S. (1996) Recent Advances in the Theoretical Studies of Low Temperature Plasma for Gas Lasers, *this volume*

3. C. Gorse, (1990) Non equilibrium excimer laser kinetics. *Nonequilibrium Processes in Partially Ionized Gases* (Ed. by M. Capitelli, J.N. Bardsley) Plenum Press, New York, 411

4. Gorse C., Capitelli M., Longo S., Estocq E., and Bretagne J. (1991) Non-equilibrium vibrational, dissociation and dissociative attachment kinetics of HCl under high electron density conditions typical of XeCl* laser discharges, *J.Phys.D* 24, 1947-1953

5. Longo S., Capitelli M., Gorse C., Dem'yanov A.V., Kochetov I.V., and Napartovich A.P. (1992) Non-equilibrium vibrational, attachment and dissociation kinetics of HCl in XeCl* selfustained laser discharges, *Appl.Phys.B* 54, 239-245

6. Kannari F., Kimura W.D., and Ewing J.J. (1990) Comparison of model prediction with detailed species kinetic measurements of XeCl laser mixtures, *J.Appl.Phys.* 68, 2615-2631

7. Estocq E., Dolouya G., Bretagne J. (1993) Self-consistent modeling of X-ray preionized XeCl*-laser discharges, *Appl.Phys.B* 56, 209-221

8. Luck H., Loffhagen D., Botticher W. (1994) Experimental verification of a zero-dimensional model of the ionization kinetics of XeCl* discharges, *Appl.Phys.B* 58, 123-132

9. Riva, R, Legentil, M., Pasquiers S. and Puech V. (1993) Ionization-attachment balance in Ne-HCl pulsed discharges, *J.Phys.D* 26, 1061-1066

10. Coutts J. and Webb C.E. (1986) Stability of transverse self-sustained discharge-excited long-pulse XeCl* lasers, *J.Appl.Phys.* 59, 704-710

11. De Angelis A., Di Lazzaro P., Garosi F., Giordano G.,and Letardi T. (1988) XeCl Laser Diagnostic by Holographic Interferometry, *Appl.Phys.B* 47, 1-6

12. Longo S., Comunale G., Gorse C., and Capitelli M. (1993) Simplified and Complex Modeling of Self-Sustained Discharge-Pumped, Ne-Buffered XeCl Laser Kinetics, *Plasma Chem. Plasma Proc.* 13, 685-700

13. Kushner M.J., (1991) Microarcs as a Termination Mechanism of Optical Pulses in Electric-Discharge Excited KrF Lasers, *IEEE Trans. Plasma Sci.* 19, 387-399

14. Dem'yanov A.V., Kochetov I.V., Napartovich A.P., Longo S., Capitelli M. (1995) Theoretical Studies on Microscopic and Macroscopic Non-Uniformities in Electric-Discharge-Excited XeCl laser, *Plasma Chem. Plasma Proc.*, in press

15. Van Kampen N.G. (1981)*Stochastic Processes in Chemistry and Physics*, Elsevier Science Publisher B.V.

16. Nicolis G., Prigogine I. (1987) *Exploring Complexity. An Introduction.* Piper GmbH & Co KG, Munich

17. Dab D. and Boon J.P. (1990) Cellular Automata Approach to Reaction-Diffusion Systems, in *Cellular Automata and Modeling of Complex Physical Systems*, Ed. P.Manneville *et al.*, Springer Verlag, 257-273

OPTICAL PUMPING AND FERRITE FLASH DISCHARGES

S.V. MITKO*, F.A. VAN GOOR, W.J. WITTEMAN
*Department of Applied Physics, University of Twente,
P.O. Box 217, 7500 AE, Enschede.
The Netherlands.*

V.N. OCHKIN
*Lebedev Physics Institute, Russian Academy of Sciences,
Leninsky pr.53, 117924, Moscow.
Russia.*

* Permanent adress: Lebedev Physics Institute, Russian Academy of
Sciences, Leninsky pr. 53, 117924, Moscow, Russia.

Abstract

Non-coherent optical pumping of gas lasers benefits over various electron
excitation techniques because of the absence of electron quenching and transient
absorption by molecular and atomic ions. Another advantage is the absence of different
discharge instabilities deteriorating the laser performance.

The absorption bands of the overwhelming majority of molecules, however, are
located in the vacuum ultraviolet spectral region where conventional flash lamp
pumping methods are not applicable.

Therefore the most important problem encountered in the design of optically
pumped gas lasers is the development of a pulse periodic, reliable VUV light source
with a geometrical length of about 1m.

In this paper a short review of different optically pumped gas lasers is presented.
It is pointed out that the blue-green XeF(C→A) laser is the most promising
representative of this class of lasers due to its great potential for spectroscopic
applications, and particularly for the amplification of ultra short pulses.

Two advanced techniques for optical excitation of this laser are considered in
detail.

In the first device a Formed Ferrite Plasma Source (FFPS) powered by a 5kJ
energy pulse is used to produce the VUV pumping radiation. A laser output energy of
41mJ is obtained on the XeF(C→A) transition in a 1μs (FWHM) pulse with 4%

W. J. Witteman and V. N. Ochkin (eds.), Gas Lasers - Recent Developments and Future Prospects, 185–204.
© 1996 *Kluwer Academic Publishers.*

outcoupling at 485nm. More than 9% of the energy stored in a capacitor bank is emitted by the FFPS in the absorption band of the parent XeF$_2$ molecule.

The second apparatus utilises a Sliding Discharge on a Conducting Surface (SDCS) powered by the same energy. An output energy of 0.22J at 485nm is obtained in a 1μs pulse with 5% outcoupling.

A comprehensive analysis of the electric, gas-dynamic and radiative properties of these pumping sources is given.

A comparison of the results, obtained with different optical excitation techniques, is carried out.

1. Introduction

Optically pumped gas lasers with excited species produced by photo dissociation or photo excitation of parent molecules have extensively been studied and developed since the early 1970's [1]. Compared with alternative techniques like electron beam or discharge excitation, optical pumping has some important advantages related with the absence of free electrons and ions in the pumping scheme. Because of this there is no electron quenching and mixing of excited states, no transient absorption of the laser radiation by atomic and molecular ions and there is no limitation of the pulse duration and laser output energy related to discharge instabilities. So, in principle, it is possible to generate laser output with unlimited long pulse duration on wavelengths from the infrared to the ultraviolet. Figure 1 shows some typical parameters of laser radiation that has been obtained up to now with optically pumped gas lasers.

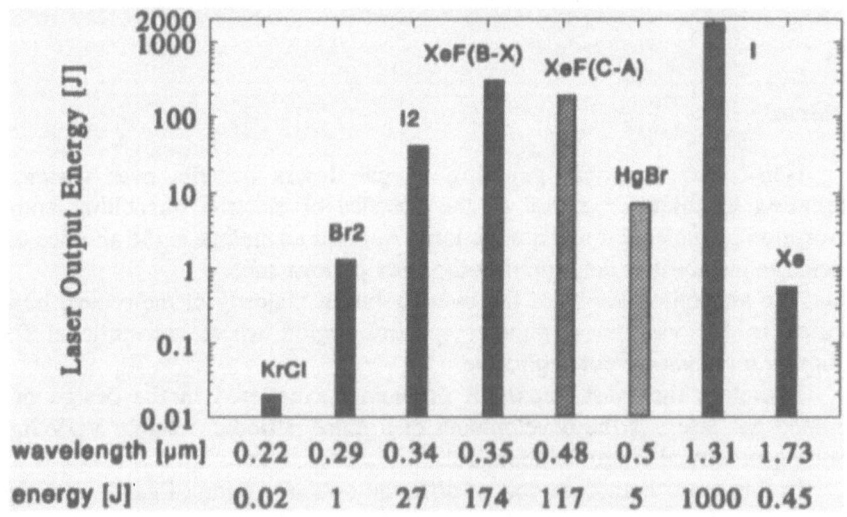

Figure 1. Output laser energy available with optically pumped gas lasers.

One of the most developed laser of this class is the iodine laser emitting on a wavelength of 1.3µm. Pulsed laser energies up to 1kJ with an average power of 500W and an efficiency of 2% have been generated with this laser. The characteristic feature of the iodine laser is that an absorption band of the active medium (230-310nm) lies in the near UV spectral region. So, optical pumping of this laser can be realised using Xe flash lamps [2]. Another example of a 'soft' excited medium is the UV (342nm) molecular iodine laser which has an absorption band near 200nm. Pulse-periodic operation of this laser, using flash lamp pumping, has also been demonstrated [3]. However, unlike the iodine laser, the absorption bands of the majority of molecules and atoms are located in the VUV spectral region. In this respect, the infrared (1.7µm) atomic Xe laser excited by direct ionisation of Xe followed by recombination is an example of a system pumped in the extreme ultraviolet spectral region with wavelengths shorter than 103nm [4]. This example demonstrates that it is possible to create photo-ionisation plasma lasers with recombination pumping excited by the broad-band thermal plasma radiation. Obviously large photo-ionisation bands of atoms and molecules combined with the low temperature of the photo-electrons provide beneficial conditions for fast electron-ion recombination. This results in a pumping scheme surpassing that of the conventional plasma creation by electric discharges or e-beam excitation techniques.

Among the lasers optically pumped with photons in the VUV spectral region, the XeF(C→A) laser has become an object of continuously growing interest [5]. In figure 2 a simplified energy level diagram of the XeF molecule is shown.

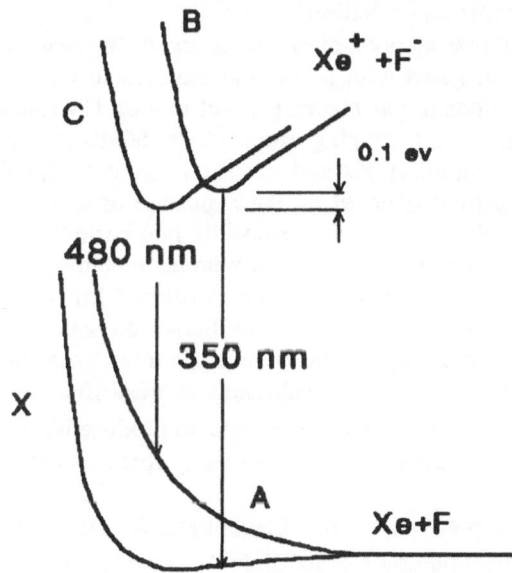

Figure 2. Simplified energy level diagram of the XeF molecule.

Excited XeF can be produced by photolyse of a parent XeF_2 molecule by photons with wave lengths between 140nm and 180nm. There are two lowest Coulomb states, $XeF(B_{1/2})$ and $XeF(C_{3/2})$ separated in energy by nearly 0.1eV. The quantum efficiency of the B-state production is 0.85 and the cross-section for stimulated emission is for the B→X transition (351nm) more than 10 times that of the C→A transition (480nm). Furthermore, the B and C states are coupled by collisions with molecules of the buffer gas. Therefore, in thermal equilibrium at 300K, more than 95% of the combined population of the B and C states is in the lower C state. Thus, although the cross-section for stimulated emission of the C→A transition ($\sim 10^{-17} cm^2$) and the quantum efficiency for its production are small compared to the competitive B→X transition, this drawback is compensated for a great deal by the much higher inversion.

The blue-green XeF(C→A) laser is very attractive for spectroscopic applications because of its tunability. This is due to the bound-free transition to a highly repulsive A state resulting in a homogeneously broadened gain profile with a band width of about 70nm. Also, the broad band width of the gain allows the amplification of ultra-short pulses with durations down to a Fourier transform limitation of about 10 femtoseconds that is short compared with the ~100 femtoseconds limitation for conventional excimers caused by their much shorter gain band-widths. At the same time, the XeF(C→A) medium has a relatively high saturation flux of about $10MW/cm^2$, so it can be used for the generation of high power, tuneable radiation.

In particular due to these properties, the XeF(C→A) laser has become a touchstone for testing various excitation techniques.

It has been found that an open discharge is one of the most suitable light sources for optical excitation of gases with absorption features in the VUV spectral region. During early investigations it was shown [6] that intense UV and VUV photon fluxes with brightness temperatures ranging from 30 to 50kK can be produced in gas mixtures with various compositions and pressures using exploding wire techniques. However, since this method is based on the explosion of a thin metal wire for each shot, it has an inevitable problem concerning its non-periodic operation in practical applications. To avoid this problem other techniques including sliding arc-discharges on various type of surfaces and materials were developed [5]. At present it is clear [7] that the radiative properties of open surface-discharges do not depend on the method of initiation as long as the discharge channel develops in a dense surrounding gas. Thus one of the main trends in modern investigations of VUV line-sources is the extensive search for the most reliable and simple technique to produce high-brightness (≥25kK), long extended (~1m) discharges in dense gases (~1bar) using relative low voltages (<50kV).

In this paper we report two advanced techniques for discharge initiation and their application for the optical pumping of the XeF(C→A) laser. The first method utilises a Formed Ferrite Plasma Source (FFPS) in which the surface discharge is initiated by a thermal explosion of a treated surface of high resistive ferrite [8,9]. The second one deals with a Sliding Discharge on a Conducting Surface (SDCS) [10], which is the resistive analogy of a discharge sliding on a dielectric surface, but with some essential advantages compared to the latter method.

2. Experimental arrangement I

Two experimental devices were used in our investigations. The first one utilised a FFPS line source as shown in detail in figure 3.

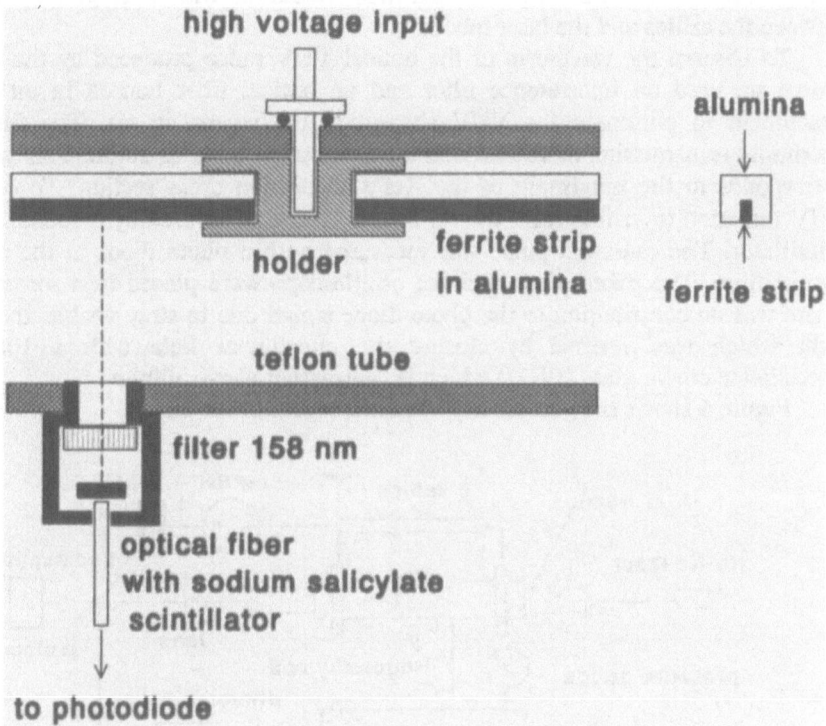

Figure 3. Design of the FFPS discharge device.

The high-brightness surface discharge was excited with an assembly of separate elements each consisting of a 1.7x4x140mm long $(NiZn)Fe_2O_4$ ferrite strip making up a discharge channel with a total length of 55cm. To create a conductive path on the surface of the strip, first a carbon line was drawn with a pencil. Than, this carbon line was burned by applying a voltage of 220V AC from the public electric network. This process led to the formation of a high-conducting amorphous layer on the ferrite surface [11]. After connection of the elements to a high voltage capacitor discharge circuit, an electrical surface discharge develops due to the thermal explosion of the thin amorphous filament. To prevent broadening of the conductive filament due to the influence of the discharge plasma the ferrite strips were glued into 8x8x140mm alumina blocks. An assembly of five elements was mounted near the wall of a 50mm inner diameter, 100cm long laser tube that was made of Teflon. Each pair of elements was held by a 50mm long stainless steel holder. The high-voltage connection was through the wall of the laser tube.

The light source was operated with energy supplied by five $1\mu F$ capacitors in parallel charged to a voltage up to 45kV. A bundle of six 2.5m long high-voltage coaxial cables connected a triggered spark-gap, mounted on the capacitor bank, to the laser tube. The discharge voltage was measured as a function of time with a resistive voltage divider, the discharge current was measured with a Rogowsky coil fitted between the cables and the laser tube.

To observe the waveform of the optical VUV pulse produced by the FFPS line source we used an interference filter and an optical fiber housed in an evacuated attachment to eliminate the VUV absorption by oxygen in air. The filter had a maximum transmission at 158nm and a transmission band of 20nm. This wavelength corresponds to the maximum of the XeF_2 absorption cross section. To convert the VUV radiation to visible light the tip of the fibre was covered by a sodium salicylate scintillator. The radiation pulse was measured with a photo diode at the exit of the optical fibre. The photo diode and the oscilloscope were placed in a screened room. There was no contribution to the photo diode signal due to stray visible and near UV light which was verified by closing the attachment hole with a 10mm thick Borosilicate crown glass (BK-7) which is transparent above 300nm.

Figure 4 shows the general experimental layout of the device.

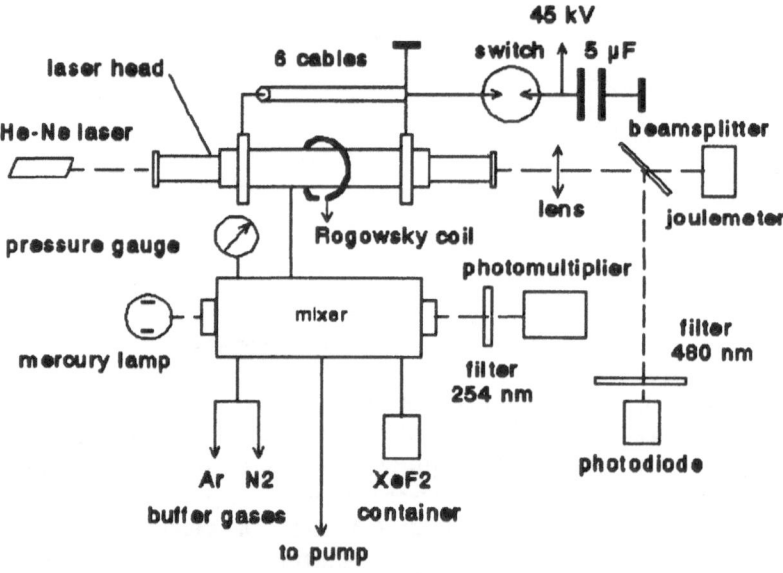

Figure 4. General layout of the FFPS driven laser.

For the experiments described in this section we used a combination of the laser tube and a gas mixer. The mixer was made of a 20cm inner diameter, 100cm long glass tube sealed with stainless steel flanges. Before each shot the whole system was evacuated to a pressure less than 0.5Torr. The XeF_2 vapour was produced from solid crystalline XeF_2 kept in a separate container attached to the mixer. The XeF_2 vapour was admitted to the evacuated mixer first followed by the addition of argon and nitrogen. The XeF_2 concentration inside the mixer was verified by the absorption of radiation from a mercury lamp at a wavelength of 254nm where the XeF_2 absorption cross section is $1.15x10^{-17}cm^2$. It was found that the life time of the XeF_2 vapour inside our mixer was about two hours (at 50% reduction of the initial concentration), so we did not care about preliminary passivation of the apparatus. The working mixture was admitted from the mixer to the evacuated laser tube about five minutes before the shot. When the system was operated as a laser, we used a mixture of 435Torr argon, 125Torr nitrogen and varying amounts of XeF_2. All experiments were performed at room temperature.

A stable cavity was formed using a 10m radius mirror and a flat out-coupler with 4% transmission mounted on the flanges of the laser tube and separated 218cm. Both 50mm diameter mirrors had dielectric coatings for lasing on the C→A transition of XeF with >85% transmission for the competing B→X transition near 350nm. A Joule meter measured the energy of the laser pulse and a photo diode behind a 480nm interference filter measured the time dependency of the laser power. The clear laser aperture was 30mm in diameter with a distance between the optical axis of the cavity and the ferrite surface of 13.6mm.

In additional experiments we investigated the dynamics of the bleaching wave driven by the FFPS light source in the active medium. During these experiments the laser mirrors were replaced by BK7 glass windows. The photo diode was placed behind a diffusor and an image of the diffusor surface, determined by a diaphragm was built inside the laser tube using a 50cm focus length lens. The axes of the FFPS assembly and that of the photo diode, diaphragm and lens were parallel. The latter sytem could be shifted in a direction perpendicular to the ferrite surface over the entire laser aperture. With this method we could observe the wave form of XeF(C→A) luminescence at different distances from the ferrite surface. It was also possible to determine the speed of the discharge expansion. The spatial resolution of the system was estimated to be 2.5mm in the direction perpendicular to the optical axis.

In the course of the experiments performed with this set-up we soon observed that large amounts of dust were produced caused by the ablation of the inner wall of the Teflon tube by the intense VUV radiation emitted by the surface discharge as well as by the discharge plasma covering the whole laser tube at the latest stage of the discharge development. As a result this effect led to undesirable contamination of the laser mirrors. It was also found that the erosion of some of the ferrite samples had a non homogeneous character as long as the ferrite operated close to the breakdown conditions. Pieces of ferrite material of 0.2-0.3mm size were splashed from the ferrite surface due to strong thermal tensions during the ~1ms duration pre-breakdown stage

of the discharge formation. In general it should be noted, that the remaining $(NiZn)Fe_2O_4$
samples have demonstrated an acceptable performance with homogeneous surface erosion of about $0.4\mu m$ per pulse at an input energy level of 70-100J per cm discharge length. No changes of the radiative properties of the light source were observed during about 10^3 pulses [9].

To avoid trouble with tube-wall ablation and to improve the performance of the element initiation a second apparatus was designed and tested.

3. Experimental arrangement II

In the second apparatus we used another technique of surface discharge based on a Sliding Discharge on a Conducting Surface (SDCS). SDCS develops on the surface of a conducting material in the form of a plasma leader that moves from the negative to the positive polarity electrode. Once the plasma leader covers the entire distance from cathode to anode, a high current arc discharge is initiated on the surface. It was shown in reference [10] that the mechanism of SDCS is of a non-thermal nature and is related with the acceleration of electrons by a strong electric field ahead the tip of the plasma leader. These electrons ionise the cold surrounding gas creating the conditions for the further plasma expansion.

The design of the device is shown in detail in figure 5.

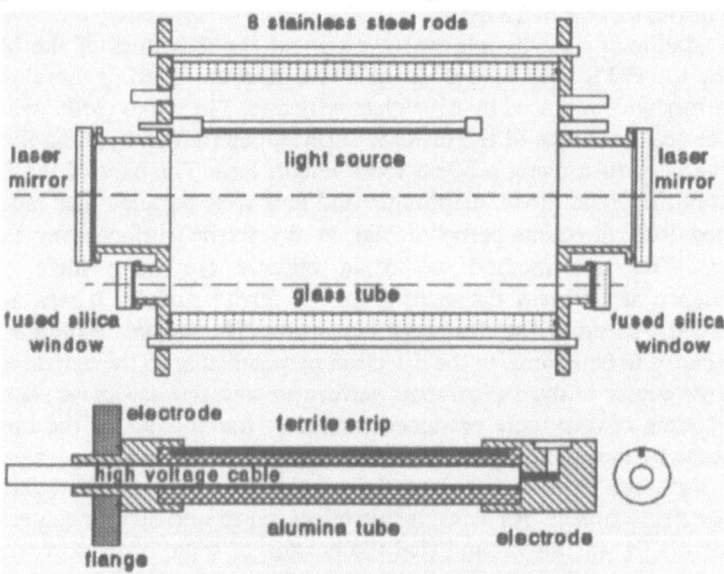

Figure 5. Design of the SDCS discharge device.

A 3x4x900mm long rod assembled with several elements of NiMn ferrite with a specific resistance of $500\Omega cm$ was used to excite a high current surface discharge. The ferrite elements, placed end-to-end, were glued in a slab machined in a 100cm long alumina tube with a 25mm internal and a 10mm external diameter. Stainless steel electrodes were glued on the alumina tube carrying the ferrite elements. To obtain a coaxial geometry of the electrical circuit the inner conductor of a coaxial cable was passed inside the alumina tube to the high-voltage electrode. The grounded electrode was connected to the cable shield through an O-ring sealed flange at the end of the alumina tube. The alumina tube assembly was mounted on one of the two flanges sealing a 150cm long, 30cm inner diameter glass tube with a wall thickness of 1cm. These flanges also contain adjustable holders for the laser mirrors, fused silica windows for monitoring the XeF_2 partial pressure, as well as connections for the gas and vacuum lines. All vacuum sealing was performed using Viton O-rings.

The general experimental layout is shown in figure 6.

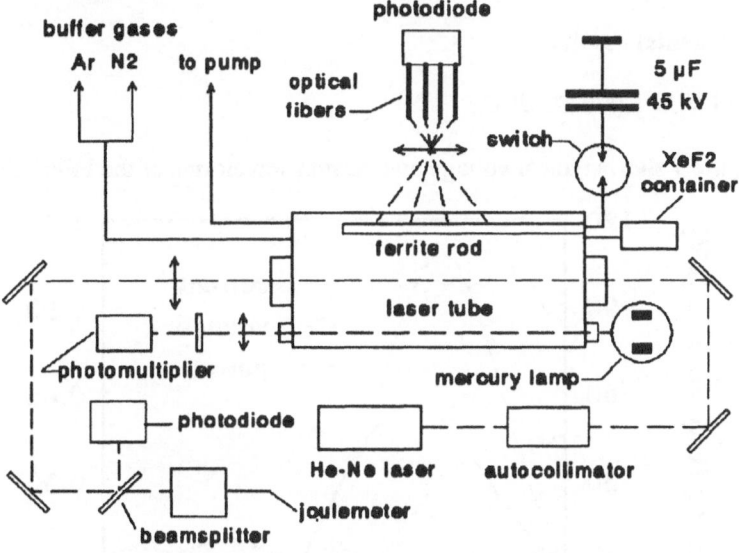

Figure 6. General arrangement of the SDGS driven laser.

All further details concerning the capacitor-spark gap assembly, XeF_2 vapour density verification inside the laser tube, measurements of the laser energy and time dependency of the laser power were the same as in the first experimental arrangement described above. The XeF_2 vapour was admitted directly to the laser tube which was evacuated prior to the experiment down to a pressure of 5×10^{-2} Torr. The life-time of the XeF_2 vapour inside the laser tube was found to be about 16hours (at halve of its initial magnitude) so there was no need to passivate the present device. After the admission of XeF_2, the buffer gases, argon and nitrogen, were added.

A cavity attached to the laser tube consisted of two flat dielectric mirrors separated by a distance of 174cm. The clear laser aperture was 8.5cm in diameter. One of the mirrors had a reflection of 99.9% and the other, out coupling mirror, had a transmission of 5%, both at the XeF(C→A) emission wavelength. The transmission of the mirrors at the wavelength of the competing XeF(B→X) wavelength at 350nm was larger than 85%. Prior to each shot the alignment of the mirrors was checked with an autocollimator having an accuracy better than 3".

During additional experiments the dynamics of the development of the SDCS along the ferrite surface was investigated. For this purpose, the images of the entrance tips of four optical fibres were built on the ferrite surface by a lens. The optical signal from the exit fibre ends was measured by a photo diode. Step-like changes in the optical signal could be observed when the plasma leader was crossing the image positions. In this way we were able to determine the moment upon which the plasma leader was at a definite position. The spatial resolution of this detection method was equal to the size of the image of the optical fibre entrance tip, about 5mm.

4. Experimental results

4.1. FFPS VUV LINE SOURCE

Figure 7 shows typical voltage and current waveforms of the FFPS discharge.

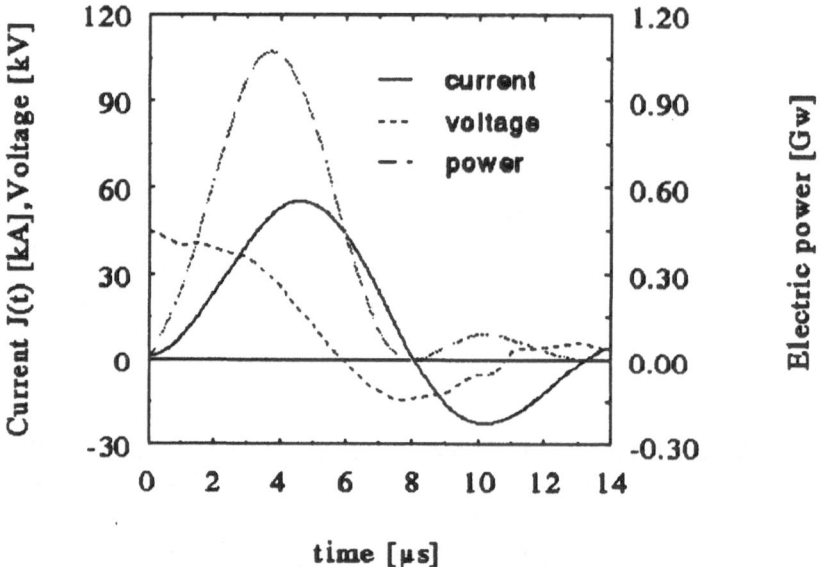

Figure 7. Current-voltage waveforms of the FFPS discharge.

The duration of the main current pulse is about 8µs followed by a second highly damped half period with a duration of 5µs. A maximum current of 55kA was measured at 45kV charging voltage of the 5µF capacitor bank. The power of the electric pulse, calculated from these data, has a duration of 4µs (FWHM) and a peak value of about

1GW at 4µs after the ignition of the discharge. The energy delivered to the plasma during the first half period of the current pulse is 4275J which is 85% of the total energy stored in the capacitor bank. There were no changes in the electrical characteristics when the XeF_2 partial pressure was varied up to 2.5Torr. The waveform of the VUV light pulse, observed in the direction perpendicular to the ferrite surface, is shown in figure 8.

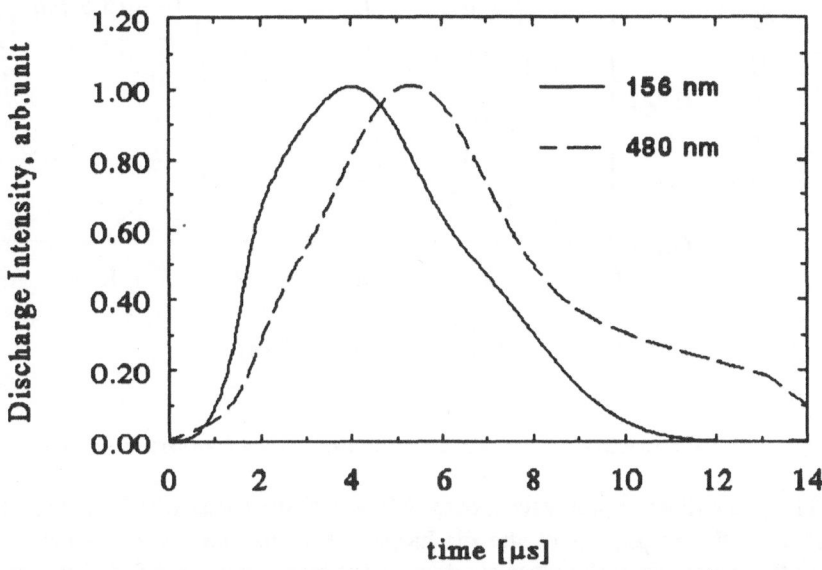

Figure 8. VUV pumping pulse produced by the FFPS discharge.

For comparison, the waveform of the visible light, emitted at 480nm, is also shown in figure 8. It can be seen that the waveform of the VUV pulse is close to the waveform of the electrical power, while the visible light pulse has been shifted toward a later stage of the discharge.

The light intensity measured along the discharge axis at different distances from the ferrite surface is shown in figure 9. During these measurements the XeF_2 density was kept at a level of $6.8 \times 10^{16} cm^{-3}$ (2Torr at 20°C). The first peak in the observed waveforms corresponds to the $XeF(C \rightarrow A)$ luminescence. The subsequent increment of the signal is related to the light emitted by the expanding discharge column. It is clearly seen, that the dissociation of the parent XeF_2 molecules proceeds in the form of a photo dissociation wave as reported by Zuev et al [5]. Space-time dynamics of the discharge and the photo dissociation wave, deduced from these data, are shown in figure 10. It is observed that the speed of the discharge expansion does not change during the main current pulse and equals to $2kms^{-1}$.

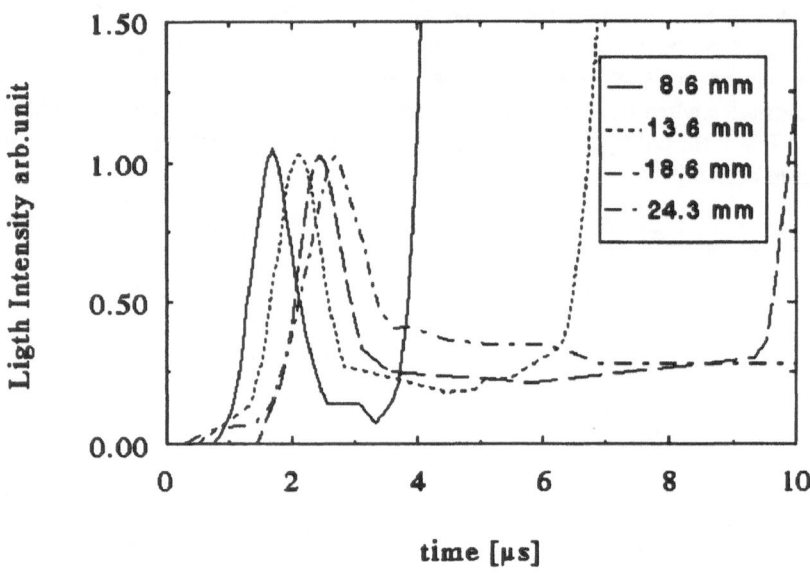

Figure 9. Side light intensity at four different distances from the ferrite surface.

The photo dissociation wave moves with a variable speed which is about 18kms^{-1} at 2µs after the beginning of the discharge. At 2.7µs the photo dissociation wave reaches the boundary of the clear aperture of our laser tube. This fact shows strikingly that only a small part of the light energy, emitted by the FFPS line source in the pumping band of XeF_2, can be used effectively. This motivated us to increase the laser aperture in our second experimental set up. Combining the space-time dynamics of the discharge and the photo dissociation wave with simple considerations of the photo dissociation process of the XeF_2 molecule [12] we could determine the absolute light intensity in the VUV spectral region near 158nm coming from the FFPS discharge.

Figure 11 shows the time-dependent brightness temperature of the FFPS plasma deduced by this method. The brightness temperature reaches its peak value of 30kK at 2µs after the beginning of the discharge and than slowly decreases. The space-time performance of the photo dissociation wave, calculated with that brightness temperature, is shown in figure 10 by a solid curve. It is seen that our FFPS discharge can produce a complete photolyse of XeF_2 in a half cylinder volume with a radius of 8cm. Using these data the total amount of light emitted in the absorption band of XeF_2 was found to be 476J or 9.4% of the energy stored in the capacitor bank.

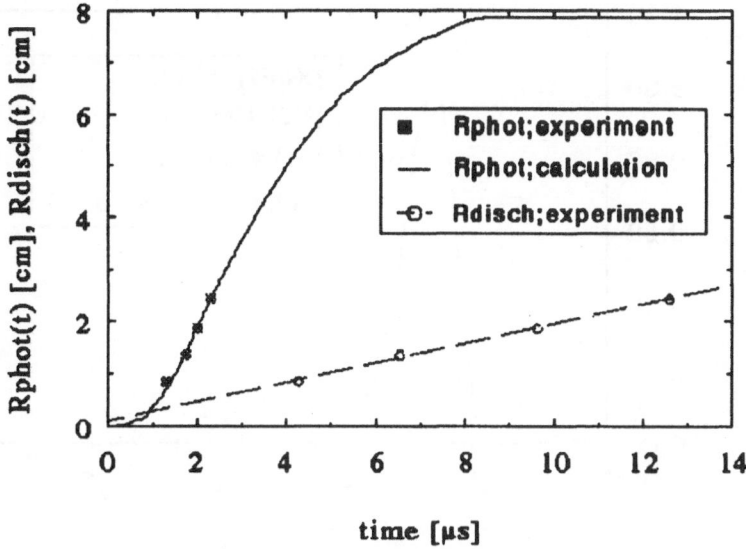

Figure 10. Time dependent radius of the photo dissociation wave Rphot, and the discharge radius, Rdisch.

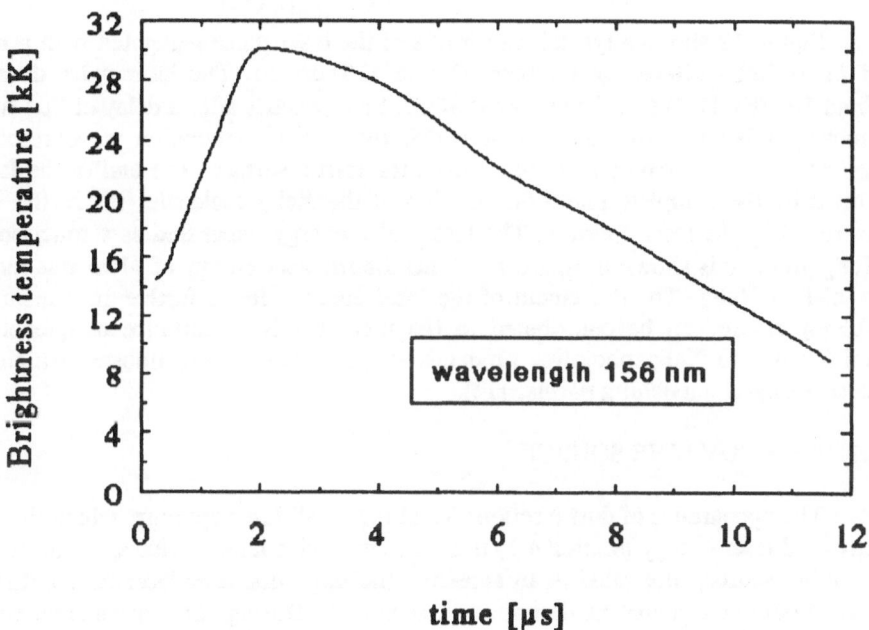

Figure 11. Brightness temperature of the FFPS discharge.

198

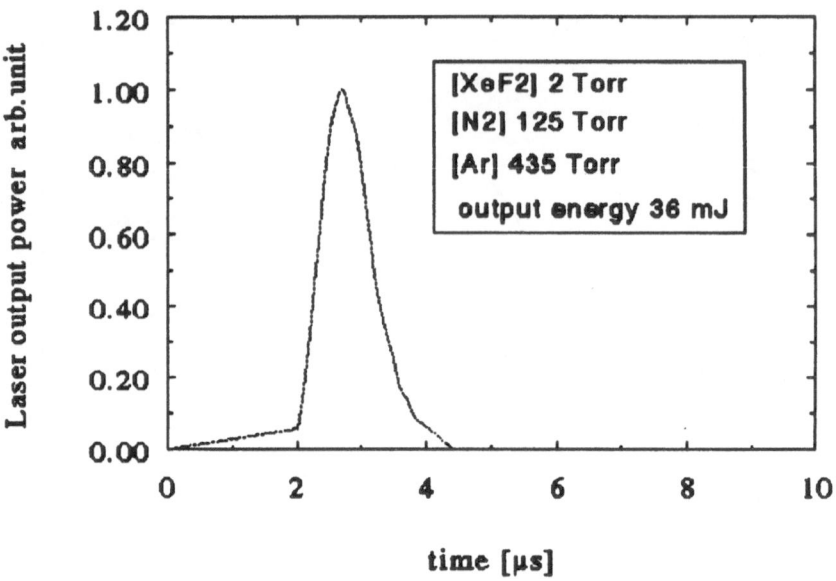

Figure 12. Output laser pulse of the FFPS driven XeF(C→A) laser.

Figure 12 shows a typical waveforms of the laser pulse generated with a mixture of 2Torr XeF_2 diluted by 125Torr N2 and 435Torr Ar. The laser pulse duration is about 1μs (FWHM) and the energy is 36mJ. Lasing starts after a delay of 2μs which is probably related to the moment at which the photo dissociation wave reaches the optical axis of the cavity at 13.6mm from the ferrite surface. The end of the lasing is caused by the complete photo dissociation of the XeF_2 molecules inside the volume restricted by the laser aperture. The laser pulse energy, measured as a function of the XeF_2 pressure is shown in figure 13. A maximum laser energy of 41mJ was measured at 2.3Torr XeF_2. The decrement of the laser energy after a further increment of the XeF_2 pressure can be contributed to the increasingly importance of quenching of XeF(C) due to XeF_2 photolyse products, in particular atomic fluorine which is the most strongest quenching partner [12].

4.2. SDCS VUV LINE SOURCE

The appearance of dust (section 2) and the small laser aperture, which limited the extracted laser energy (section 4.1) from the small-size laser device applying the FFPS pumping source, motivated us to construct the large diameter laser tube utilising the SDCD pumping source as described in section 3. During the experiments with this laser tube we used a buffer gas mixture of 114Torr N$_2$ and 342Torr Ar. When the device was operating as a laser, the buffer gas was added to 2Torr XeF$_2$.

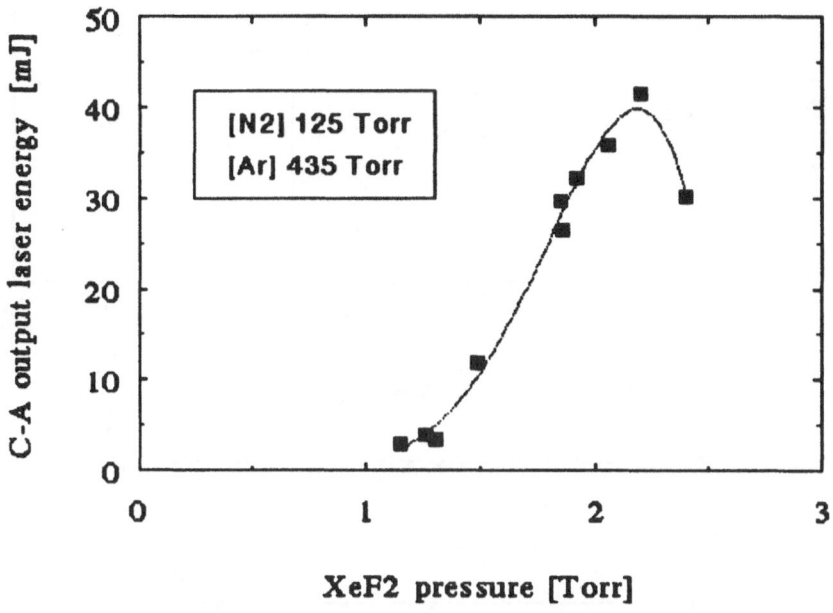

Figure 13. Dependency of the XeF(C→A) laser energy on the XeF$_2$ partial pressure. Output coupler T=4%.

Because the properties of the SDCS are not well understood up till now, we investigated the dynamics of the initiation of the discharge during preliminary experiments. Figure 14 shows the dependency of the plasma-initiation delay time on the initial voltage applied to the 90cm long discharge gap. The delay time is about 1ms at a voltage of 38kV and decreases gradually with increasing voltage. To understand the reason of this behaviour, the space-time dynamics of the discharge development was measured applying the optical fibre method described in section 3. The results of

these measurements are shown in figure 15. It is obvious from this figure that, in opposite to the FFPS, the discharge plasma appears simultaneously with the application of the voltage to the SDCS.

Initially, the plasma leader moves with a low speed, spending about 75% of the total delay time needed for the completion of the formation of the surface discharge channel, to proceed a distance of only 10cm. After this an explosive increment of the speed of the plasma leader is observed, so it takes only 25% of the total delay time to complete the low-current stage of the discharge and to cover the next 80cm of the channel with plasma. A polarity sensitivity of the discharge development, as well as t

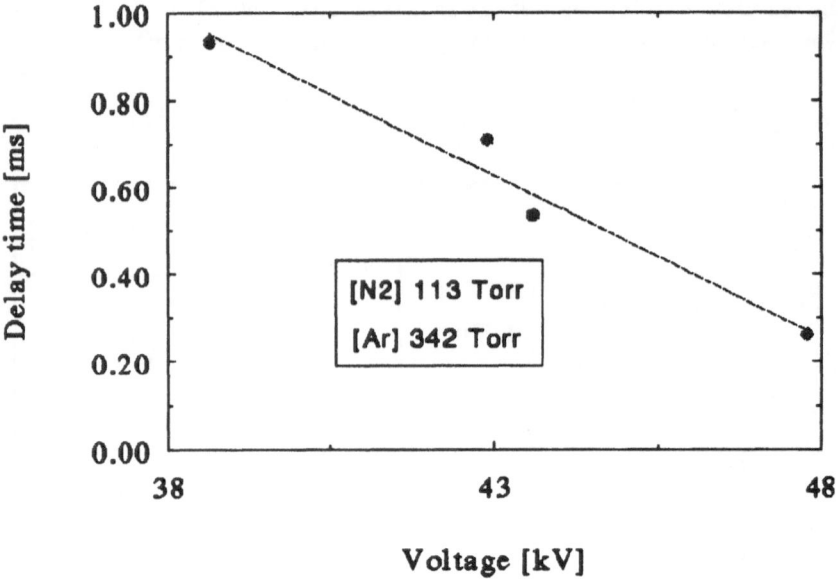

Figure 14. SDCS delay time dependency on the initial voltage at the 90 cm discharge gap.

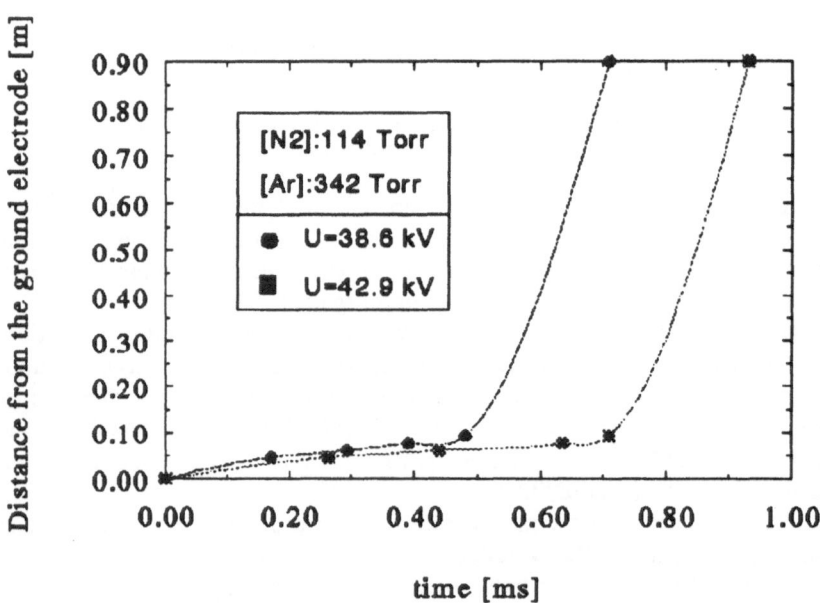

Figure 15. Space-time dynamics of the SDCS discharge.

he dependency of the plasma leader speed on the penetration depth into the discharge gap, are qualitatively supporting the mechanism of the SDCS discharge initiation as suggested in [10]. When the plasma leader reaches the positive polarity electrode, a high current arc discharge on the ferrite surface appears. Two half-periods were observed in the current pulse of this discharge at the application of the 45kV charging voltage. The duration of the first half period was 11μs with a maximum current of 30kA. The power of the electrical pulse was calculated from the current-voltage oscillograms and had a peak value of 800MW and a duration of 5.7μs. The electrical power pulse delivered an electrical energy of 4560J to the plasma or 90% of the total energy stored in the capacitor bank.

The significant decrement of the discharge current as well as its increased duration compared with the FFPS driven device is explained by the longer geometrical length of the first one.

Up to now a maximum laser output energy of 225mJ was obtained with this device. The laser output pulse, with a duration of 1μs(FWHM), appeared after a delay of about 3μs relative to the start of the high-current discharge.

A further increase of the laser outcoupling and optimisation of the working gas mixture should enlarge the laser output up to several Joules.

5. Discussion

Output pulse energies of the optically pumped XeF(C→A) laser, as reported in the literature, are shown in figure 16. The XeF(C→A) laser was pumped with segmented surface discharges [13,15,16,17] and exploding wire techniques [14]. Results, obtained in this work, are presented too. In spite of various discharge excitation techniques and physical dimensions of laser devices, it is certainly true that these results show a clear trend. A roughly linear dependence on the log-log plot shows that the main factor that increases the output laser energy is an increment of the input energy, but an efficiency of the laser device, even at the highest reported input energy of 90kJ, is still about 0.1% [17]. This is not very surprising because the most of these results were obtained under quite similar conditions considering the composition of the laser gas mixture. Typically Ar:N2 mixtures of buffer gases with a ratio close to 3:1 were used in most of these works. The only variable parameter was the pressure of the gas mixture. However, it follows from the kinetic considerations, given by Beverly [12] ,that an increment of the buffer gas pressure above about 400Torr doesn't change the rate of the XeF(C) formation. It is also known that the brightness temperature of the electric discharge weakly depends on the input energy and electric circuit parameters if the electric power is larger than 10MW-20MW per cm discharge length [6]. So, the observed growth of the output laser energy with input energy is due to the application of larger gas volumes and not to an increment of the kinetic efficiency or significant improvements in the pumping rate.

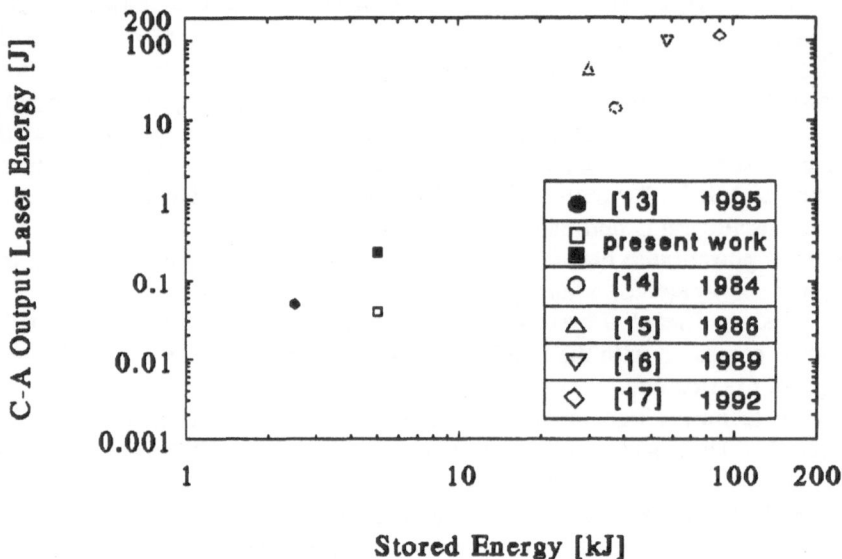

Figure 16. Results obtained with the optically pumped XeF(C→A) laser.

One of the most important processes that leads to the restriction of the laser efficiency to an unacceptable level is the atomic fluorine production. According to [12], fluorine produced as a result of XeF_2 photolyse is the most strongest quenching partner of XeF(C). The small recombination rate of fluorine favours to its accumulation in the working gas mixture during the operation of the laser. Combination of this poisoning influence with the low gain of the XeF(C→A) laser (0.1-0.6%/cm) leads to large losses of laser-active XeF(C) molecules due to quenching.

So, there are two possible ways to improve the XeF(C→A) laser performance. The first is the use of the XeF(C→A) active medium as an amplifier for short light pulses. In this case an external light pulse with energy density of about $0.1 Jcm^{-2}$, which is close to the saturation energy of the XeF(C→A) transition, should be used. The quenching effect will not manifest itself essentially due to the large initial photon flux in the active medium. Therefore, one should expect a significant increment of the kinetic efficiency of this laser. Light pulses with femtosecond durations can be amplified up to energies of 1kJ with this active medium [18]. One of the most important problems that must be solved here is to provide a coincidence between the moving excitation region and the spatial profile of the amplified light pulse.

Another research topic is the addition of chemical reagents for quick binding of atomic fluorine in working mixtures to reduce quenching of XeF(C). In our mind, the addition of several Torrs of C_2H_6 or C_2H_2 used as a hydrogen donor for the reaction

F+H→HF should solve this problem. However, this question requires further investigations.

6. Conclusions

Two techniques for optical excitation of the XeF(C→A) laser have been developed and tested. Laser output energies up to 0.22J were obtained with a laser device utilising a SDCS technique at 5kJ input energy. The main reasons of the small laser efficiency have been discussed and possible methods to improve the laser performance have been outlined.

7. Acknowledgements

One of the authors, S.V. Mitko, would like to thank Dr. Yu.B. Udalov and Dr. S.N. Tskhai for fruitful discussions and the Foundation for Fundamental Research on Matter (FOM) for support of his work at the University of Twente.

8. References

1. Rautian, S.G. and Sobelman, I.I. (1961) Photodissociation of molecules as a way to obtain the medium with negative absorption coefficient, *Sov. JETP* 41, 2018.
2. Danilov, O.B., Artemov, A.A., Zhevlakov, A.P (1994) Problems of efficiency and scaling for iodine pulse-repetitive flash-lamp pumping laser, *Digest CLEO*, 151.
3. Zuev, V.S., Mikheev, L.D., Startchev, A.V, Shirokikh, A.P., (1989) *Sov. J. Kwantovaja Elektronika* 6, 2033 (in Russian).
4. Kamrukov, A.S., Kozlov, N.P., Opekan, A.G., (1989) *Sov. J. Kwantovaja Elektronika* 16, 1333 (in Russian).
5. Zuev, V.S. and Mikheev, L.D. (1991) *Photochemical Lasers*, Harwood, Academic Publishers, Switzerland.
6. Borovich, B.L., Zuev, V.S., Katulin, V.A., Mikheev, L.D., Nikolaev, F.A., Nosach, O.Yu., and Rosanov, V.B. (1978) High-current Radiating Discharges and Optically Pumped Gas Lasers, *Radiotechnika VINITI SSSR* 15 (in Russian).
7. Mitko, S.V., Paramonov, A.V., and Shirokikh, A.P., (1991) Radiative properties of ferrite induced discharges as compared with expoding wire technique, *Proc. 2 All-Union Simposium on Radiative Plasmadynamics* 1, 35 (in Russian).
8. Watanabe, K., Kashiwabara, S., and Fujimoto, R (1987) Development of a formed-ferrite flash plasma light source for gas laser applications, *J.Appl.Phys*, 50 629.
9. Mitko, S.V., Goor, F.A. van, Holst, A.R. van der, Witteman, W.J., and Ochkin, V.N. (1994) Formed Ferrite Plasma Source for Gas Laser Applications, *Digest CLEO*, 61.
10. Mitko, S.V., Paramonov, A.V., Goor, F.A van, Ochkin, V.N., and Witteman, W.J. (1994) Sliding Surface Discharge on a Conducting Surface, *Digest ESCAMPIG* 18E, 264.
11. Iluchin, B.I., Mitko, S.V., Ochkin, V.N., Paramonov, A.V., and Shirokihk, A.P. (1991) Ferrite-Induced Breakdown Mechanism of Induction by Thermal Surface Explosion, *Sov. J. Laser Reseach in USSR* 12, 64.

12. Beverly, R.E. (1993) Kinitic Modeling of the Photolytic XeF(C→A) Laser, *J. Appl. Phys.B* 56, 147.
13. Knecht, B.A., Fraser, R.D., Wheeler, D.J., Zietkiewicz, C.J., Mikheev, L.D., Zuev, V.S., and Eden, J.G.

   (1995) Compact XeF(C→A) and iodine laser optically pumped by a surface discharge, *Opt. Letters* 20, 1011.

14. Zuev, V.S., Mikheev, L.D., and Stavrovskii, D.B. (1984) Efficiency of an optically pumped XeF laser, *Sov. J. Quantum Electron.* 14, 1174.

15. Zuev, V.S., Kashnikov, G.N., Kozlov, N.P., Mamaev, S.B., Orlov, V.K., Protasov, Yu.S., and Sorokin, V.A.
    (1986) Characteristics of a XeF(C—>A) laser emitting visible light as a result of optical pumping by surface-discharge radiation, *Sov. J. Quantum Electron.* 16, 1665.

16. Zuev, V.S., Kashnikov, G.N., Kirilenko, V.V., Mamaev, S.B., Sorokin, V.A., and Sukhorukov, V.F. (1989)
    Photodissociation XeF laser emitting visible and ultra violet radiation when pumped with radiation from a
    sectioned surface discharge, *Sov. J. Quantum Electron.* 19, 748.

17. Zuev, V.S., Kashnikov G.N., Mamaev S.B. (1992) Investigations of XeF laser optically pumped by radiation
    of surface discharge, *Preprint FIAN #23* (in Russian).

18. Zuev, V.S. and Mikheev, L.D. (1990) Fellow travellers waves of different nature in photo chemical lasers and
    their possible applications, *Preprint FIAN #190* (in russian).

PULSED PERIODIC IODINE LASER

A.A. ARTEMOV, O.B. DANILOV, A.S. GRENISHIN,
N.A. GRYAZNOV, V.M. KISELEV, A.P. ZHEVLAKOV
*Institute of Laser Physics, Vavilov State Optical Institute,
199034, St.-Petersburg, Russia*

1. Introduction

The most important problem of optical rangefinding is the enhancement of a carrier-to-noise ratio to increase the range and detection reliability. It must be emphasized, that the possibilities of such standard methods as radio frequency filtering or optical heterodyning are thoroughly studied, widely used, and practically exhausted.

The application of a quantum optical amplifier for lidars high gain, low inherent noise, and narrow amplification band [1]. The Iodine photodissociation laser, satisfies all these demands and appears to be the most suitable candidate for a power source in the location system with an optical preamplifier [2]. The high amplification of the iodine laser makes it possible to achieve small signal gain of 10^3 and higher [3] without essential increase in background due to metastable state of the upper laser level.

The repetitively pulsed mode of the iodine laser, required for the location system, is well studied also, both with optical [9-11] and chemical [12,13] pumping.

Here, we consider two key problems: precise control of master oscillator radiation frequency by magnetic field and high efficiency of the output amplifier. Besides, we analyze some problems of scaling.

This determines the low rate of the inversion decay and, therefore, long life time of gain at low pressure of an active medium. The low active medium pressure is necessary also for the receiving of the most narrow amplification bandwidth, that can be easily obtained in gaseous active media. The narrow amplification bandwidth of the optical preamplifier severely restricts a range of the tolerated optical frequencies in a reflected signal. The frequency of transmitting laser radiation, calculated with a preliminary information about a radial component of the object velocity, is to be easily changed within a required range by simple and low inertial devices. Interval between the strongest transitions of the iodine laser 2-2 and 3-4 equal to 0.46 cm^{-1} defines the natural and achievable range of tolerated Doppler frequency shifts.

The well known magneto-optical properties of the iodine laser active medium [4], which have been demonstrated in numerous investigations for free-running [5,6],

W. J. Witteman and V. N. Ochkin (eds.), Gas Lasers - Recent Developments and Future Prospects, 205–220.
© 1996 *Kluwer Academic Publishers.*

continuous [7], and single-pulse lasing [8], can be used to ensure the required radiation frequency sweeping. It is reasonable to control spectral parameters of the radiation transmitter by a low power master oscillator, which radiation is enforced up to the required level of output power by an amplification stage like described in [9].

The repetitively pulsed mode of the iodine laser, required for the location system, is well studied also, both with optical [9-11] and chemical [12,13] pumping.

Here, we consider two key problems: precise control of master oscillator radiation frequency by magnetic field and high efficiency of the output amplifier. Besides, we analyze some problems of scaling.

2. Master Oscillator

2.1. TECHNICAL REALIZATION

The construction of a laser head schematically shown in Figure 1, is based on a coaxial cavity pumping lamp (with active length of 300 mm). Laser active medium is inside a quartz tube with the internal diameter of 10 mm. An internal cylinder of cavity lamp is cooled by distillate water between the laser tube and the lamp. Outer cooling of the lamp and its electrodes is realized by a tap water.

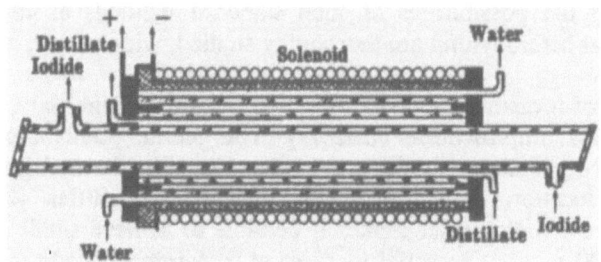

Figure 1. Laser head schematic.

The solenoid, which creates a homogenous longitudinal magnetic field in the active medium of the laser spectrum is spooled over an external surface of the cavity lamp outside the cooling loop.

The reverse pumping lamp discharge current conductor manufactured from aluminum foil is placed right under the solenoid. It has a cylindrical shape with a narrow longitudinal gap for removal of circular whirl currents, caused by the solenoid field in the cylinder body. These currents reduce magnetic field in the active medium. To exclude near-electrode zones of the active medium, where spatial homogeneity of solenoid magnetic field is distorted by ring electrodes, from the lasing area, the electrodes of the pumping lamp (with the diameter of 3 mm), pressed into a ring base (24 for every side), have lengths (24 mm) sufficient to exclude these zones from the pumping area.

Coaxial design of the pumping lamp has been chosen because it is rather compact and suitable for minimization of a solenoid diameter and decrease in energy required to generate magnetic field. It also reduces the magnetic field created in active medium by the pumping discharge current which can eventually affect the laser radiation spectrum.

High requirements imposed the parallelism of the cavity lamp elements and the homogeneity of its discharge current in transverse direction results in negligible value of magnetic field within the internal lamp cavity. [14]

The windows of the laser tube are tilted at small angle (about 10°) to the axis. That prevents a non-uniformity of losses, caused by reflection. Outer surfaces of the windows are covered with an anti-reflecting coating. The laser resonator with the length of 120 cm is formed by two plane-parallel mirrors with reflection coefficients 0.99 and 0.20. Resonator mirror reflection coefficients are 0.99 and 0.20. Two diaphragms with diameters of 5 mm are placed inside the cavity to suppress high order transverse modes. They also exclude the lasing in near-wall area, distorted by gas dynamic waves [15,16], which noticeably worsen angular divergence of laser radiation.

A cuvette with a dye (mdrk 1067) is used as a passive Q-switch [17]. It is placed in front of the high reflection mirror. The selection of the dye is determined by its relaxation properties and high resistance to heating, typical for the most of the phthalocyanine group of dyes [18]. Initial transmission of the shutter is 1.5 %. The suitable modulator for the laser is the shutter, based on a crystal of YAG, activated by vanadium (V^{3+}) ions [19], operation properties of which have been checked in our investigations as well.

Perfluoralkyliodide n-C_3F_7J, circulating in a closed loop using the principle of thermal pump [10,11,20], is used as the laser active medium. The pressure and speed of gas flow are controlled by change in heater temperature with a stable temperature of a cooler.

The capacitor with capacity of 12 μF and charging voltage of 25 kV is used to store the energy in power supply system for the pumping lamp. Commutation of the discharge circuit is performed by an ignitron. The pulse repetition frequency determined by gas flow speed in the laser active volume, in the described construction is up to 15 Hz. Further increase of the pumping pulse frequency results in saturation of the average power of and its subsequent decrease. The duration of the pumping cycle half period is 10 μs.

The homogeneous longitudinal magnetic field, required for the control of the laser radiation spectrum, is formed by a thyristor circuit, forming a current pulse with period of 600 μs, which provides a stationary value of the magnetic field intensity during the evolution of the single-pulse laser radiation. The scheme of the laser radiation frequency control based on the solenoid current change, operates under a predetermined algorithm and is able to follow Doppler shift, defined by the independent radio frequency channel of the receiving system.

2.2. EXPERIMENTAL STUDIES OF MASTER OSCILLATOR PARAMETERS

2.2.1. Measuring technique

During the investigations of the laser characteristics and their mutual influence the following parameters have been measured: light pumping energy in every pulse and the pulse frequency, radiation energy in a single-pulse and the radiation pulse period, spectrum and average laser radiation power, the iodide pressure and gas flow speed in the active medium, magnetic field strength and the radiation angular divergence.

Single pulse energy and average radiation power were measured by a standard calorimeter. Besides, radiation energy was monitored by coaxial photodetectors. Integrated signal from the detectors was subsequently measured by oscilloscope. The oscillograms made it possible to measure not only the radiation energy in each pulse but the pulse frequency too. The samples of such oscillograms or, more exactly, histograms are shown in Figure 2, where the reproducibility of the radiation energy dynamics is clearly seen. The radiation pulse duration measurement was realized also by photodetector connected to a high frequency oscilloscope and without integrating circuit. The radiation spectrum was recorded by Fabri-Perot interferometer with the 10 mm inter-mirror distance.

Interferograms were visualized using an infrachomatic film. Radiation density distribution registered in a focal plane of the lens with focus of 160 cm gave the information about the angular divergence of the output laser radiation. Gas flow speed in the active volume was measured by the methods described in [20]. The flow speed was varied within the range of 1.5...3.0 m/s. Electric power of the light pumping was determined by a charge voltage and a capacitor value. Temporal shape of the pumping current pulse was measured by Rogovsky coil, a signal from which was measured by oscilloscope. The longitudinal magnetic field strength in the active medium was measured by the induction sensor, pre-calibrated by a standard magnetometer.

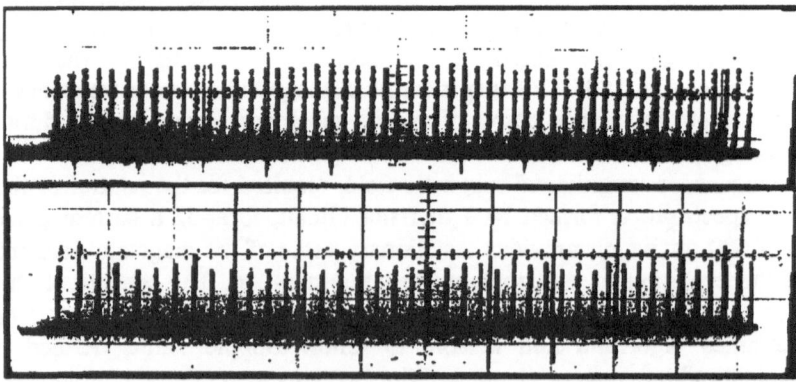

Figure 2. Histograms of repetitively pulsed laser radiation :a) in a free lasing; b) with a passive Q-switch.

2.2.2. Results of laser parameter measurements

In free-running mode without Q-switch and diaphragms in the cavity, the laser with charge voltage of 25 kV and pulse frequency of 10 Hz gives an average radiation power of about 10 W. Under these conditions, life time of the laser was about 10^4 pulses. Increase in the number of pulse, up to 10^5 results in reduction of the output power down to 30% from the initial value. Disassembling of the pumping lamp after its prolonged exploitation has shown that the output power decrease is due to the reduction of the lamp wall transmission. It occurs because of the precipitation of the sputtering products of the electrodes made from thorium tungsten. Operating gas pressure in the active volume during the specified test cycle was about 60 Torr, xenon pressure in the pumping lamp was 75 Torr.

Special attention must be paid to the influence of the distilled water layer between the laser tube and the pumping lamp on lasing. As it was found, the presence of this layer led to noticeable (about 20%) increase in the output power. The reason of the increase was most likely connected with the cutting of the ultraviolet radiation with the wavelengths shorter than 180 nm. It is known that these radiation can deteriorate the active laser medium [21]. Another reason for the increase can be related to a decrease in Fresnel of pumping radiation losses due to reflection from both quartz tube surfaces.

During laser operation with passive Q-switch the laser radiation with the pulse duration of 15...20 ns at level of $0.5I_m$ was observed. The energy of a single pulse, with the diaphragms inside the cavity was equal 30 mJ. It is necessary to note, that a single-pulse generation realization with passive Q-switch is determined by a proper relationship between the cavity losses and the gain. In the case of the iodine gas laser it can be achieved by varying operating gas pressure and optical pumping power. In this case however the optimal conditions for high amplification in the iodine active medium result in certain decrease in both laser pumping power and operating gas pressure. It was the reason for a relatively low average output power, which was equal to 300 mW with the pulse frequency of 10 Hz.

The stochastic development of a single pulse in the laser with a passive Q-switch and certain instability in the laser operating conditions affect reproducibility. They are caused mainly by the variation in flowing gas pressure typical for thermal pumping which results in noticeable, up to 15%, variations in the pulse intensity (see histogram in Fig. 26)

However, these variations can be decreased by an improvement of the gas circulation systems and more stable vaporization at the gas in the heater.

For these reasons, the reproducibility of the single-pulse temporal location with respect to the pumping pulse was relatively low. Under the laser operating conditions it was equal to ~ 250 ns, that is, however, negligible compared to the time interval between the two subsequent pulses (T=100 ms).

Application of YAG, activated by V^{3+}, as the passive Q-switch results in the decrease of pulse duration to 10-15 ns on the level of $0.5I_m$ and a single-pulse energy of 35-40 mJ. The modulator parameters were as follows: diameter of 20 mm, thickness of 5 mm, and initial admission of 1.5%. The iodide pressure in the active medium was 40 Torr added. The last circumstance was the reason for low initial shutter transmission. More detailed information about this shutter type is given in [19].

Angular divergence of the radiation in Q-switch mode is equal to $\sim 5 \cdot 10^{-4}$ rad (by the level of $0.1/_m$), moreover, its value in a pulse-periodic mode is slightly higher than the value, typical for to the laser operation without gas flow. The beam sizes in a far-field zone measure of for these two cases are comparable, however, the radiation density in the side lobs of the angular distribution under the condition of gas flow is noticeably higher. The specified radiation angular divergence remains constant after the long-term laser operation.

The measurements of the spectral parameter of laser radiation as a function of magnetic field strength.

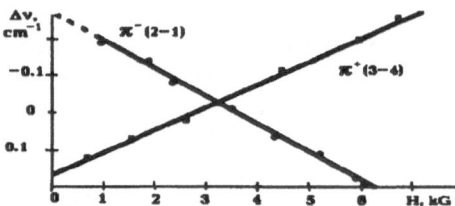

Figure 3. Radiation spectrum of the iodine laser in the magnetic field.

According to the commonly accepted terminology and the character of the used magnetic field the spectral lines, observed in the experiment, were identified as two groups of Zeeman components: π^+-components of the transition F=3→F'=4 and π^--components of the transition F=2→F'=1 between the superfine substructures of the iodine atom ground state [4]. The mutually orthogonal circular polarizations (left and right) are easy realized in an experiment no resonator inhomogeneity. The presence of these spectral components in radiation (simultaneously or one by one) is closely connected to the character of the amplification cross section change for each of them during magnetic field strength variations.

It is clear that such arbitrary fluctuations in laser output spectrum are unacceptable for rangefinding system. Therefore a single spectral line with a certain (positive or negative) frequency shift should be selected.

We select the line, corresponding to the π^+ Zeeman components of transition group 3→4, because transition 3→4, unperturbed by magnetic field, is the most convenient for realization of the backscattered signal amplification in the case of an approaching object. The gain for this transition is noticeably higher. In the opposite case, with a moving away target, the amplification of the reflected signal in the quantum amplifier is to be realized on the unperturbed transition 2→2, where the gain is lower with respect to the transition 3→4, but for moving away object in some cases it is acceptable.

The selection of the line with certain frequency shift was achieved in a resonator with selective losses for π^- and π^+-components, that (as it was shown in [6]) was realized with a quarter wave plate, Faraday rotator, and polarizer, inserted into the cavity. For this purpose, Pockels cell with a quarter wave voltage, Faraday rotator and Glan prism were used. As a result lasing on the single spectral line corresponding to π^+ or π^- Zeeman component groups was realized. However, while operating on one particular line, it must be taken into account, that the gain cross section for them varies as magnetic field changes [5]. The result of this dependence is the increase in output account. This can be observed both in the output power histogram and in the temporal delay between pumping pulse and laser pulse.

Therefore, for the output radiation stabilization, an algorithm of the spectrum control system must contain the additional program, which account for the gain is regularities and compensates them by small variations of the (capacitor charging voltage). Such correction allows us to keep stability of the laser output on appropriate level as it is shown is Figure 2b. Output power was not affected by intracavity Pockels cell, Faraday rotator, and Glan prism with an anti-reflection coatings. The loss increase was compensated by a minor increase of the pumping power and the operating gas pressure. It is necessary to note, that the polarization of the output radiation in the presence of anisotropic elements depends on their location in the cavity. The polarization is linear, when the elements are placed in front of the output mirror, and circular for location near the totally reflecting mirror. All described results were received with pure iodide n-C_3F_7J, that was satisfactory for lasing of the pulse with the above mentioned durations. For obtaining of the pulse with duration of about 1 ns buffer gases must be added to the iodide. The circulation of the mixture can be organized by thermal pumping [10] or by another technique [9]. In any case, the problem requires certain technical solution.

3. Power amplifier

The specific features of the laser described above as a prototype for the master oscillator of the range finding system determine the requirements to the amplifier, which has to increase the power up to a required value.

3.1 LAMP PUMPING EFFICIENCY AND "LIGHT POCKET" MODE

Table 1 presents key parameters of the lamps developed for high power iodine laser pumping [22]. Here $W_0 = CU^2/2$ is the energy of capacitors, $W_L \approx 0.8 W_0$ is the lamp input energy, $\eta_{\delta\lambda} = W_{\delta\lambda}/W_L$ is the lamp efficiency in spectral region of about 80 nm (270±40 nm), d is the lamp diameter, t_p is the pulse duration, and P_{Xe} is the xenon pressure. Discharge length for all lamps is 1 m. Z-pinch mode allows us to realize lamp efficiency of about 20%.

Figure 4 presents pictures obtained with a streak-camera and illustrating a temporal evolution of discharge diameter

212

TABLE 1. Parameters of the lamps operating in Z-pinch mode

W_0, kJ	d, cm	P_{Xe}, kPa	t_p, ms	$\eta\delta\lambda$, %
20	3.2	20	60	23
30	3.2	20	40	15
40	3.2	10	40	16
50	7.5	2.3	25	18
70	10.0	0.6	23	16
90	10.0	1.3	38	18

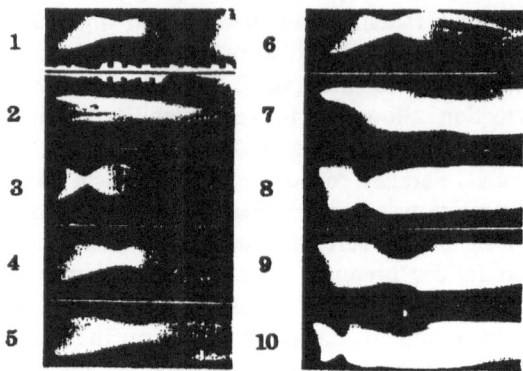

Figure 4. Discharge photochronograms in lamps with the diameter of 10 cm; input energies of 40 (1,2), 60 (3-5), 70 (7), 80 (6), 90 (8,9), and 110 kJ (10); and Xe pressures of 5 (1,3,6,8,10), 10 (4), and 25 Torr (2,5,7,9).

The selection of xenon pressure and electrical pulse energy for optimal operation makes it possible to reach the minimal discharge diameter at the moment of maximum input electrical power. Three main reasons to choose for Z-pinch mode are: 1) the best electrical energy matching of lamps with capacitors; 2) the minimum impact of gas pressure and plasma temperature on quartz lamp walls at the moment of the maximum UV lamp intensity; 3) high level of lamp efficiency (20%) in wide range of electrical parameters variation. For example, a lamp with a diameter of 7,5 cm has an electrical energy input limit of 120 kJ and operating electrical energy of 50...60 kJ. Nevertheless, the lamp efficiency of about 20% remains down to 20 kJ.

It is clear that this is the way to make pulse lamps with high radiation power. We have developed two types of pulse-periodic lamps for pumping of iodine laser. Their parameters are as follows:

type 1: average power   100 kW (10 kJ, 10 Hz)

             diameter of lamp   3.2 cm

             discharge length   50 cm

             diameter of coaxial quartz cooling tube   5.0 cm

type 2:	average power	500 kW (50 kJ, 10 Hz)
	diameter of lamp	6.5 cm
	discharge length	100 cm
	diameter of coaxial quartz cooling tube	10 cm

The lamps cooled with bi-distilled water and cooling of the electrodes is performed with ordinary water. Simultaneously, xenon flows through the lamp with velocity of about 0.1 m/s.

A schematic cross-section of the iodine laser with central lamp pumping is presented in Figure 5. The internal light diameter of the laser is 180 mm, length of its active part is 1m. Reflection coefficient of SiO_2 ceramics used for scattering reflection of UV radiation is 92% at $\lambda=300$ nm. Lamp diameter is 65 mm. In this case, the laser efficiency of 2% is measured for the laser pulse energy of 1 kJ.

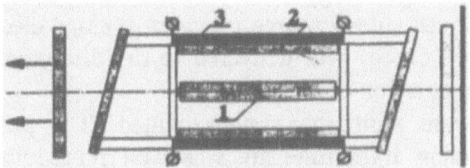

Figure 5. Iodine laser with a central lamp (1) pumping, glass-plastic cave (3), and SiO_2 ceramics (2) for scattering of UV radiation.

One of the possibilities to increase the radiation efficiency in 230...310 nm region is the application of the reflector operated in so-called "light pocket" scheme. Energy balance for this scheme is $W_L + \chi W_2 = W_T + W_1 + W_{12} + W_2$, where W_1 and W_2 are the energies of optical radiation for spectral ranges with $\lambda<\lambda_1=230$ nm and $\lambda>\lambda_2=310$ nm, respectively, W_{12} is the energy in the range between λ_1 and λ_2 (pumping of laser), χ is the efficiency which returns light back to lamp the reflector, and W_T is the energy lost on plasma heating.

For the lamp with diameter of 6.5 cm, we have experimentally obtained the following values of specified energies: W_T, W_1, W_{12}, and W_2 are about 10, 2, 9, 30 kJ, respectively. If χ is about 85%, the efficiency of iodine laser may be more than 7%.

3.2. EFFICIENCY OF MULTILAMP SYSTEM

Optical pumping of large lasing volumes may be achieved with a multilamp system. In this case, it is necessary to ensure sufficient reliability of lamp operation along with high efficiency and energy of UV radiation, typical for single-lamp pumping schemes. An analysis of a multilamp pumping system shows [23], that bending instabilities of pinch caused by ponderomotive forces deflect it to a system center. In this case, high efficiency of UV radiation for a single lamp remains the same, but limiting tolerated energy W_{lim} (destruction threshold) decreases two-fold. It was also founded that in comparison with a single pulse regime, the system lifetime in a pulse-periodic mode (number of pulses till the moment, when intensity reduces down to 80% of its original value) was the same (10^5), if storage energy W_0 did not exceed $0.5W_{lim}$.

To exclude the impact of ponderomotive forces, we used a reverse current wire as a compensating conductor. Pump source consisted of three lamps with discharge volume diameters of 4.5 cm and lengths of 100 cm. Initial Xe pressure was 30 Torr. The lamps were located in parallel and the distance between their axes was 12 cm. Brass rod with a diameter of 1.6 cm and length of 130 cm was placed between them in the center of the three-lamp system. Zero potential electrodes had a common point with the reverse current wire while high potential taps were isolated.

Discharge circuit for lamp supply included six sections of capacitors with a commutation by ignitrons. Every lamp was powered by two sections with total maximum storage energy W_0 of 5.15 kJ. Destruction threshold for one lamp was 40 kJ. A stabilized supply source provided average electric power up to 300 kW. 80% of the stored energy was delivered to the discharge volume during the first half-period with duration of 20 µs.

In this scheme, multilamp system emitted 12 µs pulses with 20 Hz repetition rate. The maximum light intensity was well reproduced from pulse to pulse with radiation efficiency of 23% in the range of λ=230...310 nm. Under such conditions, the instabilities of pinch were not observed.

Thus, radiation Z-pinch mode allows us to combine high values of efficiency and absolute flux of UV emission for any scheme of lamp operation. A circuit compensating ponderomotive interaction increases flashlamp reliability for high levels of energy and power in multilamp repetitively pulsed sources of UV pumping.

3.3. ACTIVE MEDIUM AND ITS OPTICAL QUALITY

The results presented above were obtained with two types of gas mixtures: 1) C_3F_7I + SF_6 (20 Torr + 100 Torr) and 2) C_3F_7I (C_4F_9I) + C_6F_{14} (20 Torr + 20 Torr). In both cases, laser action was free of pyrolisis and maximum volume density of laser energy equal to 140 J/L was achieved in our experimental conditions (short pumping).

One of the key problems in high power laser application is to ensure a high quality of active medium without losses in efficiency. Three processes cause optical distortions of active medium [24]: 1) shock waves from inner surfaces of the laser volume; 2) pumping inhomogeneities in laser volume; 3) self-influence (small-scale optical deformation of active medium) [25]. We assume that the self-influence followed by small-scale optical deformations of active medium leads to a disastrous situation for iodine laser, and, first of all, it is necessary to define experimentally the conditions when the self-influence is absent. We have not observed the self-influence while using pure CF_3I or C_3F_7I and two types of gas mixture: C_3F_7I (CF_3I) + SF_6 (pressure of SF_6 was 100 Torr) and C_3F_7I (CF_3I) + C_6F_{14}. To combine homogeneous active medium with highest laser efficiency it is necessary to make a proper choice of laser cave diameter and gas pressure. Use of laser cavity with an inner diameter of 420 mm give us an opportunity to realize the following laser parameters: pulse energy of 1 kJ, laser efficiency of 1.8%, and gradient of refraction index of 10^{-8} cm^{-1}. All these parameters are realized for gas mixture $C_3F_7I + C_6F_{14}$ (2 Torr + 20 Torr).

3.4. CLOSED LOOP OF ACTIVE MEDIUM

Figure 6 presents schematic of the iodine laser with a closed loop of active medium. Fan (5) forms a gas flow transverse to optical laser axis. Gas mixture goes through a microporous absorber (6), which uses certain types of active carbon compounds.

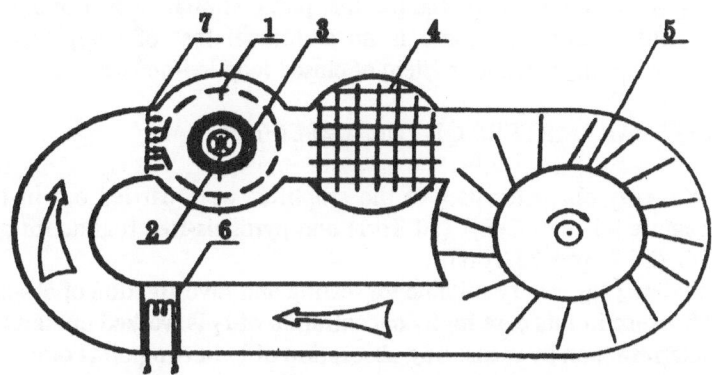

Figure 6. Schematic of closed loop iodine laser with lamp (2) in the center of laser cave (1), fan (5), absorber (6), quartz tube (3) to limit water for lamp cooling, and refrigerator (4) for gas cooling.

One of main absorber parameters is the absorption magnitude a (in g/g) (unit of substance weight per unit of absorber weight). Table 2 presents the values of absorption a for two types of absorbers and for some iodine substances: iodides (CF_3I, C_2F_5I, C_3F_7I), buffer gases (C_6F_{14}, SF_6) and for I_2 (I_2 is the main quencher of the upper laser level).

TABLE 2. Values of absorption for different absorbers and absorbents.

Substance		CF_3I	C_2F_5I	C_3F_7I	C_6F_{14}	I_2	SF_6
a, g/g	type 1	0.566	0.895	0.864	1.01	1.899	0.389
	type 2	0.758	1.378	1.388	1.63	2.896	0.541

The condition for proper operation of the absorber can be specified as
$a(I_2) < a(C_3F_7I) < a(\text{buffer gas})$

In this case:
  1) I_2 is absorbed very effectively (high absorption);

216

2) the absorber operates as very large and good buffer volume for C_3F_7I (middle absorption);

3) buffer gas goes through the absorber device with low absorption. The best buffer gas for this purpose is SF_6.

Let us consider an example of typical gas mixture and its components in the closed loop and absorber. Pressure of C_3F_7I is 25 Torr inside the laser volume (gas phase). The weight of C_3F_7I inside the absorber (liquid phase) is about 35 kg and pressure of SF_6 is 75 Torr (gas phase). The most important fact is that under these conditions, the absorber is able to absorb 7.7 kg of I_2. It means that in practice, the pressure of I_2 during all lasing period is close to zero.

C_3F_7I is also liquid inside the absorber and I_2 molecules are dissolved in liquid phase of C_3F_7I as well as inside the thermal pump similar to that one of Baker and King [10]. But in this case there is no additional loss of energy so we have a possibility to solve the scaling problem of closed loop iodine laser.

3.5 ENERGY PARAMETERS OF THE AMPLIFIER.

Studies on energy characteristics of the amplifier were carried out in two modes: pyrolisis regime for pure C_3F_7I (20 Torr) and pyrolisis-free regime for gas mixture C_3F_7I + SF_6 (25 Torr + 75 Torr).

Pyrolisis regime [26] is very suitable for testing and investigation of absorber operation because in this case high concentration of I_2 is worked out, and for gas velocity determination by measuring absorption of I_2 in two points of closed loop. Figure 7 shows radiation pulses in pulse periodic free running mode. One can see reasonable amplitude stability of laser radiation pulse intensity with gas velocity equal to 2 m/s, although maximum gas velocity in the closed loop is 5 m/s. In this pyrolisis regime, laser parameters were as follows an average power 350 W, pulse repetition rate 10 Hz, and laser efficiency 0.7%. Laser parameters in pyrolisis-free regime were: average power 500 W, pulse repetition rate 10 Hz, and laser efficiency 1.0 %.

Large aperture of the apparatus gives the opportunity to design a multistage amplifier, using telescopes for saturation regime optimization and saturable absorbers for isolation of stages. Calculations using the equations of Frantz and

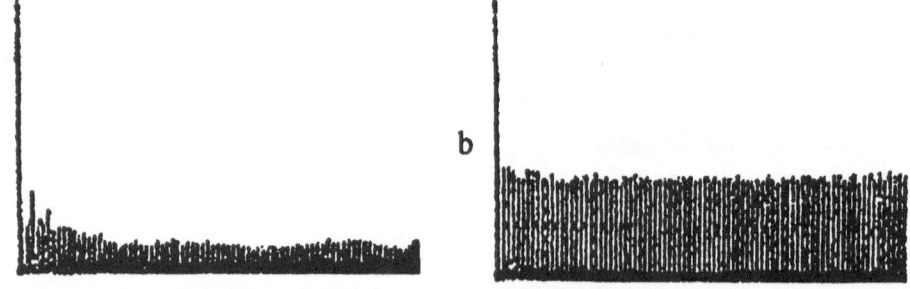

a   b

Figure 7. Oscillograms of lasing pulses at gas velocity of 0 (a) and 2 m/s (b).

Nodvik [27] have shown that energy parameters of the module at compact filling up of the aperture let us reach output pulse energy of 10 J for input energy of 30 mJ, although its design in this case is rather complicated.

It is necessary to note that the parameters of the closed loop system and laser cavity give an opportunity to use the second type of repetitive pulsed by lamp (see section 3.1) and we can expect essential improvement of performance for operation in the power amplifier mode Application of a pumping lamp with larger diameter and higher power results in significant increase of the amplifier cross-section, that allows us to enlarge the number of amplifying stages in a single module and to increase the energy of the output optical pulse. This circumstance opens additional opportunities for energy scaling of the device.

3.6. SPECTRAL PROPERTIES

Frequency variation range is determined by the possible velocities of a target. The amplification unit has to ensure efficient power extraction within the range between the most intensive superfine components of the lasing transition (0.46 cm^{-1}) for the system, operating with different spectral lines in the cases of the approaching and moving away objects. The broadening of the gain curve by a buffer gas pressure (6-8 atm) under the conditions of gas flow is technically complicated and pointless. The most logical solution of the problem is the application of magnetic field [28].

The gain curves of the iodine laser active medium, calculated for the mixture of the iodide with a buffer gas and for the inhomogeneous magnetic field, are presented in Figure 8.

Application of the inhomogeneous magnetic field allows us to obtain a uniform gain curve in the required range even at low pressure, of buffer gas. The mixtures of iodide with buffer gas studied in pulsed periodic regime (see Section 3.5) correspond to broadening of about 0.1 cm^{-1}. Nevertheless, even in this case, non-uniform character of the requires control of input signal value to compensate variations in gain due to the laser frequency shift.

To exclude action of transverse magnetic field of the discharge current in the case of central lamp pumping, it is necessary to use the moment of zero current. The presence of the transverse component in the magnetic field and axial inhomogeneity of its longitudinal component result in slight distortion of output radiation polarization [28]. Ellipticity degree varies with radius and frequency of input radiation, but it hardly affects the energy of laser output.

4. Summary

The results presented above provide us with the opportunity to design a high-efficient laser source for range-finding system with an optical preamplifier based on the studied iodine photodissociation laser apparatus.

218

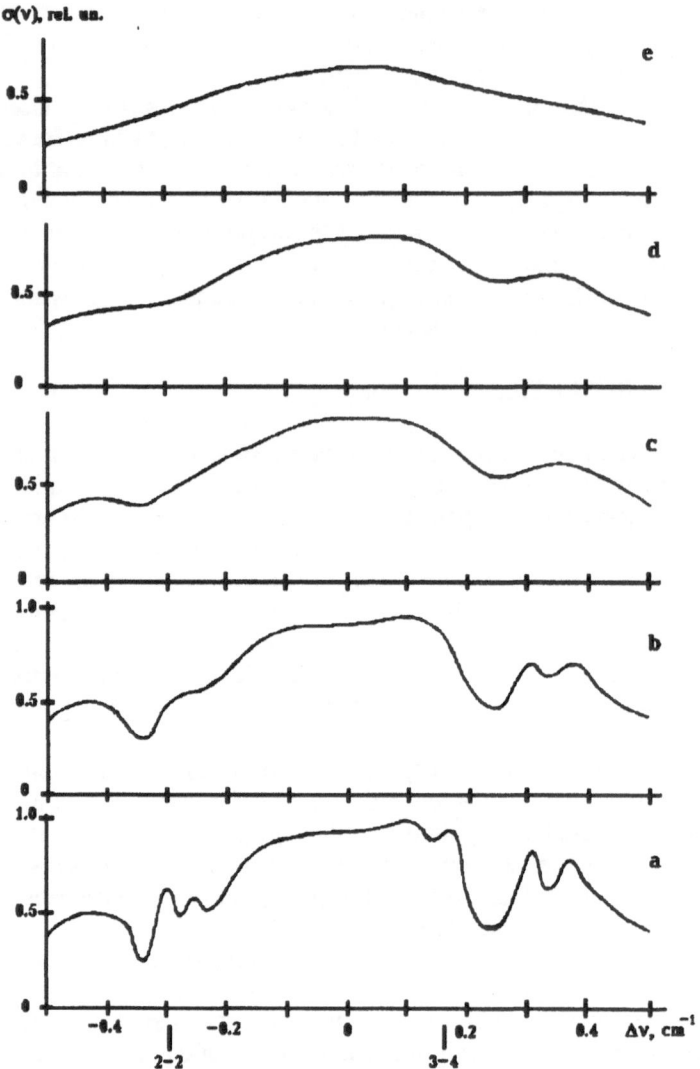

Figure 8. Spectra of the iodine laser transition in the inhomogeneous magnetic field H_m=5 kG. $\Delta\nu_L(cm^{-1})$ = 0.016 (a); 0.032 (b); 0.09 (c); 0.135 (d); 0.275 (e).

5. References

1. Grenishin, A.S., Gryaznov, N.A., and Kiselev, V.M. (1993) Repetitively pulsed iodine laser with passive Q-switch and controlled spectrum of radiation, *Proc. SPIE* **2095**, 171-179.

2. Kutaev, Yu.F. (1994) Comparison of radars on the base of energetical efficiency criterium, *Journal of Optical Technology* (rus.), n.3, 65-68.
3. Zuev, V.S., Katulin, V.A., Nosach, V.Yu., Petrov, A.L. (1980) Generation and amplification of nanosecond pulses in iodine lasers, *Proc. PIAS "Optical Pumped Gaseous Lasers"* (SU) **125**, 46-103.
4. Belousova, I.M., Bobrov, B.D., Kiselev, V.M., Kurzenkov, V.N., and Krepostnov, P.I. (1974) I^{127} atom in a magnetic field, *Opt.Spectrosc.* **37**, 20.
5. Belousova, I.M., Bobrov, B.D., Kiselev, V.M., Kurzenkov, V.N., and Krepostnov, P.I. (1974) Photo-dissociation laser on I^{127} atom in a magnetic field, Sov.Phys. *JETP* **38**, 258.
6. Fill, E.E., Thieme, W.H., and Volk, R. (1979) A tunable iodine laser, J.Phys.D **12**, L41-L45,
7. Kelly, M.A., McIver, J.K., Shea, R.F., and Hager, G.D. (1991) Frequency tuning of a CW atomic iodine laser via the Zeeman effect, *IEEE J.Quant.Electron.* **27**, 263-273.
8. Kiselev, V.M., Grenishin, A.S., Kotlikova, T.N., and Rodina, L.I. (1984) Passive mode-locking in an iodine laser subjected to a longitudinal inhomogeneous magnetic field, *Sov. J. Quant. Electron.* **14**, 650-654.
9. Danilov, O.B., Artemov, A.A., and Zhevlakov, A.P. (1992) Iodine photodissatiative pulse repetitive laser; problems of efficiecy, *Proc. SPIE* **1980**, 41-49.
10. Baker, H.J. and King, T.A.(1981) Repetitively pulsed iodine laser with thermal gas flow cycle, *J.Phys. D* **14**, 1367-1376.
11. Jelinek, M., Trenda, P., and Hermoch, V. (1986) Iodine photodissociation laser with thermal gas flow, *Proc. Workshop "Iodine lasers and applications"*, Bechyne, Czechoslovakia.
12. Basov, N.G., Vagin, N.P., Kryukov, P.G., Nurligareev, D.Kh., Pazyuk, V.S., and Yuryshev, N.N.(1984) CH_3I and $n-C_3F_7I$ as donors of iodine atoms for a pulsed oxygen-iodine chemical laser, Sov. J. Kvantovaya Elektronika **11**, 1893-1894.
13. Schmiedberger, J., Kodimova, J., Kovar, J., Spalek, O., and Trenda, P. (1991) Magnetic modulation of gain in a chemical oxygen-iodine laser, *IEEE J.Quant. Electron.* **27**, 1262-1264.
14. Babaritsky, A.I., Balagurov, A.Ya., Bistrova, T.B., Kalachev, B.V., and Smirnov, A.N. (1973) Nonmagnetic pulse coaxial lamp, *Sov.J.Prikladnaya Spektroskopiya* **19**, 372.
15. Golubev, L.E., Zuev, V.S., Katulin, V.A., Nosach, V.Yu., and Nosach, O.Yu. (1973) Investigation of the optical inhomogeneities, arising in the active medium of photodissociation lasers during lasing, Sov. J. Kvantovaya Elektronika, n.6, 23-30.
16. Bobrov, B.D., Kiselev, V.M., Grenishin, A.S. (1983) Influence of gasdynamic shock waves on the lasing kinetics in a flashlampe-pumped iodine photodissociation laser, *Sov.J.Quant.Electron.* **13**, 1175-1178.
17. Batashev, S.P., Galpern, M.G., Katulin, V.A., Lebedev, O.L., Mekhryakova, E.A., Mizin, V.M., Nosach, V.Yu., Petrov, A.L., and Petukhov, V.A. (1979) Investigations of the new compositions characteristics for passive shutters of iodine laser, *Sov. J. Kvantovaya Elektronika* **6**, 2651-2654.
18. Gryaznov, Yu.M., Lebedev, O.L., Serebryakov, V.A., Starikov, A.D., and Shwom, E.M. (1969) Stable passive shutter for neodime lasers, *Sov.J.Prikladnaya Spektroskopiya* **10**, n. 5.
19. Grenishin, A.S., Kiselev, V.M., Krutova, L.I., Lukin, A.V., and Sandulenko, V.A. (1993) The cristalline passive shutter for iodine laser, *Summary of papers Conf."Laser Optics"*, St.Petersburg, 248.
20. Zalessky, V.Yu., Kokushkin, A.M., Polikarpov, S.S. (1982) CW-generation of photodissociation iodine laser with cyclical circulation of gaseous perfluoralkil iodide, *Sov. J. Pisma v Tekhnicheskaya Physika* **8**, n. 15.
21. Brederlow, G., Fill, E., and Witte, K.I. (1983) *The High-Power Iodine Laser*, Springer-Verlag, Heidelberg,.
22. Ageev, B.M., Artemov, A.A., Vakorin, A.A., Gromovenko, V.M., Danilov, O.B., Zhevlakov, A.P., Koryakovskii, A.M., Lapshin, V.A., Pasunkin, V.H., Petrikin, V.S., and Shustrov, N.V. (1991) Pulse periodic iodine laser with lamped pump, *Sov.Izv.Ac.Nauk., Ser.Phys.* **2**, 231-235.
23. Gavrilov, V.E., Danilov, O.B., Zhevlakov, A.P., and Tulsky, S.A. (1982), Radiation of strong-current pulse tube lamps, *Sov. Journal of Optical Technologies* **7**, 3-7.
24. Belousova, I.M., Danilov, O.B., Sinitcina, I.A., and Spiridonov, V.V. (1970) Study of optical inhomogeneities of active substance of laser, operating on photodissociation of molecules CF3I, *JETF(sov)* **58**, 1481.
25. Borovich, B.L., Zuev, V.S., Katulin, V.A., Nosach, V.Y., Nosach, O.Y., Startzev, A.V., and Stoilov, Y.Y. (1975) Parameters of iodine laser amplifier of short pulses, *Sov. Quantum Electronics* **2**, 1282.
26. Danilov, O.B., Zhevlakov, A.P., Tulskii, S.A., Yachnev, I.L. (1981) An investigation of free-running photodissociation iodine laser, *Pisma v JETF (sov)* **7**, 1160-1163.

220

27. Frantz, L.M. and Nodvik, J.S. (1963) *J.Appl.Phys.* **34**, 2346-2349.
28. Bobrov, B.D., Kiselev, V.M., Grenishin, A.S., Rodina, L.I. (1985) Spectral characteristics of power iodine laser radiation with magnetic controlled amplification, *Opt.Spectroscop.* **59**, 1085-1094.

CCRF EXCITED COPPER-ION-LASER

J. Schulze, C. Lücking, N. Reich, D. Teuner,
J. Mentel, M. Grozeva *, J. Mizeraczyk **

Ruhr-University Bochum
Department of Electrotechnical Engineering
Division AEEO
D-44780 Bochum, Germany

* Bulgarian Academy of Sciences, Institute of Solid State Physics
Tzarigradsko Chaussee 72, 1784 Sofia

** Polish Academy of Sciences, Institute of Fluid Flow Machinery
Fiszera 14, 80-952 Gdansk

1. Abstract

A capacitively coupled radio frequency excited cw Cu^+-laser was realized by evaporating CuBr of high purity into a discharge operated in He or Ne. With a tube filled with He as a buffer gas laser action on 4 infrared copper II lines (740.4 nm, 766.5 nm, 780.8 nm and 782.6 nm) was achieved. Using a Ne-CuBr gas mixture laser gains on 14 UV-lines between 240.3 nm and 272.2 nm were observed. The tube design and some operating characteristics of the CCRF excited CuBr-laser are presented.

2. The capacitively coupled radio frequency (CCRF) discharge

The discharge is excited in a cylindrical tube made of dielectric material between two electrodes facing each other along the tube axis. The walls in front of the electrodes are negatively charged by electrons oscillating in the applied RF-field. Positive space charge layers are formed in front of the walls due to the fixed spatial ion distribution. It generates together with the electrons on the wall a strong electric DC field directed from the plasma to the wall. Electrons accelerated by this field penetrate the plasma with high energy.

Special characteristics of the CCRF-discharge are:
- relatively simple technology of tube manufacturing
- high purity of the discharge due to the electrodes outside the tube

W. J. Witteman and V. N. Ochkin (eds.), Gas Lasers - Recent Developments and Future Prospects, 221–224.
© 1996 *Kluwer Academic Publishers.*

222

- homogeneity of the discharge along the tube axis
- no cataphoretic effects along the tube axis

3. RF excitation and design of the matching

The design of the matching between the RF-generator output resistance of 50 Ω and the laser discharge tube impedance is of main importance for an appropriate operation of the laser tube. The tube can be represented electrically by a small ohmic impedance in series with a large capacitive impedance for which a so called L-matching formed by two capacitors and a coil is suitable. However, with a simple L-matching the discharge spreads within the whole tube and becomes inhomogeneous within the gap between the electrodes. This problem is avoided by using a symmetric matching realised by a symmetrizing transformer. With the symmetric matching a homogeneous discharge is nearly perfectly confined within the electrode gap [1,2]. Moreover no longer electromagnetic interferences are observed in electronic devices in the vicinity of the tube which is a severe problem in the case of an unsymmetric matching.

4. Tube design

A temperature of about 600 - 800 K is sufficient to vaporize the CuBr because of its relative high vapour pressure compared to pure copper. The copper-vapour-density in a tube operated with CuBr can be regulated independently using a sidearm oven with CuBr in it. This is not possible if the copper vapour is produced by sputtering from an inner electrode [3]. The tube design with such an oven is shown in Fig. 1 [4]. The discharge is operated in an Al_2O_3 ceramic tube of 40 cm length which is enveloped by a fused silica tube. To achieve a good CuBr distribution the heated CuBr reservoir is distributed along the discharge tube as shown in Fig. 1. The reservoir has six tube connections evenly distributed along the active length of the tube.

To achieve a long lifetime of the tube the Brewster windows have to be protected from dust particles produced by sputtering at the tube wall. For this purpose the condensation chambers are equipped with heating wires producing by thermal buoyancy gas whirls within the condensation chambers. Due to these gas whirls the flow of dust particles from the tube ends to the Brewster windows is deflected onto the wall.

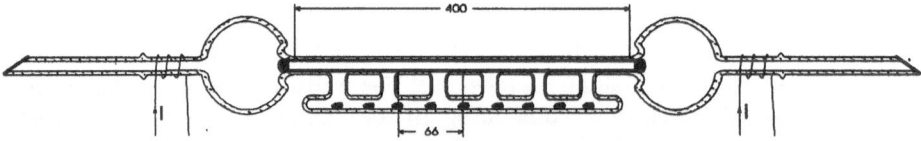

Fig. 1: Tube for a CCRF-excited He-Cu$^+$ laser operated with CuBr. It is vaporized by an oven distributed along the discharge tube into a capillary made of Al_2O_3 ceramic which is inserted into a fused silica tube.

5. Excitation of the He-Cu⁺- and the Ne-Cu⁺-lasers

The main processes involved in generating the laser action are:
- Dissociation of the CuBr
- Ionisation of helium or neon by electron-impact
- Production of excited copper ions by charge transfer

$$He^+ + Cu \rightarrow He + Cu^{+*}$$

$$Ne^+ + Cu \rightarrow Ne + Cu^{+*}$$

- Depopulation of the lower laser levels by UV- and VUV- transitions
  to the ground level of the copper ion
- Destruction of the copper ions by recombination at the wall,
  three body recombination or chemical recombination

6. Characteristics of the He-Cu⁺ (CuBr) laser

The measurements were made at a temperature of the CuBr side arm oven of 723 K. The He-pressure dependencies of the CCRF excited He-Cu⁺ laser intra-resonator powers of both infrared lines (740.4 nm and 780.8 nm) when oscillating separately are shown in Fig. 2. It has to be noted that the resonator power of the 780.8 line is four times higher than that of the 740.4 nm line.

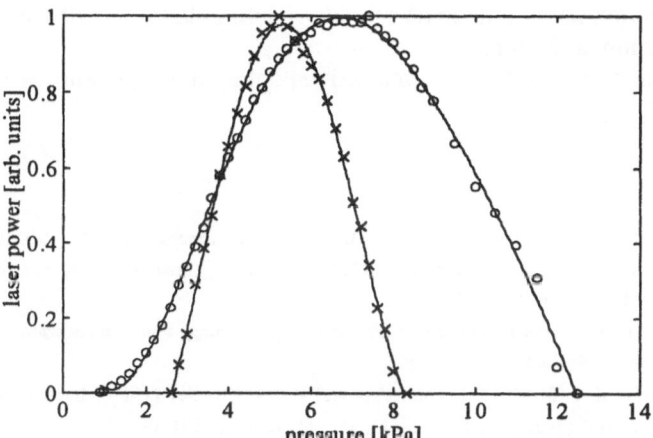

Fig. 2: Dependency of the resonator power of the Cu⁺ lines
at 740.4 nm (×) and 780.8 nm (o) on the He pressure

224

Fig. 3 shows the output coupling dependency of the laser output power at 780.8 nm for different input RF powers.

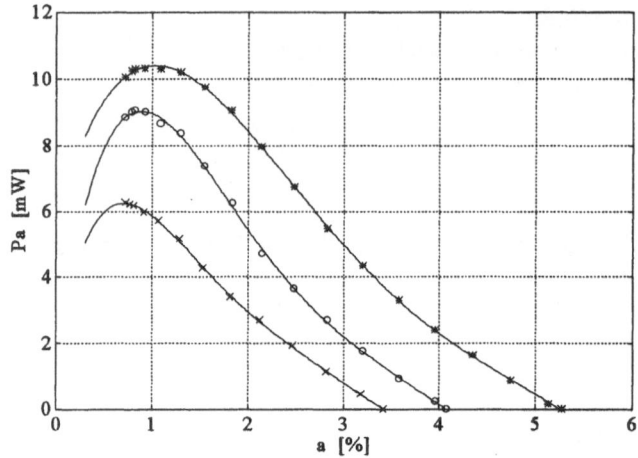

Fig. 3: He-Cu$^+$ laser output power at λ = 780.8 nm as a function of output coupling for different RF input powers (600 W (x), 700 W (o), 800 W (*)), He pressure = 7.2 kPa

7. Characteristics of the Ne-Cu$^+$ (CuBr) laser

We observed laser gains on the following 14 UV lines: 240.3 nm, 242.4 nm, 247.3 nm, 248.6 nm, 250.6 nm, 252.9 nm, 254.5 nm, 259.1 nm, 260.0 nm, 270.1 nm, 270.3 nm, 271.4 nm, 271.9 nm, 272.2nm.
The strongest line was the 248.6 nm line with an optimum Ne pressure at 1 kPa.

8. References

This work was supported by the European Commission Copernicus Programme CIPA-CT 93-0219.

[1] N. Reich, J. Mentel, G. Jakob, J. Mizeraczyk (1994), Cw He-Kr$^+$ laser with transverse radio frequency excitation, Appl. Phys. Lett., 64, no. 4, p. 397

[2] N. Reich (1994), Transversale kapazitive Hochfrequenzanregung von Gasentladungslasern, Ph. D. Dissertation, Ruhr-Universität Bochum, Germany

[3] V. S. Mikhalevskii, M. F. Sem, G. M. Tolmachev and V. Ya. Khasilev (1980), Laser action on the ionic transitions of copper in a rf discharge, Zh. Prikl. Spektrosk, vol. 32 (4), p. 591-593

[4] N. Reich, J. Mentel, G: Jakob (1993), Capacitively coupled transverse rf-discharges for multiline lasers, Proceedings II of the XXI. Int. Conf. on Phenomena in Ionized Gases, Ruhr-Universität Bochum, Arbeitsgemeinschaft Plasmaphysik (Publ.), p. 102 -103

ADVANCED TECHNOLOGY FOR HIGH AVERAGE POWER PULSED GAS LASERS

W. MAYERHOFER, M. JUNG, E. ZEYFANG
DLR Institut für Technische Physik
Pfaffenwaldring 38-40
D- 70596 Stuttgart

ABSTRACT

Repetitively pulsed e-beam controlled multigas lasers are very promising for a wide area of industrial and military applications, because they combine high power light emission with good beam quality in the near and mid infrared. We present specific data of the laser output (1.7 to 10 μm) at different gas mixtures (12 liter cavity) and of its pulsed power components ($\leq$ 200 kW power modulator). Using a power supply with 300 kW/m^3 power density (high frequency IGBT switched) a power modulator was realized with a power density of more than 30 kW/m^3 at up to 90 % efficiency. In this way pulse to pulse reproducebility of the loading voltage was improved from 10 % up to 0.1 % in burst mode operation for up to 1000 pulses.

1. Introduction

Pulsed gas lasers (high intensity light) as well as pulsed power components (high energy electron beam) can be used for many applications[1] (paint stripping, surface modification, pollution reduction). This has generated a demand for a laser concept with a compact high power unit volume which can function reliably and savely to realize customer specified requirements. We present basic laser data of our multigas laser concept in a single pulse operation. A breakthrough for repetitively pulsed operation was possible after a new advanced technology for pulsed power supply became available. Using this technology we realized a spark gap switched power modulator with a power density of more than 30 kW/m^3. A next step to improve the system is to replace the spark gaps by advanced high power semiconductor switches. First experiments show promising results with respect to di/dt, blocking voltage, and reprate operation. For the high pressure discharges we utilized high energy electrons from a large scale accelerator (window area A= 120 cm^2) with cold cathode (carbon felt) emitters. It has been able to run for up to 1000 pulses at 5 Hz respectively 20 pulses at 60 Hz reprates limited by strong pressure rise in the e-gun enclosure. Further investigations and selection of new emitter types are under way.

W. J. Witteman and V. N. Ochkin (eds.), Gas Lasers - Recent Developments and Future Prospects, 225–229.
© 1996 *Kluwer Academic Publishers.*

2. Multigas Laser Concept

The laser is a sealed off testbed unit with an e-beam sustained transverse flowed discharge[2] (gas flow velocity up to 100 m/s). For operation at the different laser wavelengths (1.7/ 5/ 10 µm) we changed only the gas mixture for pressures up to 1 at and the output optics with optional stable or unstable resonator. In addition we used a separate high pressure cavity (ArXe/2.8 at/1.2 liter) with respect to the pressure limitation of the closed cycle concept. Specific laser data (pulse energy E_L[J/lat] versus electrical discharge energy E_d[J/lat] are presented in figure 1 for the different laser gases , different types of resonators (T: transmission percentage of the stable resonator output mirror, M: magnification of the unstable resonator) and the laser efficiency [%] as a parameter. The maximum measured laser pulse energy E_L[J], specific pulse energy E_L [J/lat.], laser efficiency [%] and averaged laser power P_{Lav}[kW] are shown in table 1 for a burst mode operation (limited by the e-gun operation time).

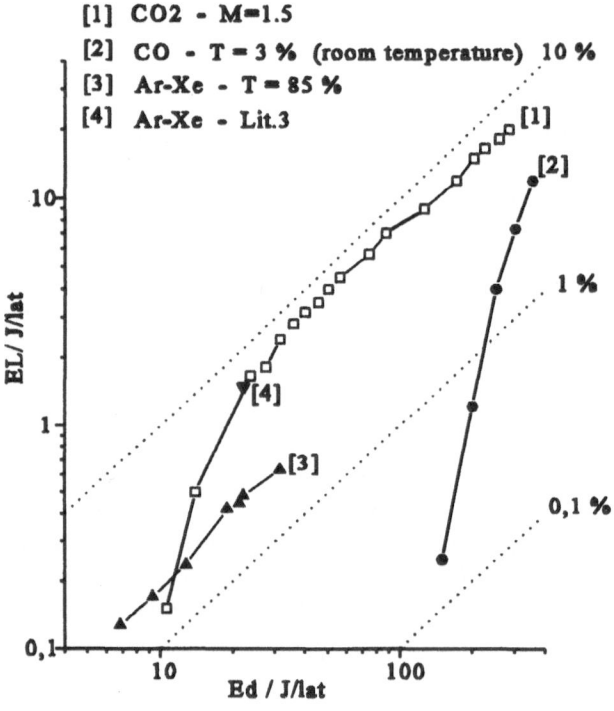

Figure 1. Specific laser pulse energy EL vs. discharge energy Ed

Table 1: Maximum measured laser data

mixture :	$CO_2/N_2/He =$ 10/20/70	$CO/N_2/He =$ 7/43/50	$Ar/Xe=$ 1/150
pressure [at]	0.5	0.6	2.8
volume[liter]	12	3,4	1,2
E_L [J]	138	24	2
E_L [J/lat.]	23	12 (Ref .2)	0.6 (1.5; Ref.3)
η [%]	8	4,6	2 (8; Ref. 3)
P_{Lav} [kW]	3	1	0.1

3.Specific data of the laser and its power supply

Design objectives are to provide components with high power per unit volume which can function reliably and savely. For the main pulsed power components scaling relations of our closed cycle laser concept (10 kW-class) are shown in figure 2. The power modulator consists of a 150 kJ/s power supply (2.27 m³) main discharge PFN (1.73 m³) and e-beam-PFN (1.04 m³) resulting in 30 kW/m³ in total. Considering in addition the laser closed cycle unit (3.33 m³) and the vacuum system for the e-gun (0.64 m³) we realized a laser system at a power density of 1 kW/m³ in total. Regarding this scaling relations the high pressure atomic xenon laser is becoming very promising, because the gas cooling (and corresponding to this fact the system size) can be reduced drastically for this noble gas laser.

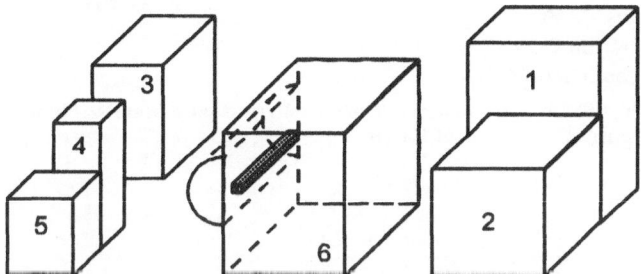

Figure 2.   Volume scaling of the laser and it's components: power supply[1], main discharge-PFN [2], e-beamPFN [3], vacuum system [4/5], laser closed cycle unit [6]

4. Advanced power supply

The design for the power supply is based on the use of 30 kJ/s standard units (commercial available) utilizing high frequency (50 kHz) resonant current switching by parallel IGBT's. Advanced thermal management and high frequency switching results in a power supply unit with a fundamental power density (30 kJ/s versus 0.1 m³). For operating the multigas laser experiment an array of 5 units was integrated into a single enclosure and tied together through a parallel network to combine and protect their outputs during parallel power supply charging.

The highlights of this advanced technique can be summarized by the following topics:
- efficiency: the power supply efficiency is better than 90 %.
- stored energy: low stored energy (10 Joule) results in low device stress and high controllability.
- regulation: the system provides a much better regulation (0.1 %) than any other approach. This is due to the use of one small power supply at the end of charge.
- soft failure and system safety: in the event of a single component failure the system continues to operate at slightly reduced output power.
- cost effectivity: a failed power supply is simply replaced.

With respect to the operational limits of the e-gun we used high power liquid electrolyte resistors as dummy loads and supersonic gas flow spark gaps (R.E.Beverly III and Associates, Ohio, USA) as switches to test the power modulator. We achieved full spezification , that means : 1000 shots at 100 Hz reprates [4]. A typical voltage at the e-beam load in sequence mode is shown representatively in figure 3.

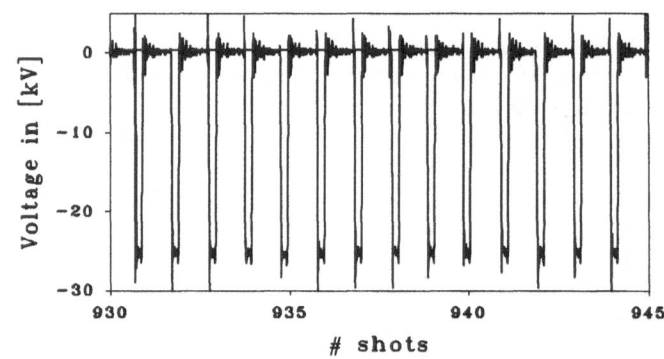

Figure 3. Typical voltage at the e-beam load in sequence mode at 100 Hz

5. Conclusion

An electron-beam sustained multigaslaser offers interesting aspects for a wide area of applications with respect to it´s high power light emission and beam quality in the near and mid infrared. We presented measured laser pulse energy of a 12 liter cavity using different gas mixtures for different laser wavelegths (10/5/1.7 μm) in single pulse and repetitively pulsed (≤ 60 Hz) burst mode operation (up to 1000 pulses).To realize customer specified requirements we need a laser concept with a high power unit volume which can function reliably and savely. Reduction of system size requires size reduction of many different components. The first goal on this way was to install an advanced power supply. Preliminary analysis showed that IGBT technology offers the highest power per unit volume (300 kW/m³) and per unit cost of presently commercial available power supplies. By using this technology we designed a power modulator concept (30 kW/m³) and operated the laser in burst mode. With respect to the operational limits of the e-gun we tested the power modulator at dummy loads. We achieved full specification at 100 Hz rep rates. Further improvement can be realized by optimization of the e-gun itself and by introducing high power semiconductor switches [5].

6. References

1. M. Kristiansen, digest of technical papers: 9th IEEE Pulsed Power Conference (1993), pp. 46.
2. A. Ionin, W. Mayerhofer, S. Walther, E. Zeyfang, Roomtemperature repetitively pulsed e-beam sustained carbon monoxide laser, SPIE Vol. 2502, pp.44 (1994).
3. W.J. Witteman, P.J.M. Peters, H. Botma, S.N. Tskhai, Y.B. Udalov, Qi-Chu Mei and V.N. Ochkin, High power atomic xenon laser, SPIE Vol. 2502,pp.497 (1994).
4. M. Jung, W. Mayerhofer, Th. Schweizer, E. Strickland, Spark gap switched PFN's operating at 100 Hz for a multi kW pulsed CO_2 laser, abstract for 10th IEEE Pulsed Power Conference (1995).
5. M. Jung, W. Mayerhofer, M. Edele, O. Gstir, Th. Schweizer, E. Zeyfang, Test of fast SCR's as spark gap replacement, abstract for 10th IEEE Pulsed Power Conference (1995).

MEASUREMENT OF THE OPTICAL LOSSES OF CO$_2$-LASER COMPONENTS

G. JAKOB, J. TEICHMANN, U. BERKERMANN, G. SCHIFFNER
Ruhr-University Bochum
Department of Electrotechnical Engineering
Division AEEO
D-44780 Bochum, Germany

Abstract - The optical losses in the resonator critically determine the efficiency of a CO$_2$-laser. These losses have been calculated by analysing the circulation of radiation power, reflected or transmitted, when passing through a Scanning-Interferometer. The system consists of a stable optical resonator. One mirror of this resonator is movable by a circular arrangement of piezo elements, so it is possible to change the length of the resonator by several wavelengths of the laserbeam coupled into the interferometer. Different optical elements such as waveguides or Brewster windows can be analysed. The reflectivity of laser mirrors and the optical losses from slab waveguides, made of aluminia ceramics or copper, are determined for different wavelengths between 9.2 μm and 10.8 μm. Furthermore, optical losses of ceramic-waveguide resonators have been measured for different mirror-waveguide distances and for varying mirror tilt-angles. The results are compared with different theoretical models of optical losses, calculated by several other authors.

1. Introduction and Experimental Setup

This paper reports a method to measure the reflectivity of laser mirrors and to determine the optical losses of laser elements such as windows, capillaries and especially waveguides and slab waveguides (hereafter referred to as 'slabs'). Measurements have been carried out by a Scanning-Fabry-Perot-Interferometer, which forms a stable optical resonator and creates thereby laser-like conditions. Subsequently we describe the system with the main elements and we report some typical results. The structure of the system is shown in figure 1 and the main parameters are described in table 1.

The interferometer consists of two mirrors S_1 and S_2, which form the stable optical resonator. Both mirrors can be adjusted by the motors $M_1...M_4$. The beam of a 5 W CO$_2$-laser L (with supporting equipment K and Hf) is coupled into the interferometer. The mirror on the opposite side of the laser L is movable by a circular arrangement of

231

232

piezo elements $PZ_1...PZ_3$, so it is possible to tune the length of the resonator parallel to the optical axis by several wavelengths of the enclosed laserbeam. The piezo elements with stretch sensors, applied at their surface, are driven by a piezo driver PT in a automatic control loop. So the output of the signal generator SG is converted into a piezo movement without hysteresis effects. Passing the polarizer P and the Fresnel-rhomb FR, the laserbeam is coupled into the interferometer. The polarizer and the Fresnel-rhomb are used as an optical diode, which avoids a modulation of the laser power by the reflected signal of the interferometer. Because of fitting in the aperture B, the optical lowest order mode is mostly excited. The detector D measures the light transmitted by the interferometer. The measured signals are amplified by the amplifier A, observed by the oscillograph O, and are registered in a computer by a AD-converter with a high conversion rate. The beam of the HeNe-laser LJ is congruent to the CO_2-laser beam and is used to adjust the arrangement. The laser power is controlled by the power meter LM and the head MK. Using the dotted drawn detector D next to the polarizer one can measure the beam reflected by the interferometer. This enables the determination of losses in optical slabs in a hybrid-resonator with one cylindrical copper mirror.

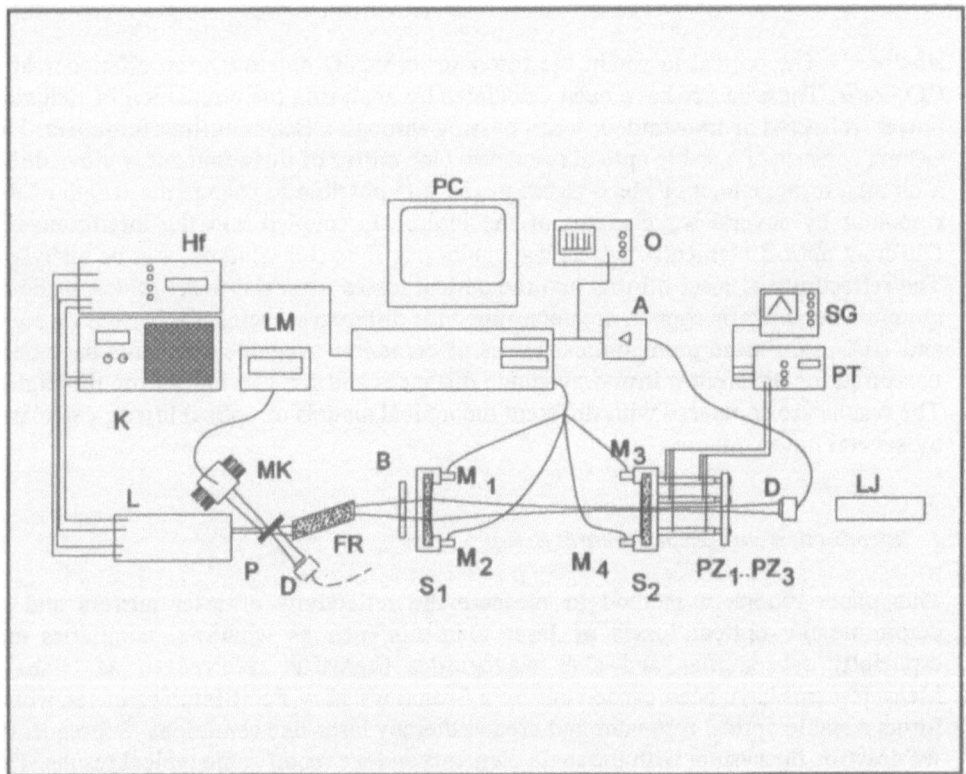

Figure 1. Scanning-interferometer for the measurement of optical losses of CO_2-laser components

2. Interpretation

When monochromatic radiation is coupled into a Fabry-Pero-Interferometer, the power of the transmitted and reflected radiation can be described by means of Airy-functions [1] as a function of the interferometer length and the losses of one cycle in the interferometer. By evaluating the typical peaks of the measured airy-curves one can calculate the internal losses in a interferometer. The peak width depends on the losses. These losses are described by the entire reflectivity (hereafter referred to as 'Rg') of the interferometer. This reflectivity consists of the geometrical mean value of the reflectivity of the mirrors and the transmission of the medium in between. By moving one mirror in the direction of the resonator axis, the transmitted power becomes maximal when the resonance condition of the resonator mode is fulfilled. When the resonator losses are very low, the peakwidth becomes small and it is necessary to detect the signal with a very high temporal resolution. Rg is computed by the quotient of the peakwidth and the distance between the peaks, which is $\lambda/2$. To measure the optical losses of waveguides and windows, one must first determine Rg of the 'empty' interferometer, which is the geometrical mean value of the mirror reflectivities. Next, the optical element must be inserted into the interferometer. Then, the optical losses of the elements can be calculated using Rg.

3. Experimental results

The reflectivity of different laser mirrors, the optical losses of ceramic and copper waveguides, and coupling losses were determined for different wavelengths between 9.2 μm and 10.8 μm. The standard deviation of the measurements is about 1 % of the losses of one cycle in the resonator.

The reflectivity of laser mirrors was measured in the interferometer with free beam propagation. To determine the losses of waveguides and slabs e.g. for CO_2-lasers, the elements were measured in an interferometer with well-known mirror reflectivity.

Several authors (e.g. [2],[3],[4]) have calculated coupling losses for different mirror-waveguide distances and for varying mirror tilt-angles using several theoretical models. The executed measurements indicated, that the real losses are smaller than calculated. Measuring the losses of mirrors with large radii or plane mirrors we detected mirror-distances, where the waveguide losses decrease as the distances grew. This behaviour corresponds with the results of calculations carried out by Gerlach et al.[5]. It is also necessary to consider higher order modes in the waveguide and in the area of free beam propagation between waveguide and mirror. Using an aperture between waveguide and mirror, the measured losses have been smaller than without an aperture.

Also, measurements of losses of Al_2O_3-waveguides with square profiles have been carried out between 9.3 μm and 10.8 μm. Contrary to the assumption that the losses decrease for lower wavelengths, there is a considerable increase below 9.8 μm. This result agrees with the calculation of Barker [6] regarding the dispersion and absorption lines of Al_2O_3.

The Scanning-Fabry-Perot-Interferometer can be used for a precise determination of the reflectivity of laser components. Also, the arrangement is a powerful tool for the measurement of optical losses from Brewster windows, waveguides, slabs and coupling losses, existing in real laser resonators. Table 1 shows the main parameters of the system and some typical results of measurements of CO_2-laser components and waveguides.

Table 1. Measurement of the reflectivity and the optical losses of CO_2-laser components and waveguides

parameters:

wavelength	10.6 μm
laser	CO_2-Waveguide
radiation power	5 W
piezo-stretching	60 μm
scan-frequency	5 Hz
sensor	HgCdTe-Detector
sample-frequency	2 MHz
storedepth	128 k samples
range of loss parameter	0...0.3

typical measurements:

a) Reflection of a ZnSe concave mirror, radius 1 m, nominal reflectivity: 0.99

result: 0.990189 ± 0.00014

b) Coupling loss in a square ($\square = 2.25 \times 2.25 mm^2$) waveguide-resonator, distance Al_2O_3-waveguide - mirror: 30 mm

result: 0.0233 ± 0.00014

c) Losses in a copper-waveguide, length: 300 mm, distance between the electrodes: 2.5 mm

result: 0.00124 ± 0.000068

4. References

[1] G. Hernandez (1988), *Fabry-Perot-Interferometers*, Cambridge University Press, Cambridge.

[2] J.-L. Boulnois and G.P. Agrawal (1982), Mode discrimination and coupling losses in rectangular waveguide resonators with conventional and phase conjugate mirrors, *J. Opt. Soc. Am.*, vol. 72, no. 7, p. 853.

[3] C.A. Hill, P.E. Jackson and D.R. Hall (1990), Carbon dioxide waveguide lasers with folds and tilted mirrors, *Appl. Opt.*, vol. 29, no. 15, p. 2240.

[4] C.A. Hill and D.R. Hall (1985), Coupling loss theory of single-mode waveguide resonators, *Appl. Opt.*, vol. 24, no. 9, p. 1283.

[5] R. Gerlach, D. Wei and N. M. Amer (1984), Coupling effiency of waveguide laser resonators formed by flat mirrors, *IEEE J. Quantum Electron.*, vol. QE-20, no. 8, p. 948.

[6] A.S. Barker (1963), Infrared lattice vibrations and dielectric dispersion in Corundum, *Physical Review*, vol. 132, no. 40, p. 1474.

MULTIKILOHERTZ REPETITION RATE CW DISCHARGE CO_2 LASER WITH AVERGE POWER UP TO 2,5 KW

G.N. Grachev, A.G. Ponomarenko, A.L. Smirnov and
V.B.Shulyat'ev

Institute of Laser Physics, Sibirian Division of the Russian
Academy of Sciences, Prosp. Lavrent'eva 13/3
630090, Novosibirsk, Russia

Abstract

A line-tuned pulse-periodic CO_2 laser with a Q-switched resonator is described. The laser is pumped by a continuous discharge in the cross gas flow. The tunability range covers about 60 lines of P,R branches of transitions 001 - 100 and 001 - 020. The repetition rate can be set from 1 up to 100 kHz . A maximum peak power of about 700 kW (at frequencies <5 kHz) and an average power up to 2,5 kW (at frequencies > 25 kHz) have been obtained at line 10P20.
The aim of this work is to extend the applications and to increase the efficiency of CO_2 lasers with continuous pumping in selected areas such as infrared laser chemistry, isotope separation, laser pyrolysis and thermal technologies.

1. Laser design

The laser is created on the basis of a continuous electric discharge CO_2 laser: LOK-3M [1]. The general discription of the construction of the laser head and the basic technical solutions are given in [2]. The active medium is exited by a DC discharge in a CO_2 : N_2 : He gas mixture. The discharge is realised in the gas flow in a rectangular gas dynamic channel, the flow velocity in the input of the discharge gap being 45 m/s. The direction of the flow, electric current and radiation propagation direction are mutually perpendicular. An electrode configuration has 2 discharge gaps, wich are formed by a common plane anode with a width along the flow of 100 mm being placed in the central part of the gas dynamic channel and 2 cylindrical cathodes of 16 mm diameter and 800 mm length which are set from both sides of the anode at a distance of 50 mm from it.

W. J. Witteman and V. N. Ochkin (eds.), Gas Lasers - Recent Developments and Future Prospects, 235–240.
© 1996 *Kluwer Academic Publishers.*

236

Nonsectional, water cooled single element electrodes are made of copper. In the laser there is a ballastless power supply on the basis of the inductive-capasitive transmitter with independent current adjustment in the discharges [3].

To obtain a pulse-periodic generation mode we have reconstructed a self-filtering unstable resonator of the LOK-3M laser [1] with the addition of an optical Q-switching sheme and line generation tuning (QSLT) in the feedback arm.
The main advantages of such optical system are as follows:
- achivement of high beam quality close to the diffractional limit;
- pulse-periodic mode with power (average and pulse) by an order of magnitude greater than in stable Q-switching resonators as the radiation power in the feedback arm makes up about 0,1 from the output laser power [4].

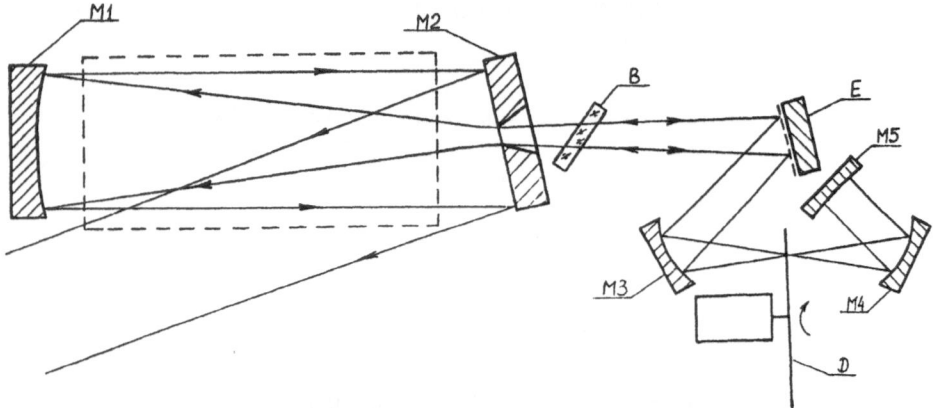

Figure 1. Resonator sheme. M1 - concave mirror, R = 14,6 m; M2 - output mirror with filtering aperture; B - KCl output window; E - echelette; M3,M4 - concave mirrors, R = 0.15 m; M5 - concave mirror, R = 3,2 m; D - chopper disc.

Fig. 1 presents a resonator scheme. A resonator comprises a self-filtering unstable resonator (SFUR) [4], which has an optical shceme of Q-switching and line tuning (QSLT) in the feedback arm. In active medium arm a beam makes 3 passes in ever discharge gap. In the common resonator frame there is a feedback arm with the optical scheme QSLT, separeted by the KCl window. It is set aligned at Brewster angle and creates plane polarised beam.

The optical scheme QSLT consists of a echelette with 100 lines/mm and 2 short-focus concave mirrors to create and correct astigmatism of the incklination beams. Such scheme allowed us to obtain 2 mutually perpendicular waists of the feedback beam of 2 X 0,15 mm. The radiation power in the feedback arm makes up about 10% of the output power (SFUR magnification is 4,5 [4]). In such a way we managed to avoid the optical discharge on the modulator surface. This discharge limited the development of a Q-switched pulse. The Q-switching was accomplished by a rotating copper disk with slits in the waist plane. The linear slit velocity (up to 150 m/s) provides switches of the resonator Q-factor for the period comparable with the generation development time.

The pulse repetition rate is regulated by both the change of rotational speed of the modulator motor (50 - 200 rot/s) or replacement of the discs with different number of the slits placed on the circle (from 6 up to 360 slits).

2. Results

The laser tuning was carried out in the vibrational-rotational branches 9,4 and 10,4 microns (about 60 lines in whole). Maximum power was up to 2,8 kW in the continuous mode and up to 2,5 kW (average power) in the pulse-periodic one at line 10P20.

In Fig. 2 one can see a spectrum and average laser power at pulse repetition rate 43 kHz, pumping power 50 kW and gas mixture ratio CO_2:N_2:He as 2,5:6:12,5 Torr respectively.

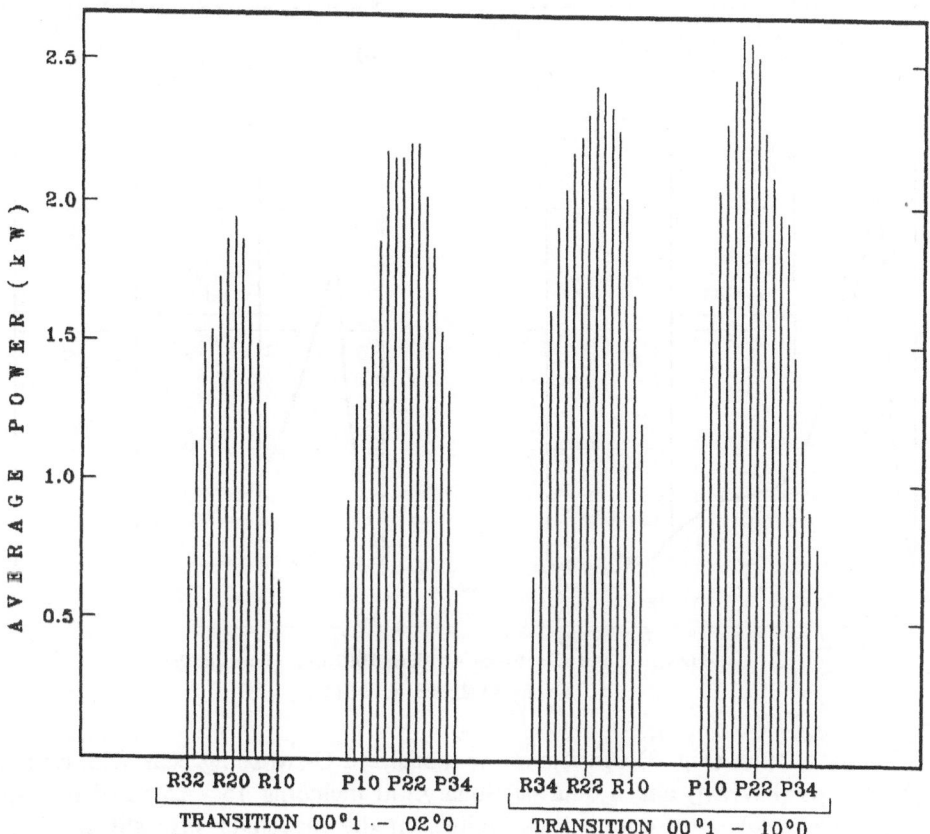

Figure 2. Tunability range and average laser power at pulse repetition rate 43 kHz, pumping power 50 kW, gas composition CO_2:N_2:He = 2,5:6:12,5, total pressure 21 Torr.

238

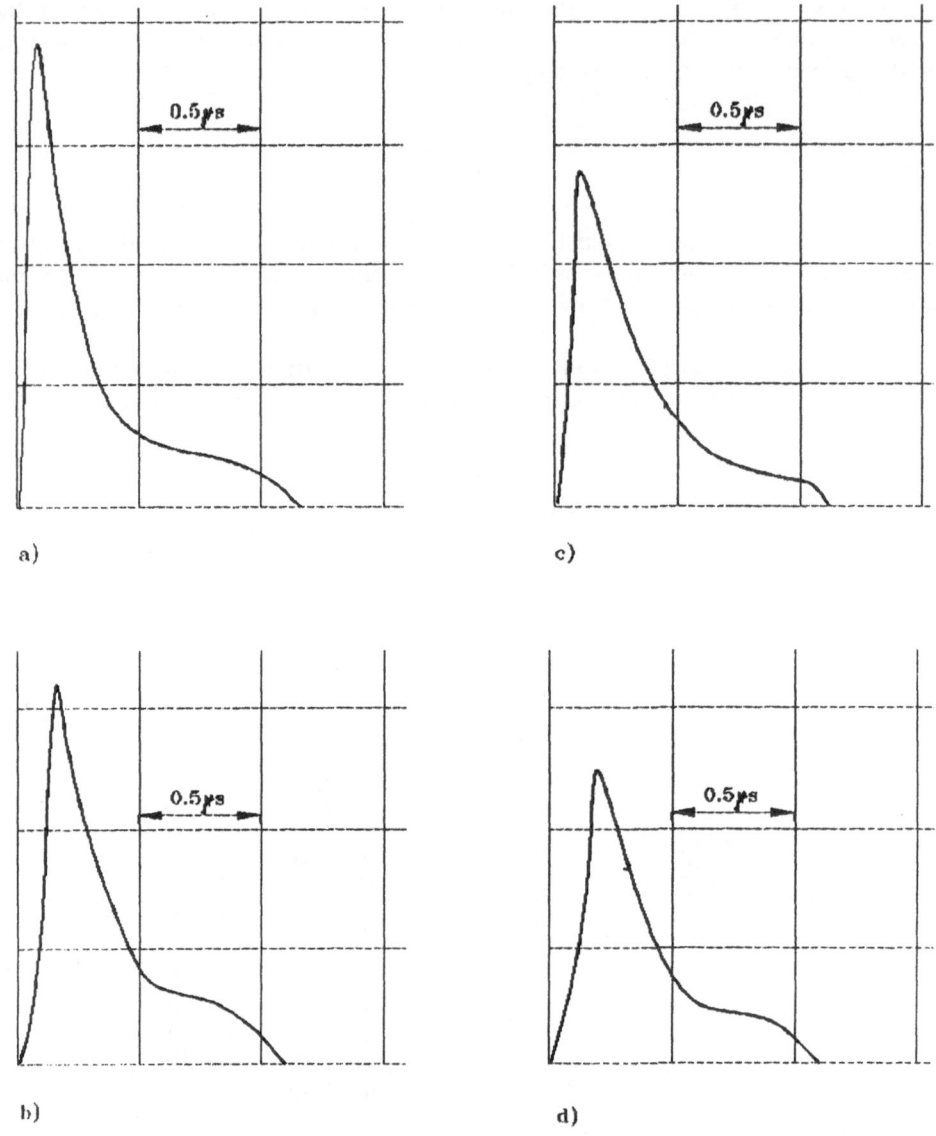

Figure 3. Laser pulse traces. a,b - lines 10P20, c,d - lines 9R20;
a,c - f = 3 kHz, b,d - f = 96 kHz.

The laser pulse (Fig. 3) consists of the modulated Q-switshing peak with duration 0,2 - 0,3 μs (FWHM) and a quasi-stationary tail including 15 - 25 % of the pulse energy. The total pulse duration dapending on the modulator disc slit width and modulator rotational speed makes up from 1 to 3 μs.

We have measured the average and peak laser power in the Q-switched mode at "strong" lines and some lateral one in P,R branches of spectrum in the pulse frequency range 1 - 100 kHz. Fig. 4 shows a characteristic plots.

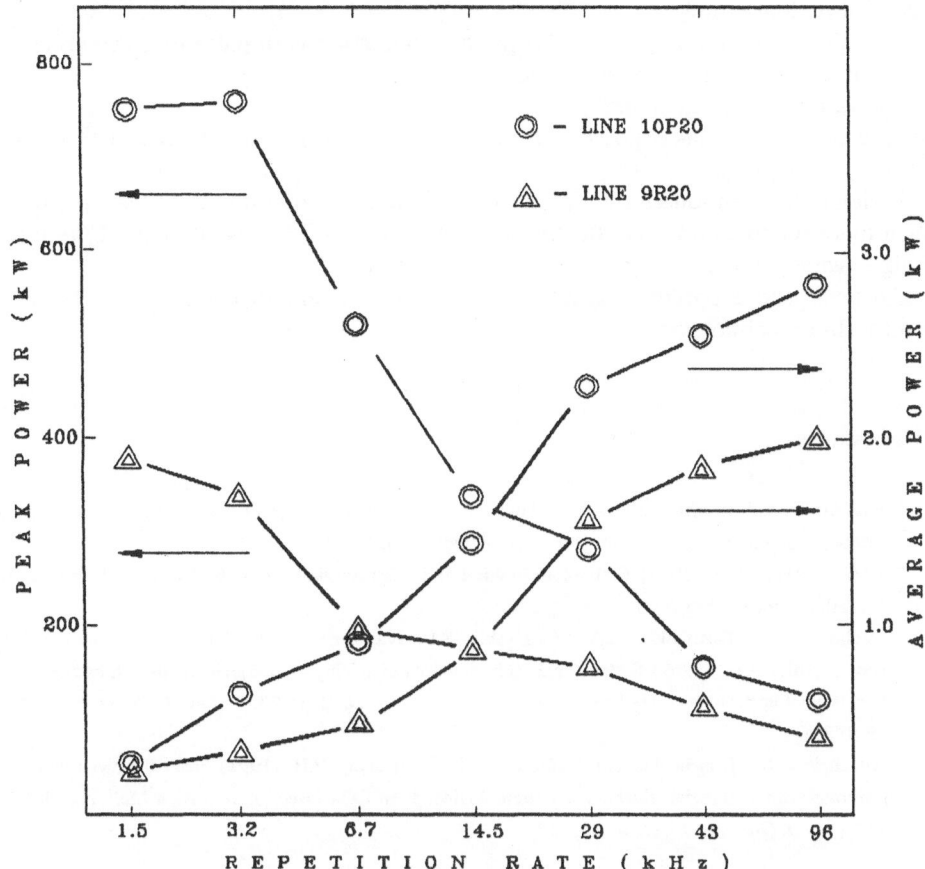

Figure 4. Average laser power and peak laser power versus pulse repetition rate.

up 0,8 mrad at the beam d Maximum peak power about 700 kW (at frequences < 3 kHz) and average power up to 2,5 kW (at frequences > 25 kHz) have been obtained at line 10P20. The decrease of the relation between the peak and continuous powers from 250 to ~150 (at frequences < 3 kHz) at tuning from the "strong" line to the lateral lines were registered. Such tuning gives also the Q-switched pulse duration from 0,3 up to 0,5 μs. This effect is, apparently, connected with the decrease of the relation of the unsaturated gain before the pulse and saturated gain at the Q-switched pulse rising.

The full divergence angle (on power level 0,86) makes iameter of 30 mm and the peak radiation intensity in the lens focus is 0,2 - 2 GW/sm^2 (at frequencies 5-100 kHz) at the generation at line 10P20.

3. Conclusion

CO_2 laser is created with the following features:
- generation line tuning;
- a possibility of continuous and pulse-periodic operation with pulse repetition rate and pulse duration control at a broad range;
- high peak and average output power;
- high beam quality makes possible a fine focusing (with power density 2×10^8 - 2×10^9 W/cm^2) at lens focus.

It should be mentioned, this lasers with continuous pumping and Q-switching are much more reliable and cost effective as compared to TEA pulse-periodic CO_2 lasers of high average pover.

All listed above provide a good grounds for a wide employment of these lasers in selective laser technologies.

4.References

1. Ivanchenko, A.I., Krasheninnikov, V.V., Smirnov, A.L., Shulyat'ev, V.B. (1994) Industrial 3 kw co_2 laser emitting high-quality radiation, *Quantum electronics* 24, 591-594.

2. Abil'siitov G.A. (Ed) (1991) Tekhnologicheskie Lasery: Spravochnik. (Industrial Lasers: Handbook Vol. 1, Mashinostroenie, Moscow).

3. Volkov, V.V., Denisenko, A.A., Zakrevskii, S.I., Ivanchenko, A.I, Koba, A.P., Lysenko K.L., Ponomarenko A.G. (1989) Systema nakachki s vysokimi udelnymi energeticheskimi characteristikami dlya technologicheskogo CO_2 lasera, *Kvantovaya Elektron.* **16**, 2234-2236. (*Sov. J. Quantum Electron.* **19**, 1437).

4. Grachev, G.N., Ivanchenko, A.I., Smirnov, A.L., Shulyat'ev, V.B. (1991) Neustoichivyi resonator s prostranstvennoi filtraziei islucheniya v technologicheskom CO_2 lasere, *Kvantovaya Elektron.* **18**, 131-133. (*Sov. J. Quantum Electron.* **21**, 118).

LASER BEAMS WITH HELICAL
WAVEFRONT DISLOCATIONS AND THEIR
APPLICATIONS IN THE DIAGNOSTICAL
AND METROLOGICAL SYSTEMS

P.V.KOROLENKO
Moscow State University, Faculty of Physics
Moscow, 119899, Russia

Abstract

The paper presents the results of the investigations of the conditions when in gas lasers' cavities the beams with helical wave front appear. It has been affirmed that in lasers' cavities a regular helical wave as well as a complex set of HDs may be generated. In highly degenerated resonant cavities, in wide aperture ones and in co-axial lasers helical waves are frequently realized as so called "multy-pathing modes" (M-modes) having circular arrangement of the light spots. For experimental data explanation and for the description of helical waves' properties the wave as well as geometric models have been applied. The mechanism of M-modes spatial synchronisation has been studied. The method allowing to produce the sharply directed radiation beams in wide-aperture lasers is suggested. The divergence of the beams produced is near to the diffraction limit. The method of ingomogeneity degree estimation of the propagation media and reflecting surfaces has been put forward. It bases on calculation of dislocations number in the radiation received. The method suggested to sense the power of ingomogenity of transmitting media and reflecting surfaces. It bases on estimating of HD number in the received radiation.

1. Introduction

Recently the experimental data have detected the existence of processes causing the helical deformation of wavefront. The distortions of this type considerably changing the wavefront topology appear in speckle fields [1], laser's radiation [2], optical light guides [3], laser beams propagating through the turbulent media [4]. Beams with helical wavefront structure have in their cross-section one or several special points where intensity is equal to zero and phase varies to $2\pi\ell$, while passing around these points. Here ℓ is

W. J. Witteman and V. N. Ochkin (eds.), Gas Lasers - Recent Developments and Future Prospects, 241–248.
© 1996 *Kluwer Academic Publishers.*

a whole number not equal to zero; it determines the topological charge of helical surface.

The unique properties of helical fields allow to raise the problem of their applications to improve characteristics of different optical arrangements. The spatial geometry of helical fields as a rule is characterized by intensity decreasing near the axis; this characteristic property may appear to be optimal when carrying out the technological operations, connected with influence of radiation on materials [5], in the case of diffractionless propagation of beams in laser gas media [6]. There is an information about helical fields application for the correlation photon spectroscopy [7] and improvement of metrological arrangements' characteristics [8]. However, the lack of theoretical and particularly of experimental data about the processes of forming and propagation of beams with wavefront helical dislocations (HD) narrows the possibilities of their practical usage.

The work presented is based on the study of helical fields formed in laser resonators and randomly inhomogeneous media. The problems of such fields applications for the diagnostical and metrological arrangements are considered.

2. Laser Systems

It is known [9], that the Helmholtz's equation for field amplitude U of opened stable resonator admits the solution as mode $TEM^*_{p\ell}$ with helical wavefront structure

$$ u_{p\ell} \sim f_{p\ell}(r/w) \cdot \left[a_{p\ell} \cdot \exp(i\ell\varphi) + b_{p\ell} \cdot \exp(-i\ell\varphi) \right], \qquad (1) $$

where r, φ - are the polar coordinates in the mode cross-section; p,ℓ are whole-numerical indexes, w is the parameter characterizing the distribution width; $f_{p\ell}(t) \sim t^\ell \cdot L_p^\ell(2t^2) \cdot \exp(-t^2) \cdot \exp(\frac{i\pi r^2}{\lambda R})$; $L_p^\ell()$ is Lager's polynomial; R is the wavefront curvature radius; $a_{p\ell}$ and $b_{p\ell}$ are arbitrary complex constants.

Presence of the exponential multiplier $\exp(\pm i\ell\varphi)$ in (1) points to helical form of phase surface. The sign in index of the exponent determines a direction of azimuthal phase rotation. When the amplitudes of field components having phases, rotating in opposite directions ($a_{p\ell}=b^*_{p\ell}$) are equal the classical for stable resonators' modes $TEM_{p\ell}$ will take place; they contain nodal lines in azimuthal direction.

Experiments were performed using He-Ne, Ar and CO_2-lasers. They included the ordinary piping, gasodynamic, TEA and coaxial lasers. The holographic contrivances and shift interferometers were used for registration of helical dislocations on laser beams' wavefronts [4]. The existence of helical dislocations and their locations were fixed by points of fringe's branching on holograms and interferograms. Investigations have shown, that the features of mode structure transformation and character of helical fields formed in resonators are closely approximated by (1) in the case of not very large apertures. For example if the laser confocal resonators are used the increasing of their aperture (starting from the threshold value) leads to successive changes of states with following set of parameters: (p=0, ℓ=0) → (p=0, ℓ=1, a≠0, b=0) → (p=0, ℓ=1, a>b>0) → (p=0, ℓ=1, a=b) → (superposition of $TEM_{p\ell}$ modes with different values of p and ℓ). The second of these states corresponds to "pure" regular helical field, the third one describes the helical field, in which the uniform azimuthal phase change is disturbed by additional field with opposite phase rotation.

Experiments show, that it is efficient to form helical fields in wide-aperture lasers by excitation of multi-passing mode (M-mode) with a circular location of light spots in their resonators [11]. It was made either by simple adjustment of the mirrors or by inserting the masks of appropriate configuration in optical resonator. Each M-mode excited in a planospherical cavity is usually described by 2 indices: N and K. The index N represents the number of light spots on each of the mirrors and the index K is the number of return passes of a beam along the azimuth needed to close the path [10]. These indices are determined uniquely by the geometric parameters of the cavity:

$$\frac{K}{N} = \frac{1}{\pi} \cdot \arccos\sqrt{1-(L/R)} \, , \tag{2}$$

where L is the cavity length and R is the radius of curvature of the spherical mirror. Because of strong competition of oscillation types, forming M-mode with fixed traces of creating beams, each M-mode has specific distribution of light oscillations phase over light spots. At the same time competition between M-modes with different beam traces is more weakly expressed. The condition of spatial synchronizing is characteristic for the generation of such modes superposition. These properties of M-modes permit to form highly coherent beams with stable parameters.

Without inner-resonator's mask the geometry of M-modes, amplitude and phase distributions of light oscillations were determined by random distribution of active medium inhomogeneities and defects of reflectors. But the inserting of mask in the resonator permitted to regulate the amplitude-phase distribution of field and to control it easily by changing of mask configuration. The latter was accomplished by its shifting or rotating.

It follows from experimental results [12] that when M-mode beams are coupled out through a uniform semitransparent cavity mirror, the output radiation has exactly the same angular divergence as in the case of generation of high-order axial modes. For this reason the M-mode divergence is very high in wide-aperture laser systems.

However, there is a simple way of reducing the divergence of the M-mode output radiation. It involves excitation in the laser cavity of an M-mode with a high index N and coupling out of the radiation energy from that part of the exit mirror where one of the local light spots is located. Such a spatially non-uniform energy extraction has been achieved in a TEA CO_2-laser by a mirror with a radially located semitransparent zone or with the aid of a coupling plate near a mirror [20].

Another way of reducing the divergence of the radiation generated in M-mode cavities is based on the formation, in a certain part of the cavity of system, of parallel beams and local energy extraction from this region [11]. Experiments in the case of gas dynamic CO_2 laser have shown that the divergence of such a system of parallel beams is close to the diffraction limit. Such allow angular divergence can be explained by the fact that special locking of the M-modes occurs in the process of generation and that the optical oscillations in the emerging beams are phase-locked.

The property of M-mode degeneracy relatively to sizes of light spots circle gives the possibility of auto-tuning of radiation intensity maximums to the gain profile of the active substance. This fact is very important for the energy extraction in coaxial lasers with small slit between the walls of coaxial active camera, since the use of inner-resonator fields of the other kinds often leads to serious difficulties in the tuning of the optical elements of laser system.

Experiments with coaxial laser the active medium of which was excited by HF field showed that the application of M-mode resonator permits not only to extract radiation of narrow-angle TEM_{00}-mode, but to realize single frequency generation mode by the absolute attaching of the frequency to the centre of the working line as well. It may be explained by the fact that for energy extraction of M-mode with large effective length through the resonator's slit a mode near to a super luminescent one is likely to be generated. In this case generation as a rule takes place at a single frequency at the line centre. The experiments carried out showed, that the coaxial laser with the helical M-mode fulfils all the requirements for laser-heterodyne and may be used in precise laser measuring systems.

3. Randomly Inhomogeneous Media

As it is known [14], light beams with HD of wavefront can be formed when radiation propagates through the randomly inhomogeneous layers or media. As a result the light structure acquires the speckle-like character. HD is formed in the points at the speckle borders, where light intensity is equal to zero. Such effects come into being when light propagates through the turbulent atmosphere [15]. Near the points of zero intensity the field can be written as follows:

$$u = C_x \cdot x + i \, C_y \cdot y, \qquad (3)$$

where x, y are coordinates in the plane of beam's cross-section; C_x, C_y are the arbitrary constants.

Since the number of HD in coherent beams is related to the magnitude of refractivity fluctuations along the path there is a possibility to estimate the degree of medium inhomogeneity by the investigation of wavefront dislocation's structure. In this work, this possibility has been accomplished by the estimation of structure characteristic C_n^2 variations on the atmospheric paths due to the turbulent motions of air mass.

The different methods of C_n^2 estimations are well known. They use the results of radio- and optical soundings of the atmosphere. Some of these methods use the fluctuations' intensity estimation [16,17], the others are based on the densitometric analysis of interferometric or diffraction patterns [18]. These methods, as a rule, used dependencies, followed from the "2/3 law" and are limited by the ranges of inertial interval. This is a disadvantage of these methods. Moreover, they usually are not very precise in the case of strong intensity fluctuations and require comparatively large exposition time. The last fact does not allow to investigate the fine-scaled processes in the atmosphere.

The method of C_n^2 estimation on the base of HD registration has been verificated on horizontal and slant tropospheric paths, located near the MSU main building on

Lenin Hills. The length of the path in one direction was approximately 300 m. (More detailed description of the paths' geometry and transmitting and receiving devices is given in work [4]). As the wave front of the gas laser's beam gauge the Mach-Zehnder's shift interferometers were used.

In the case of strong fluctuations the number of HD, appearing on the wavefront, can be described with a good approximation by the expression:

$$N \approx r_n^2 (C_n^2 k^2 L)^{6/5}, \qquad (4)$$

where r_n is a beam radius.

Magnitude N is defined by ratio of beam cross-section to the correlation region. The dependence (4) permits to use the relation

$$C_n^2 \approx (N/r_n^2)^{5/6} (k^2 L)^{-1} \qquad (5)$$

to estimate the C_n^2 value.

Pertinent estimations have been made for different seasons. Some common results are presented in the table 1.

Table 1

HD number	C_n^2 meteo	C_n^2 optic	HD number	C_n^2 meteo	C_n^2 optic
1	$1 \cdot 10^{-14}$	$2 \cdot 10^{-14}$	4	$1 \cdot 10^{-15}$	$2 \cdot 10^{-14}$
2	$2 \cdot 10^{-14}$	$4 \cdot 10^{-14}$	4	$2 \cdot 10^{-14}$	$7 \cdot 10^{-14}$
2	$1 \cdot 10^{-14}$	$2 \cdot 10^{-14}$	5	$3 \cdot 10^{-15}$	$3 \cdot 10^{-14}$
2	$1 \cdot 10^{-15}$	$4 \cdot 10^{-14}$	5	$8 \cdot 10^{-16}$	$9 \cdot 10^{-15}$
2	$3 \cdot 10^{-15}$	$7 \cdot 10^{-15}$	6	$1 \cdot 10^{-13}$	$3 \cdot 10^{-14}$
3	$3 \cdot 10^{-14}$	$8 \cdot 10^{-14}$	7	$2 \cdot 10^{-15}$	$3 \cdot 10^{-14}$
3	$2 \cdot 10^{-15}$	$2 \cdot 10^{-14}$	7	$4 \cdot 10^{-15}$	$3 \cdot 10^{-14}$

These data allow to reveal a good agreement between the C_n^2 values defined from meteo data and optical measurements. In the same time, almost in all cases the effective C_n^2 value is found to be greater than that defined from the meteo data.

4. Application of HD for Phase Objects Investigation

Dislocations, as a pure phase phenomenon, can be used for the exploration of phase objects, such as unhomogeneous medium, reflecting surfaces, transparent objects and so on. The dislocations are considered to be a measure of the phase distortions; so they may appear useful for obtaining the information containing in the phase of wave. It is suggested [19] to use dislocations of wavefront (not helical, but linear ones) for super-resolution of phase object with a priori information. Parameters of phase micro object,

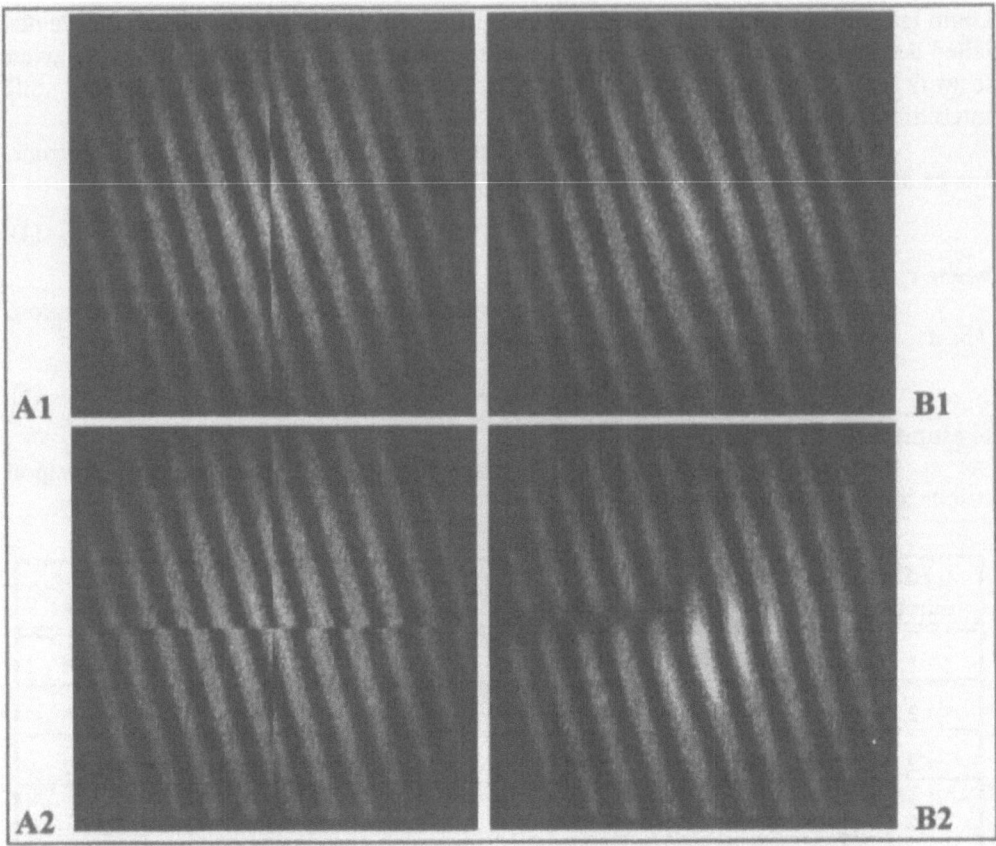

*Fig.1. Demonstration of better detection of HD relative to LD. (A) - sharp dislocations,
(B) - smooth phase tilt; (1) - linear dislocations, (2) - transformed to HD.*

with the linear dislocations (LD) can be defined by analysis of object interferogram even beyond the limits of Raleigh resolution.

In such cases we suggested to use the helical dislocation instead of linear one. LD on a reflected wave front can be transformed into HD by adding of the wave with its copy in both space and phase quadrature. This is more advantageous because it is easier to observe HD on the interferogram. Figure 3 shows the model interferograms. Fig.A presents sharp dislocations (jumping phase), fig.B - characterizes smooth phase tilt. In the second row LD is transformed to HD, using the method described above. As it is seen, the scrap fringe for the sharp LD transforms into its bent and is getting less noticeable, whereas the branching of the fringe presenting HD is easy to detect. In addition, HD is more stable than LD under the diffraction conditions.

There is one more application of HD to be proposed. Let us suggest, that the exploring phase object has the profile of known configuration, but its other parameters are unknown, for example, the modulation depth. In this case the presence (or absence) of HD on the wavefront can give an information about initial phase distortions.

The number of theoretical investigations and numerical experiments have been carried out to explore phase dislocations appearance under the diffraction conditions, as a result of escalation of initial wavefront distortions when the amplitude is constant. It was concluded that in this case the dislocations appear only when initial perturbations exceed some critical value. This value is different for each field configuration and varies 0.5π to 1.5π.

For example, the calculations of field distribution behind the one-dimensional sinusoidal phase grating $m \cdot \cos(2\pi \cdot x/L_x)$ show that dislocations appear when the modulation depth $m > \pi/2$. Numerical calculations of diffraction of two-dimensional fields with the phase modulated as $m \cdot \cos(2\pi \cdot x/L_x) \cdot \cos(2\pi \cdot y/L_y)$ give the critical value of modulation depth equal to 1.32π.

5. Conclusions

The studies accomplished show that the peculiarity of the processes of helical field forming and propagating not only broadens our understanding of structure and salient features of the light diffraction but gives the new possibilities for the improvement of optical devices for metrology and diagnostics as well. Primarily it is regarding to the laser systems. In this case the using of helical fields allows to improve the parameters of the radiation up to the level admitting to use such systems when providing the precise measurements.

In the cases of wave propagation through the media containing random inhomogeneities and phase objects the helical fields are realized as HD systems on the wave front. It allows to estimate the degree of medium inhomogeneity and phase object structure by the registration of HD number and location.

The properties of HD such as the insensitivity to the diffraction energy redistribution and their manifesting on the interferograms allow to improve the ways of registration of the wave objects structure to get the super Raleigh resolution. One may confirm that the further study of HD characteristics will widen the field of their application in different precise devices. Stability to diffractive redistribution of energy and single-valued appearance on shift interferograms allow to improve the registration methods of phase objects structure.

Acknowledgements

This work is supported by Russian Foundation of Fundamental Researches.

References

1. Pilipetsky N.F., Shkunov W.W., Zel'dovich B.Ya. (1985) *Wave Front Conjugation* Moscow, "Nauka".
2. Korolenko P.V., Tikhomirov V.N. (1991) *Kvantovaja elektronika*, **18**, N9, 1139-1141.

3. Bajenov V.U., Soskin M.S., Vasnetsov M.V. (1990) *JETF Letters*, **52**, N8, 1037-1039.

4. Arsenyan T.I., Fedotov N.N., Kaul S.I., Korolenko P.V., Ubogov S.A. (1992) *Radiotekhnika i elektronika*, **37**, N10, 1773-1777.

5. Buchroeder R.A., Ring of light laser optics system. Pat. #4623776 USA.

6. Askaryan G.A., Chisty I.L. (1970) *JTF*, **58**, N1, 34-38.

7. Pusev P.N., Vaughan J.M., Willetts D.V. (1983) *J.Opt.Soc.Am.*, **73**, N8, 1012-1017.

8. Vaughan J.M., Willetts D.V. (1983) *J.Opt.Soc.Am.*, **73**, N8, 1018-1021.

9. Ananiev Yu.A., Anikichev S.T. (1989) *Optika i spektroskopija*, **67**, N3, 693-696.

10. Degnan J.J., Ramsay I. (1970) *Appl.Opt.*, **9**, N2, 385-392.

11. Korolenko P.V., Novoselov A.G., Sharkov V.F., Stepina S.A. (1986) *Kvantovaja elektronika*, **13**, N12, 2562-2564.

12. Korolenko P.V., Shulga A.G., Vasiliev A.B. (1989) *JTF Letters*, **15**, N22, 91-94.

13. Troitsky U.V. (1974) *Kvantovaja elektronika*, N1, 124-128.

14. Khmelevtsov S.S., Gurvich A.S., Kon A.I., Mironov V.L. (1976) *Laser Radiation in Turbulent Atmosphere*. Moscow, "Nauka".

15. Semenov A.A., Arsenyan T.I. (1978) *Electromagnetic wave fluctuations at near-the-ground paths-* Moscow, "Nauka".

16. Arsenyan T.I. (1982) *VI USSR Symposium on Radio Meteorology (Tallinn)* 148.

17. Kravtsov Yu.A., Orlov U.I. (1980) *Geometric Optics at Non-uniform Media*. Moscow, "Nauka".

18. Abdullaev S.S., Zaslavsky G.M. (1981) *JETP*, **80**, N2, 524-536.

19. Tavrov A.V., Tychinsky V.P. (1990) *Kvantovaja elektronika*, **17**, N3, 264-266.

20. Vasiliev A.B., Kornienko L.S., Korolenko P.V. (1987) *Opt.Spektrosk.*, **63**, 214 [*Opt.Spectrosc. (USSR)* **63**, 125]

RESEARCH OF SOME NEW WAYS TO IMPROVE THE EFFICIENCY AND OPTICAL QUALITY OF INDUSTRIAL CO$_2$ - LASERS

V.S. GOLUBEV
Research Centre for Technological Lasers of Russian Academy of Sciences,
140700 Shatura, RUSSIA

1. SUMMARY

Several types of high-power industrial CO$_2$-lasers are under development at the NICTL; they include waveguide diffusion-cooled lasers, fast-transverse flow and fast-axial flow gas-discharge lasers. A review of results of some new R&D aimed to improve the energetical efficiency and optical beam quality of these lasers is presented. Most important of these results are: high-efficient AC-excited monobeam waveguide laser; "crossed-electrodes" fast transverse flow laser; phase-locking of multichannel lasers using CAD-optical components; a special type of optical resonator which enables to obtain a compact output beam of a fast-flow laser.

2. INTRODUCTION

The NICTL had developed several models of high power industrial CO2-Lasers of 0,5-10 kW range. These models include diffusion-cooled waveguide type, fast transverse and fast axial flow lasers [1]. The progress in technical and economical characteristics of these lasers has been made and continues due to certain R&D activity aimed to find and realize new ways in improvement of energetical and optical parameters of lasers.

In the field of problems of energetical efficiency of lasers several proposals have been realized:

- a concept of high-power waveguide diffusion-cooled CO2-laser was realized using the electrodeless AC-discharge (10-20 kHz) [2]. The absence of active ballast loading (the capacitive ballasting was used) allowed to achieve 15-20% value of laser electro-optical efficiency. The optical schemes of the laser included multibeam 2 kW MTL-2 Laser [2] as well as monobeam arrangement (500 W MTL-500 laser (Fig.1.))
- a new concept of "crossed-electrodes" discharge in fast-transverse flow laser [3] using current source for power supply allowed to avoid active ballasting of the discharge and to increase the electro-optical efficiency of laser (Fig.2)

W. J. Witteman and V. N. Ochkin (eds.), Gas Lasers - Recent Developments and Future Prospects, 249–255.
© 1996 *Kluwer Academic Publishers.*

250

- a detailed investigation of the discharge inhomogeneities allowed to identify the zones of excessive energy losses and to recommend the optimal schemes of optical resonator, thus increasing the electro-optical efficiency of the laser in fast-transverse flow laser [4]

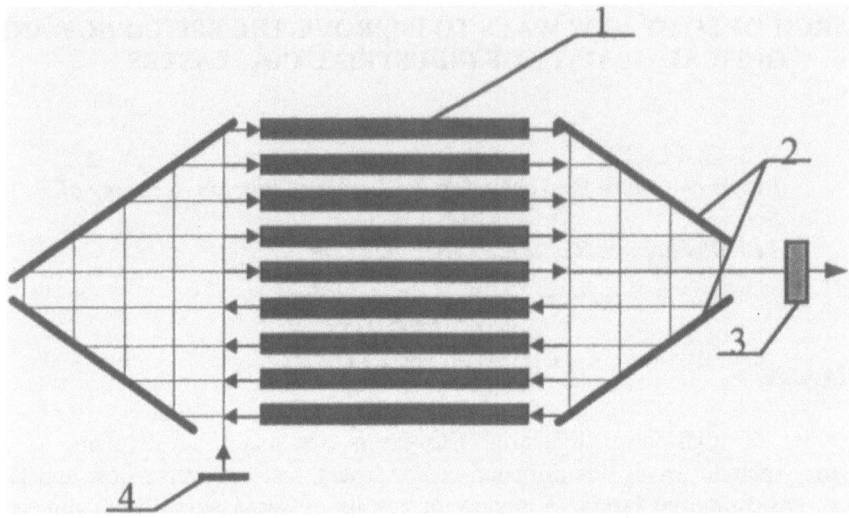

Fig.1.   The optical scheme of the resonator with 9 discharge tubes.
For M=500:
1-discharge tube; 2-flat full reflection mirror; 3-ZnSe output mirror;
4-end resonator mirror.

- an investigation of the role of molecular and turbulent diffusion of excited molecules in diffusion-cooled and in fast-axial flow lasers [5] gave some recommendations to improve their energetical efficiency and optical beam quality.
In the field of problems of laser beam quality some new proposals have been introduced:
- the multichannel diffusion-cooled lasers were provided with phase-coupling optical arrangements based upon different principles (Talbot-effect, focal-space-filter etc.). A 70% single-lobe efficiency has been achieved [6] in 1 kW power range.
- a nonlinear interaction of the beam with optical components of a 1,5 kW fast-transverse flow laser has been investigated, giving some recommendations of correction of these components [7]

251

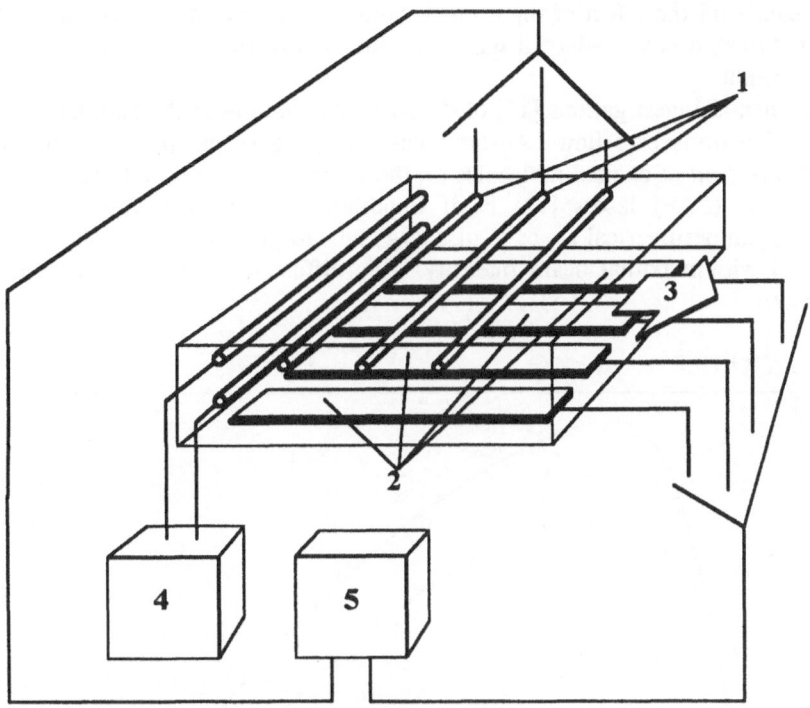

Fig.2. Gas discharge geometry and power supply performances of the "crossed electrodes" CO_2 - laser pumping system [3]:
1- cathodes; 2- anodes; 3- gas flow; 4- power supply for preonizer;
5- thyristor inverter.

- it was shown that the beam quality of high-power (more than 5-10 kW) fast-transverse flow lasers depends upon the resonator scheme [8] and gas flow turbulence [9] enhanced by a nonlinear interaction of gas-refraction inhomogenetives with discharge and active laser medium. Some special resonator schemes have been proposed including that of generator-amplifier and of compactization space transformations of the beam, allowing to achieve M2=1,2 beam quality factor of a 5-10 kW laser [10].

3. An investigation of the influence of the diffusion of excited molecules on energetical efficiency of CW-CO₂ lasers.

The role of molecular and turbulent diffusion of excited molecules in CO_2-lasers has been investigated in [5]. This phenomenon is of importance in all types of active media cooling geometry: diffusion, fast-axial and fast-transverse flow. The volume filled with laser electromagnetic field practically always is less than the volume of active media, and the effect of the excited molecules energy transfer by molecular or turbulent diffusion can produce a noticeable increase of the energetical efficiency of laser generation.

Experimental investigations [11] of the space distribution of the gain coefficient in a 1.5 kW fast-transverse flow CO_2-laser have shown a strong positive effect of the turbulent diffusion of excited molecules on the value and homogeneity of the gain. In fast-axial-flow (FAF) lasers [12] is of importance turbulent diffusion of excited molecules from peripherical zones of discharge tube to the central zone; [Fig.3] which is provided with maximal beam intensity. This diffusion gives an increase of laser output power.

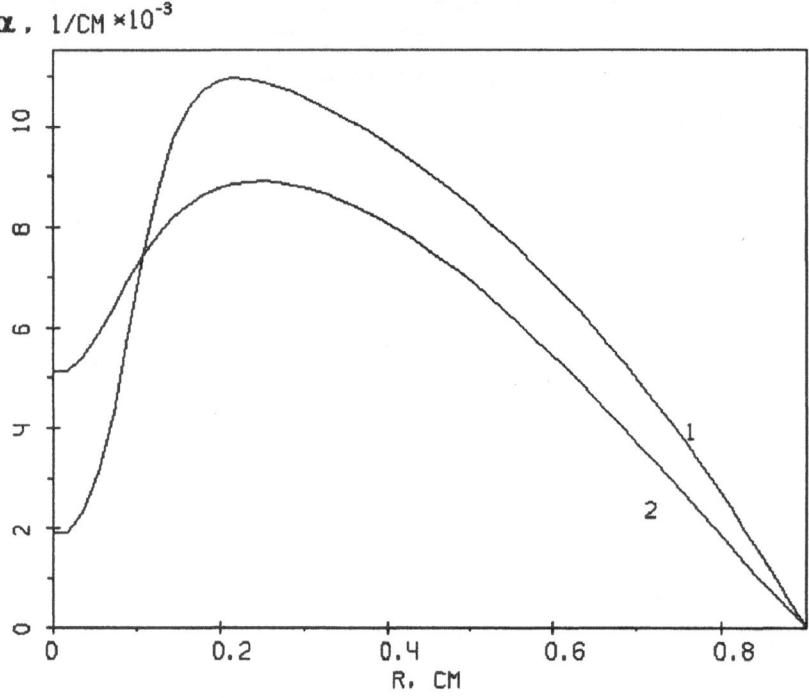

Fig. 3. Calculated radial distribution of longitudinal-ly-averaged saturated gain coefficient α in fast-axial-flow laser [12]. The beam radius equals to the turbulent diffusion length. Curve 1-in abcsence, curve 2-in presen-ce of turbulent diffusion.

Two adjacent kinetic phenomena in FAF-lasers are the longitudinal gain inhomogeneity in gas flow and additional gain of active media zones in the relaxing gas flow, outside the discharge zone [13]. An analysis of influence of all these factors, including the turbulent diffusion, has allowed to understand the reason of high electrooptical efficiency (20%) of the FAF-lasers.

4. Investigation of new resonator schemes of FTF-lasers

Promising results in obtaining high beam quality gives a semiconfocal stable resonator with diffractional beam output. The properties of such a resonator are very similar to properties of the self-filtrating resonator, but the first one is more compact. However, both of these resonator schemes require sufficiently long optical paths (> 8m).

In the case of high (more than 5kW) laser powers, the oscillator-amplifier scheme provides good level of the beam quality. We have fulfilled some investigations of this scheme with a 5 kW FTF amplifier, using a 1.5 kW master laser with stable semiconfocal resonator, and have obtained $M^2 = 1.2$ quality of the amplified beam.

5. Improvement of beam quality with a "stable-unstable" optical resonator.

One of important laser parameters determining the quality of laser material processing is the beam far-field intensity distribution. Most laser applications in Laser Material Processing can be well satisfied with gaussian beam. In the case of a confocal unstable resonator with an annular output beam, the far-field intensity distribution is presented by a peak with a pedestal. The authors [10] have proposed and realized in a 5 kW fast-transverse flow CO_2-laser a new resonator scheme of "stable-unstable" type [Fig.4]. The resonator contains only reflecting optical components. The near-field beam intensity distribution has rectangular geometry (18x20mm, corresponding to horizontal stable plane and vertical unstable resonator planes, along and transverse to the gas flow, respectively).

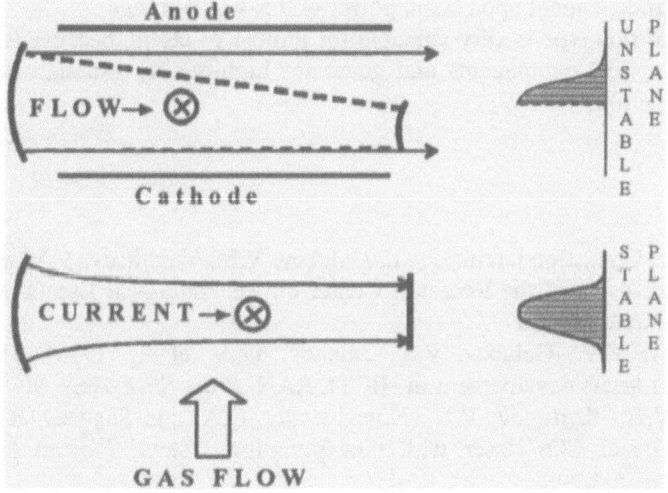

Fig. 4 Scheme of "stable-unstable" resonator [10] used in a fast-transverse flow CO_2- laser.

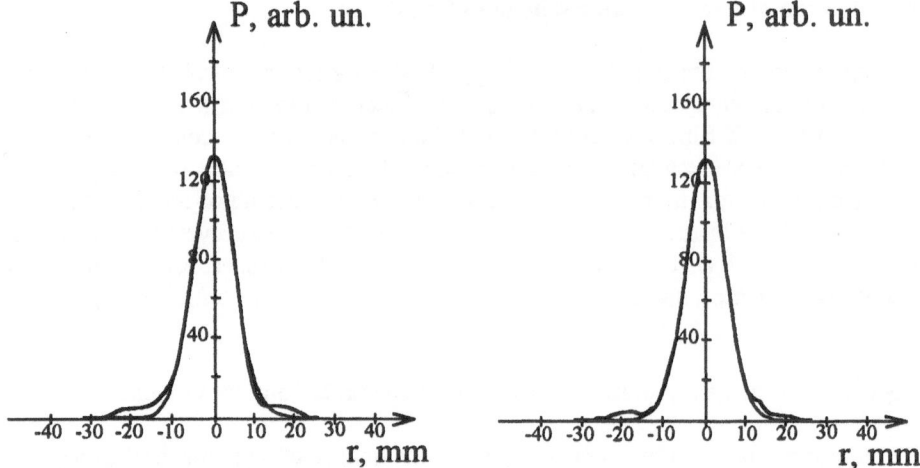

Fig. 5. Far-field (F=705mm) beam intensity distribution of "stable-unstable" resonator. Laser output power P=4.6 kW. Beam near-field aperture 18x18mm^2. Left: stable plane distribution, right: unstable plane distribution. Thin line: gaussian distribution.

The measured far-field beam intensity distribution is gaussian with 10% accuracy in the power range 0.5-4.6 kW [Fig 5]. Unstable resonator beam distribution was substantially non-gaussian.

The calculated full-angle beam divergence (1 mrad),corresponding to 86% energy, was practically independent upon beam power in 0.5-4.6kW range.

The resonator of this type is very valuable for multi-kW lasers, because it consists of only reflective optical components and generates high-quality gaussian beam in full laser power range (0.5-4.6 kW).

6. References

1. Abilsiitov, G.A., Bondarenko, A.I., Golubev, V.S., Vassiltsov, V.V. et al. (1990) Industrial lasers of the Research Center on Technological Lasers, *Kvantovaya Elektronika* **17**, 672-677.
2. Vassiltsov, V.V., Golubev, V.S., Zelenov, Ye.V. et al. (1994). Multichannel industrial lasers development in NICTL RAN, *Proc. SPIE* **2257**, 90-99,
3. Niziev, V.G., Kortunov, V.N., Novodvorsky, O.A. and Sagdeev, R.Ya. (1992) Technological CO_2- laser with new pumping system, *Plasma Devices and Operations* **5**, 89-98.
4. Golubev, V.S. (1994) Recent investigations of gas discharge and beam quality problems of fast-flow CO_2- lasers, *Proc. SPIE* **2502**, 111-119.

5. Galushkin, M.G., Golubev, V.S., Zabelin, A.M. and Zavalova, V.Ye. (1993) Influence of excited molecules' diffusion on beam energetical parameters of industrial CW-CO_2-lasers, *Izvestia AN, Ser. Fyz.* **57**, 83-89.

6. Vassiltsov, V.V., Zelenov, Ye.V. et al. (1993) Multichannel AC-excited industrial CO_2- lasers, *Proc. SPIE* **2109**, 122-128.

7. Galushkin, M.G., Golubev, V.S., Garshev, V.I. et al. (1994) Nonlinear properties of high-power industrial CO_2- lasers stable resonator, *Proc. SPIE* **2257**, 136-143.

8. Galushkin, M.G., Golubev, V.S. et al. (1993), Optical systems of high-power industrial CO_2- lasers, *Izvestia AN, Ser. Fyz.* **57**, 63-68.

9. Galushkin, M.G., Golubev, V.S., Zabelin A.M.and Panchenko, V.Ya. (1992) Light-induced small-scale optical ingomogeneities of the active medium of CW-CO_2- lasers, *Izvestia AN, Ser. Fyz.* **56**, 199-205.

10. Zabelin, A.M., Korotchenko, A.V. and Samarkin, V.V. (1995) A study of laser beam quality in industrisl CO_2-5 kW laser with the "stable-unstable" resonator, Proc. of V-th Intern. Conf. "Industrial Lasers&Laser applications'95", Shatura. To be published in "Proc. SPIE".

11. Galushkin, M.G., Golubev, V.S., Zavalova, V.Ye, Novodvorsky, O.A. and Panchenko, V.Ya. (1994) Turbulent diffusion's influence on amplification of radiation in the gas-discharge chamber of a fast-flow CO_2- laser, *Kvantovaya Elektronika* **22**, 485-487.

12. Galushkin, M.G., Golubev, V.S., Dembovetsky, V.V., Zavalov, Yu.N. and Zavalova, V.E. (1995) Optical inhomogeneities in active medium of the high-power fast-axil-flow industrial CO_2- laser, Proc.V-th Int.Conf. "ILLA'95", Shatura. To be published in "Proc. SPIE".

13. Galushkin, M.G., Golubev, V.S., Dembovetsky, V.V., Zavalov, Yu.N., Zavalova, V.E, and. Panchenko, V.Ya. (1995) Influence of excited molecules turbulent diffusion on energy parameters of fast- axial-flow CO_2- laser beam, Proc.V-th Int.Conf. "ILLA'95", Shatura. . To be published in "Proc. SPIE".

LARGE APERTURE DISCHARGE PUMPED EXCIMER LASERS AS DRIVERS FOR SOFT X-RAY SOURCES.

T. LETARDI
ENEA Dip. Innovazione, C.R Frascati
C.P. 65 - 00044 Frascati, Rome (Italy)

1. Introduction.

Few years ago the expanding interest in the development and applications of soft X rays was predicted in a paper that appeared in "Physics Today" with the title "The renaissance of X-ray optics" [1]. This prediction has been largely confirmed by the evolution of the field in these last years. First of all, we must underline that we refer to a spectral region ranging from about 100 eV to few keV. This region received small attention in the past times mainly for the lack of efficient sources, for the lack of usable optics, for the lack of well defined applications. The situation began to change first of all with the discovery of the synchrotron light as efficient source, than with the interest of the Astrophysics for the observation of the Universe in spectral regions different from the visible, and finally with the development of multilayer coatings and Fresnel plates that can be used as optical components up to photon energies of hundreds of eV. At the end, in some respect, the technologies of the optical region, utilised in the past up to about 200 nm, have been extended in these last years of two order of magnitude, with expected important impacts in many application fields as microlithography, nanomecanics, and X–ray microscopy.

It is important to underline that the synchrotron light remains largely the most bright source of X rays practically in all the spectral regions. Anyway, the laser plasma sources are receiving increasing attention, because the development of the laser sources, the investigation on the spectral emission of the produced plasma, the first successful applications are very promising, and in some cases could be competitive with synchrotron light.

The aim of this work is to compare, first of all, the brightness of synchrotron sources with the one of the laser plasma sources, then to present the work done in Frascati first of all in adapting a large aperture excimer laser (Hercules) to this specific application, then the realisation of the X-ray source, the spectra measurements and the first applications.

All this work has been done in the frame of many collaborations, namely L'Aquila Univ., National Health Institute of Rome, "Tor Vergata" Univ. of Rome, INFN of Frascati, MISDC-NPO"VNIIFRI" of Moscow, London Univ.

2. Comparison Between Synchrotron And Laser Plasma Sources.

A useful parameter for the comparison of different X-ray sources is their spectral brightness B(E), at the photon energy E, that is measured in number of

W. J. Witteman and V. N. Ochkin (eds.), Gas Lasers - Recent Developments and Future Prospects, 257–262.
© *1996 Kluwer Academic Publishers.*

photons/(s·mm^2·mrad^2·0.1%b.w.), where b.w. is the spectral bandwidth. For the synchrotron light, the data have been taken from [2] and refer to the a storage ring (ALS–Advanced Light Source, Berkeley) operating at 1.5 GeV electron energy, at a circulating current of 400 mA, using as deflecting devices bending magnets and undulators.

The estimation of the brightness of laser plasma sources requires some further considerations and approximations.

First of all, according to the measurements of Kodama et al. [3], about 50% of the impinging radiation is remitted by the hot plasma in the X spectral region, in the case of the experimental conditions of the Authors. Then, we assume that the remitted radiation has a black-body spectrum. Finally, it is supposed that the laser beam can be focused in a 10 μm diameter spot and that the beam intensity can be controlled also by means of the control of the pulse length, at constant pulse energy, in order to shift the peak emission in the wavelength of interest. The result of these considerations are reported in Figure 1 [4] where figure 1a refers to the average spectral brightness, and the estimation for the laser plasma source is obtained for a laser average power of 100 w, while figure 1b refers to the peak spectral brightness, and the estimation is at a laser peak power of 100 MW.

Some experimental measurements [5] confirm reasonably the previous estimations.

It is important to underline that, in order to heat the plasma at the temperatures interesting for X-ray emission, the laser intensity on the target must exceed values of the order of 10^{11} W/cm^2, as it can be simply estimated by the consideration that the black–body emission, at the temperature of 50 eV, is of the order of 5×10^{11} W/cm^2.

To increase the laser intensity, it is necessary to control the beam quality, to

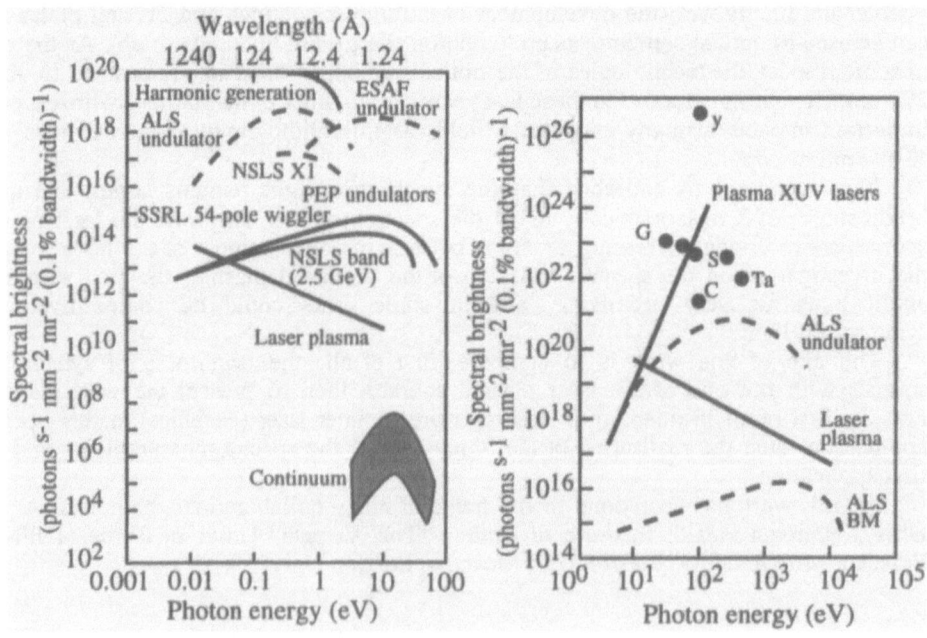

Figure 1. Brightness comparison between ALS and Laser plasma sources.

increase the active volume section, because the maximum laser output is limited by power or energy saturation, to decrease the pulse time-length. Some developments and some results obtained with the laser source Hercules will be presented in the following sections.

3. Operation Of The Laser Hercules At Low Divergence.

The laser source Hercules used for this experiment, is a large aperture XeCl laser fully developed in the ENEA Frascati Centre [6]. The active volume is $10 \times 10 \times 100$ cm^3, and it is pumped by a self-sustained discharge after X-ray preionization. The gas is recirculated, so that it can work up to a rep. rate of 10 Hz. Normally it works with a flat-flat cavity, as oscillator. In order to improve the characteristics for this type of application, the following changes have been made:
a) Operation with Positive Branch Unstable (PBUR) cavity with magnification M=5
b) Operation with PBUR cavity with magnification M=9
c) Operation as amplifier for a high-quality, 10 nanosec pulse.
    The improvements of the performances are shown in table 1.
In order to have a low divergence laser beam, the active medium must have very uniform characteristics (gas density, gain profile and so on) [7]. For Hercules, very uniform gain profile has been obtained by means of a careful control of the preionization (figure 2), while no effect of the recirculation gas has been observed on the divergence

TABLE 1. performances of HERCULES with different operating configurations

Cavity	Flat-Flat	PBUR M=5	PBUR M=9	Amplifier
En/pulse (J)	8	5	3.5	2
Pulse Width (ns)	160	120	90	10
Beam dimensions (cmxcm)	5x10	4x9	4x9	4x9
Beam div.(mradxmrad.80%b)	6x4	0.1x0.1	0.073x0.03	0.1x0.1
Av.bright.(80%beam) W/cm^{2}sr	3.3×10^{10}	9×10^{13}	1.7×10^{14}	3.5×10^{14}
Peak bright.(30%beam)W/cm^{2}sr				1×10^{15}

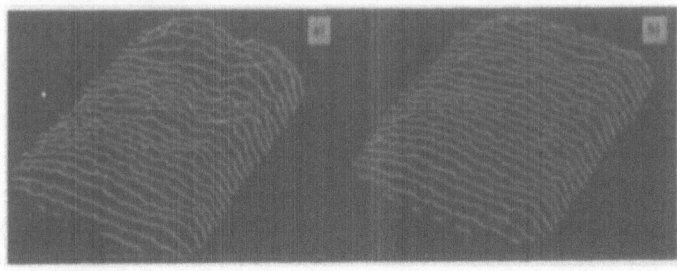

Figure 2. Uniformity of Hercules laser beam using one a) or two b) preionization

4. Emission Spectra Measurements.

The laser beam has been focused on targets with different Z (C, F, Cu, Ni, Na, Ti, Y, Mo, O in SiO_2, Mg, Sn, Ta, Al) [8]. The experimental apparatus is shown in figure 3.

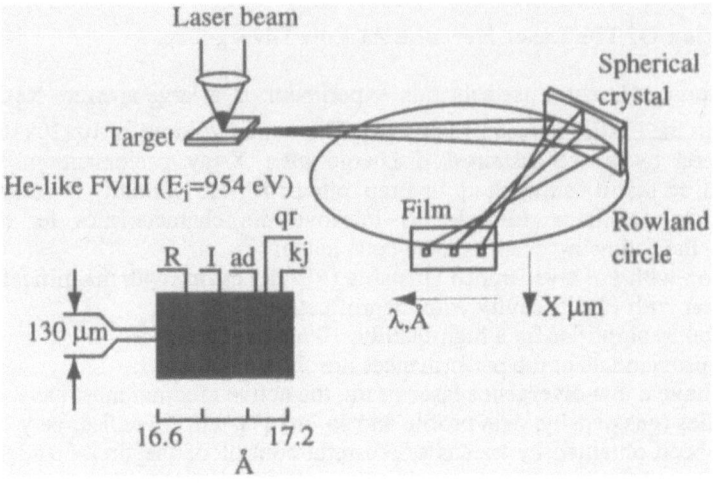

Figure 3. T he focussing spectrograph with spatial resolution (FSSR-1D) scheme for registration of spectra of multicharged ions with spatial resolution. A mica sphericallly bent crystal with curvature of R=100 mm was used.

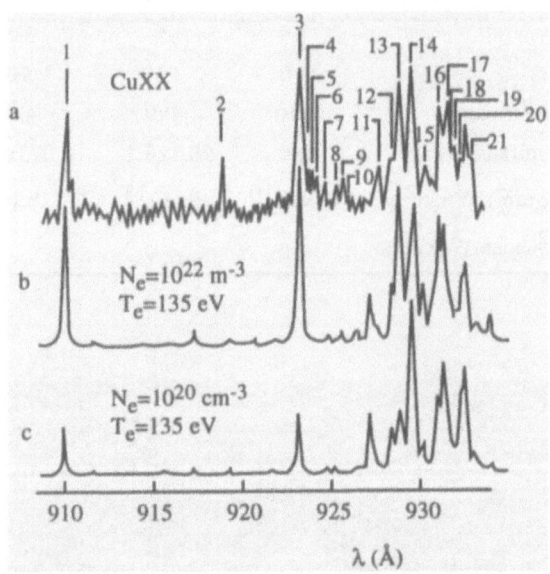

Figura 4. Densitograms of experimental spectra, obtained in plasma, heated by XeCl laser (a). Model spectra for different electron temperature and density of plasma (b,c).

The last focusing element is a triplet with numerical aperture N=3. The dimensions of the X-ray emitting region has been measured with the pin-hole technique, and a minimum spot diameter of 15 μm has been measured; this size corresponds to the estimated laser beam waist on the target, for which the laser power density is near 10^{14} W/cm^2.

The measuring apparatus consists of PIN diodes, with suitable filters for the different spectral regions, and of two crystal spectrometer: the first one (mica, 2d=19.9 Å) with high resolution ($\Delta\lambda/\lambda = 10^{-4}$), the second one (RbAP, 2d=26.1 Å) with low resolution ($\Delta\lambda/\lambda = 10^{-2}$), both operating in the spectral range 1 keV-2 keV. In figure 4 it is shown a high resolution spectrum of CuXX, and the comparison with two different simulations [8].

As a result, we have achieved a total conversions efficiency of 6% in the spectral region 283-532 eV, it is 70 mJ/shot of X-rays, and a total conversion efficiency of 0.6% in the spectral region > 1 keV, it is 7 mJ/shot of X-Rays.

5. X-Ray Contact Microscopy.

When the X-rays impinge on a biological specimen in close contact with a photoresist, a high resolution image of the specimen can be obtained after development of the photoresist. Moreover, if the X-ray are inside the so-called water window (2.2−nm 4.4 nm), where the absorption coefficients of water (it is O) and proteins (it C) are very different, high contrast picture of the constituents of a cell can be taken. This process usually damages the cells. If the X-ray fluence is high enough, so that in a single shot a picture can be taken, in times so short that the deformations have no time to develop, this is a unique method to take microscopic images of a living cell.

Experiments have been done using the Hercules X-ray source [9], and Clamydomonas cells in close contact with a photoresist to a distance of up to 5 mm

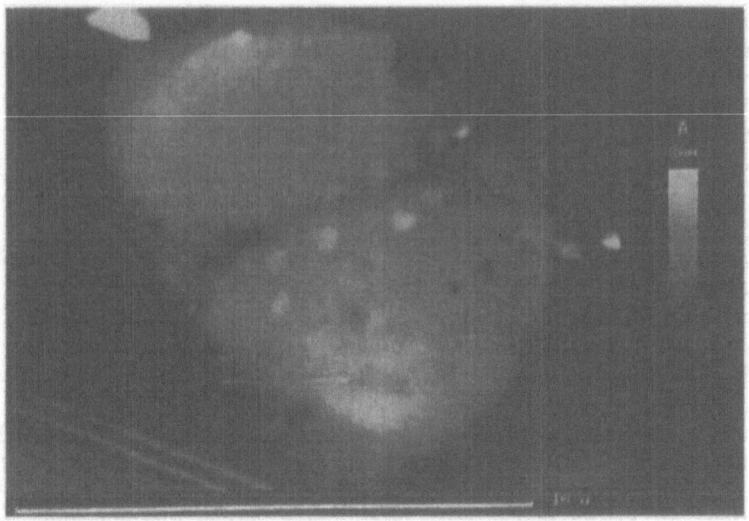

Figure 5a. Chlamydomonas in mitotic division phase.

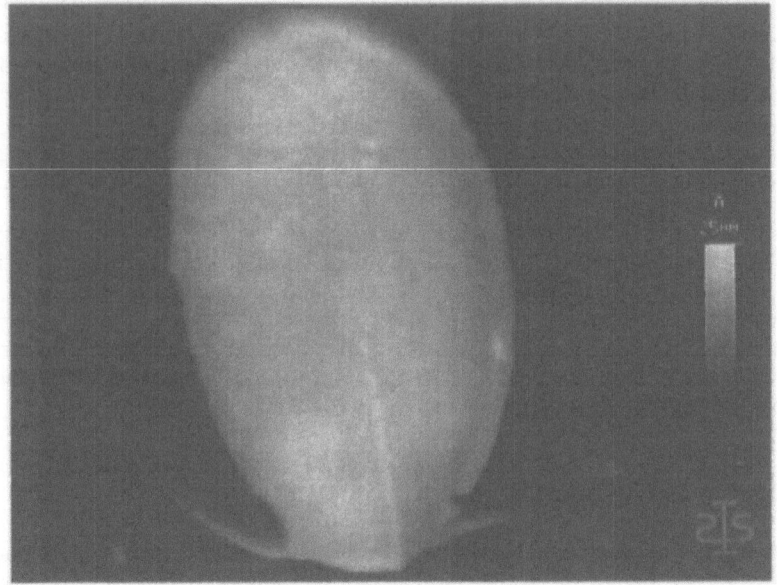

Figure 5b. X-ray microscopy of chlamydomonas-dimensions of cell: 8 μm x4.2 μm.

from the X-ray source. Some results are shown in figure 5, where some details (internal structures, flagella) can be seen. The best resolution has been estimated in 100 nm.

6. References

1. Underwood J.H and Attwood D.T., (1984) The renaissance of X-ray optics, *Physics Today*, p. 45 (April 1984).
2. O'Neill F., (1988) Laser Plasma XUV Sources, *RAL-88-101, 1988*".
3. Kodama et al., (1986) *J. Appl. Phys* **59**, 3050.
4. Dattoli G. and Letardi T.(1994) Soft X-ray sources: a comparison *ENEA Report RT/INN/94/29*
5. Turcu I. et al., (1994) SPIE Vol. 2015, paper 2015-37
6. Bollanti S., Di Lazzaro P., Flora F., Lisi N., Giordano G., Hermsen T., Letardi T., Zheng C.E., (1990) Performance of a ten liter electron avalanche discharge XeCl laser, *Appl. Phys B* **50**, 415-423.
7. Di Lazzaro P., Letardi T., Zheng C.E., (1992) Discharge Medium Uniformity Influence on XeCl Excimer Laser Beam Quality, *Il Nuovo Cimento* **14D**, 41-48.
8. Bollanti S., Di Lazzaro P., Flora F., Letardi T., Palladino L., Reale A., Batani D., Mauri A., Scafati A., Grilli A., Faenov A.Ya. Pikuz T.A. Pikuz S.A.Osterheld, (1995) Na-Like Autoionization States of Copper Ions in Plasma, Heated by Excimer Laser, *Physica Scripta* **51**, 326-329.
9. Bollanti S., Di Lazzaro P., Flora F., Giordano G., Schina G., Zheng C.E., Filippi L., Palladino L., Reale A., Taglieri G., Batani D., Mauri A., Bnelli M., Scafati A., Reale L., Albertano P., Grilli A., Faenov A., Pikuz T. ,Cotton R., (1995) Long Duration Soft X-Ray Pulses by a XeCl Laser Driven Plasmas and Applications, *Accepted for publication in "J. of X-ray Science and Technology.*

ULTRAHIGH RESOLUTION LITHOGRAPHY WITH EXCIMER LASERS

F.K. TITTEL, M. ERDÉLYI*, C. SENGUPTA, ZS. BOR*,
G. SZABÓ*, J.R. CAVALLARO, M.C. SMAYLING, W.L. WILSON
*Department of Electrical & Computer Engineering and Rice
Quantum Institute, Rice University, P.O. Box 1892, Houston, TX
77251, USA*
**Department of Optics and Quantum Electronics, JATE University,
Dóm tér 9, H-6720 Szeged, Hungary*

1. Introduction

The photolithography process is central to integrated circuit fabrication. Through this process an integrated circuit is patterned by imaging a photomask onto a layer of photoresist. The light source currently being used by the semiconductor industry in the photolithographic process for 0.35 micron feature size is the mercury lamp. This light source has wavelengths of 436 nanometers (g-line) and 365 nanometers (i-line). As the feature size for integrated circuits moves below 0.35 microns, a new source of shorter wavelength light and higher power must be found to replace the mercury lamp. Excimer lasers are capable of producing wavelengths at 248 nanometers and 193 nanometers which could allow for design features down to 0.18 microns. Accordingly, as mercury lamp technology reaches its limits with the introduction of the "shrink" version of the 64 megabit DRAM (0.32 micron) and advanced microprocessors, the semiconductor industry will shift to excimer DUV lithography starting in ~1996. Furthermore, higher performance integrated circuits will require improved spatial resolution, larger projection areas, and higher throughput lithographic strategies. Wavefront enhancement technologies such as phase shift masks and/or off axis illumination will play an important role in these new generations of integrated circuits.

Since deep-UV lithography was first demonstrated using KrF excimer lasers, [1,2] the performance and reliability of these deep UV light sources have continuously improved. Current excimer laser models satisfy the optical specification and uptime requirements necessary for pilot integrated circuit production. Attention has now been focused on the Cost of Ownership (CoO) for these excimer laser light sources. Both line-narrowed KrF lasers for stepper/scanner systems and broadband KrF lasers for scanner systems must achieve reductions in CoO in order to compete on cost basis with I-line steppers using phase shifting masks or non-conventional illumination schemes.

263

W. J. Witteman and V. N. Ochkin (eds.), Gas Lasers - Recent Developments and Future Prospects, 263–272.
© 1996 *Kluwer Academic Publishers.*

The key technological needs for excimer lasers in microlithography are: 1) lowered operating costs (longer chamber lifetimes, lower gas consumption); 2) higher repetition rates ($> = 1000$ Hz) for scanned exposures; 3) improved pulse to pulse energy repeatability; and 4) both line-narrowed (refractive lenses) and broadband (reflective/catidioptic) optics.

As feature size shrinks, there are two choices for decreasing the resolvable linewidth, W. From the relationship

$$W = k_1 \frac{\lambda}{NA} \tag{1}$$

where λ is the illumination wavelength, NA is the numerical aperture of the optical system, and k_1 is a system dependent parameter. It is obvious that a smaller W can result from either a reduction of the illumination wavelength λ or a larger imaging system numerical aperture, NA. The desire for shorter wavelengths will lead to the adoption of excimer lasers (KrF and ArF) as high brightness UV illumination sources. Achieving a larger numerical aperture is a challenge and is one of the reasons for the high cost of modern photolithographic steppers. The depth of focus (DOF) is given by

$$DOF = k_2 \frac{\lambda}{NA^2} . \tag{2}$$

Thus, decreasing λ and increasing the NA cause a serious degradation to the depth of focus and, consequently, the final image at the wafer. In particular, the inverse square relationship between DOF and NA is a serious problem. A smaller DOF requires a more stable and controlled, and hence costlier, stepper. The problem of reduced DOF becomes important as the surface of the fabricated wafer becomes increasingly non-planar as process steps increase in complexity.

An alternate approach to decreased W and increased DOF is to alter the optical system coefficients k_1 and k_2. One technique that has been shown to significantly improve both resolution limit and depth of field of the photolithographic image is that of phase shifting [4-6]. By appropriate optical manipulation, the electric field of adjacent regions of the photolithographic image can be made to be 180° out of phase with one another. This assures that when these images overlap (due to diffraction) there will be some place where the two images exactly cancel one another, resulting in near 100% contrast. The desired phase shift is usually accomplished by making some regions of the photolithographic mask optically thicker than in other regions. This is achieved by either etching regions of the mask, or by adding an additional layer to the mask, and then etching it away from unwanted regions. The finished mask and the resultant image at the silicon wafer surface are shown schematically in Figure 1. Besides the alternating phase-shift regions mentioned above, phase-shift regions can be placed at the rim of larger structures to enhance the contrast of their image. More recently, chromeless phase shift masks [6], as well as attenuating masks which achieve the desired

Principle of Phase Shift Masking

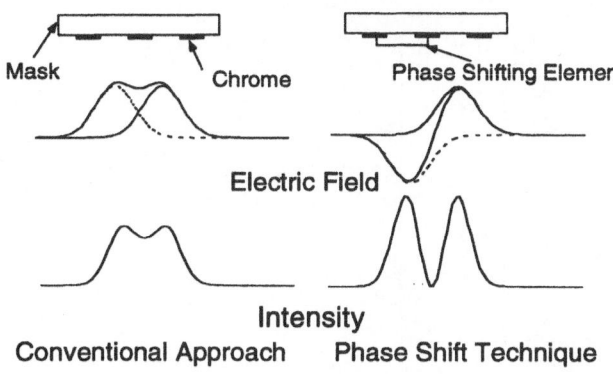

Figure 1: Comparison of Conventional and Phase-Shift Masking

phase shift effect by propagating some of the light through lossy regions in the mask, have been studied [7,8].

While any of these approaches to phase shifting have shown significant promise and offer distinct advantages over conventional photolithography, implementing the phase shifting technique to deep UV wavelengths requires tighter tolerances for the appropriate phase shift material as the wavelength is decreased. It is also much harder to find appropriate phase shifting materials which do not absorb strongly in the DUV region.

Recently, we reported progress on a novel phase shifting technique which does not require special phase shifting regions built into the mask [9-11]. By using a reflective chrome mask in a laser-based interferometric scheme, the desired phase shift can be achieved.

As outlined in Figure 2, light from a laser is divided into two beams by a beam splitter. One beam passes through the mask, while the other is reflected off the back of the mask. The two beams are then imaged onto the wafer surface. The optical path lengths of the two beams are adjusted so that their relative phase is 180° apart and the desired phase-shift effect is achieved. With this technique, we have demonstrated that it is possible to write line and space patterns with a linewidth of less than 0.3 μm using a frequency tripled Nd:YAG laser at 355 nm as the illumination source. CCD camera imaging, as well as computer simulations, obtained using DEPICT [12], show the effectiveness of this new scheme, as shown in Figure 3. One of the unique features of this approach is that both the amplitude and the phase of the two incident images are independently adjustable even after mask fabrication, which is not the case for most other phase shifting techniques.

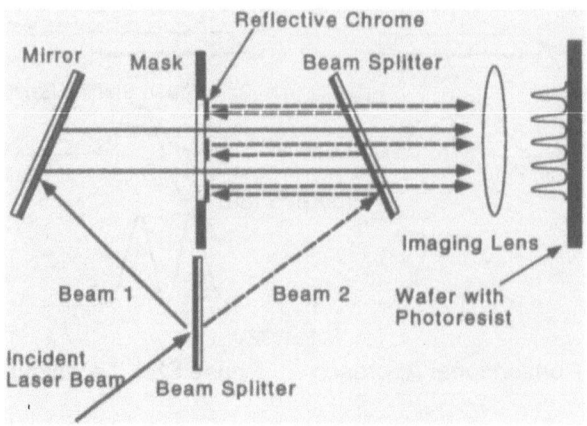

Figure 2: Interferometric Phase-Shift Technique

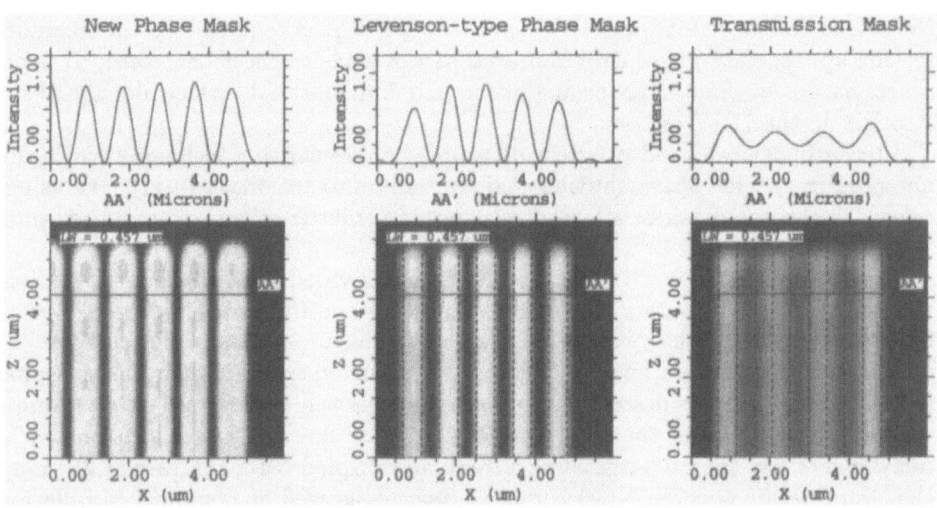

Figure 3: Simulation results for the interferometric phase-shift method, Levenson-type phase-shift mask and transmission mask

In this paper we report on recent investigations where the interferometric technique described above is combined with off-axis illumination [13]. The ability to independently control the amplitude of the two out-of-phase beams is especially helpful in this application. Improved resolution, as well as greater depth of focus, result from applying this new imaging scheme.

Phase shifting can increase the depth of focus by about 50%. Off-axis illumination can add an additional 40% of depth of focus [14]. However, since off-axis illumination uses the first order diffracted beam (amplitude $\frac{1}{\pi}$) instead of the principal or zero order beam (amplitude $\frac{1}{2}$), the image contrast is reduced by about 5%. The additional use of an attenuated phase shift mask with a transmission of about 5% can compensate for this, however, and the equalized intensity can result in a 100% image contrast [15].

2. Experimental Set-Up

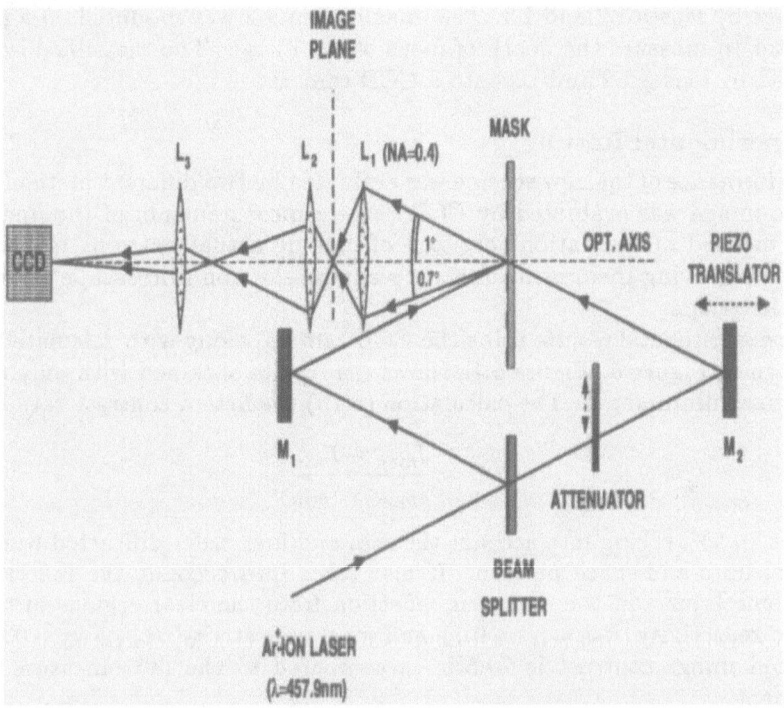

Figure 4: Experimental scheme of the off-axis illumination, combined with interferometric phase shifting. The mask is illuminated symmetrically from both sides with a beam splitter and two mirrors (M_1, M_2).

Figure 4 shows the experimental arrangement. The mask was a patterned evaporated, reflective chrome layer on fused silica substrate, forming a line and space pattern, with a spatial frequency of 16 μm.

The output from an Ar^+ laser beam operating at 457.9 nm was split into two beams and used to illuminate both the front and back surface of the mask. The intensity and the phase of the back illumination was controlled by a variable attenuator and a piezo-controlled linear translator, respectively. A microscope objective (magnification (M) = 20X, NA = 0.4) was used to image the mask onto the photoresist. The mask to objective distance was adjusted to the microscope tube length to ensure a nominal magnification ratio of 20X and high image quality. The off-axis illumination angle of the mask was 1° and the first order diffraction angle was 0.7° while all other diffraction orders were rejected by the aperture of the lens. The image of the line-space patterns formed by the lens L1 was magnified with two microscope objectives, L2 and L3, in tandem (M = 20X, NA = 0.5, and M = 40X, NA = 0.65). Special care was taken to avoid any optical degradation of the image by lenses L2 and L3. The imaging lens L2 was mounted on a precision translator to measure the depth of focus of the image. The magnified image was projected by lenses L2 and L3 onto a CCD camera.

3. Experimental Results

The performance of the new scheme was evaluated by two different methods. In the first the image was evaluated by CCD camera measurements of the image. The second method of evaluation consisted of making actual patterns in photoresist and then observing them using either a scanning electron microscope or an atomic force microscope.

The experimental results using the CCD camera, along with calculated results, are shown in Figure 5. Figure 5.1a shows the results obtained with only the front of the mask illuminated. The calculation (5.1b) predicts a contrast ratio C

$$C = \frac{I_{max} - I_{min}}{I_{max} + I_{min}} \tag{3}$$

of about 70.5%, taking into account the zero and first order diffracted beams from an equal lines-and-space pattern. It also takes into account the reflectivity for the chrome layer and the dielectric reflection from the clear regions of the mask (chrome reflectivity $R_{chrome} = 0.71$ and mask reflectivity $R_{mask} = 0.032$). The calculated image contrast is 70.5%, as compared to the 69% measured in this experiment.

Figure 5.2a shows the image of the mask when the front reflected beam was blocked and only the transmitted beams were intercepted by the imaging microscope lens L1. A careful comparison of Figure 5.1a and 5.2a shows (see the vertical dotted lines), that these images are spatially shifted by half the period of the pattern. This means that there is no transmission from the back side, where there is a reflection from the front side of the mask (see Figure 4).

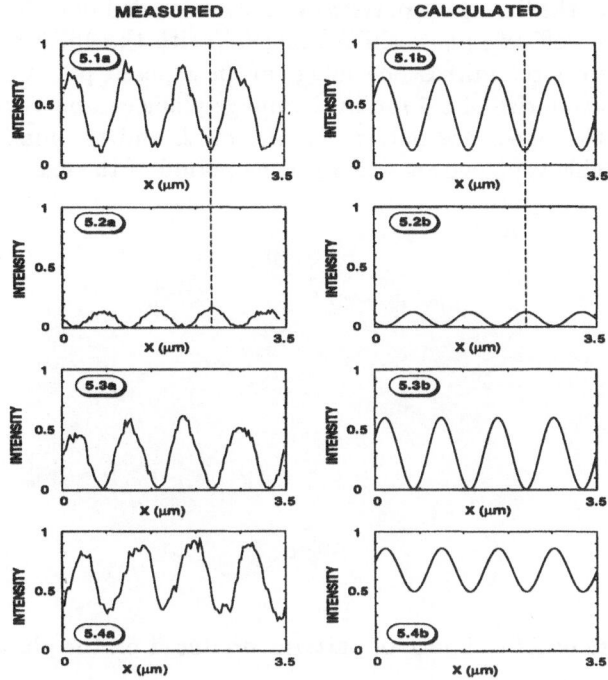

Figure 5: Intensity patterns obtained by the CCD camera. The (a) figures show the measured and the (b) figures show the calculated curves. 5.1: blocking the back illumination; 5.2: blocking the front illumination; 5.3: using both beams with 180° phase difference; 5.4: using both beams with 0° phase difference on the surface of the mask.

Figure 5.3a shows the interference pattern, when both beams were used and the phase of the transmitted image was shifted by π with respect to the reflected image, using the piezo-controlled translator of mirror M2. The intensity of the transmitted beam was adjusted with the attenuator shown in Figure 4, so that the peak intensity of the transmitted pattern was equal to the minimum intensity of the reflected pattern (see Figures 5.1a and 5.2a). Due to the π phase shift of the patterns, the electric fields are subtracted, resulting in a nearly 100% modulation depth of the image. Due to the high spatial coherence of the illuminating laser beam, the high contrast of the image remains for almost a 5 μm defocus in either direction. When the phase shift between the transmitted and reflected images is set to 0 or multiples of 2π, the electric fields of the two beams are added and the visibility decreases (Figure 5.4a) significantly.

The image of the line-space patterns was also recorded in a photoresist (Shipley 94314) using the off-axis-phase shifted scheme, with the 457.8 nm Ar^+ ion laser line. An atomic force microscope image of an exposed photoresist is shown in Figure 6. A feature size of 0.4 μm in the image plane can be calculated, knowing the period of the line-space pattern on the mask and the magnification of the objective L1. This value agrees well with the period of the exposed pattern.

Figure 6: AFM image of patterns produced on the photoresist.

4. Conclusions

The feasibility of a new high resolution photolithography scheme combining off-axis illumination with interferometric phase shifting was demonstrated. Using a laser operating at 457.9 nm, feature sizes of 0.4 μm were obtained with a DOF of 10 μm. Because of the freedom to adjust the relative phase and amplitude of the two illuminating beams, the modulation depth could be improved to almost 100%. This freedom of adjustment makes this scheme a useful test bed for studying various phase shifting and off-axis illumination schemes.

5. Acknowledgment

This research was supported in part by NSF under grant DDM-9202639 and NSF-INT 9020541.

6. References

1. Pol V. et al (1986) Excimer laser-based lithography: A deep ultraviolet wafer stepper, *Proceedings of Optical/Laser Microlithography* **Vol. 633**, SPIE, 6-16.

2. Partlo W. N., Sandstrom R. L., Fomenkov I. V., and Das P. P. (1995) Low cost of ownership KrF excimer laser using a novel pulse power and chamber configuration, *Proceedings of Optical/Laser Microlithography VIII* **Vol. 2440**, SPIE.

3. Pätzel, R., Kleinschmidt, J., Rebhan, U., Franklin, J., and Endert, H. (1995) KrF Excimer Laser with repetition rates of 1 kHz for DUV Lithography, *Proceedings of Optical/Laser Microlithography VIII* **Vol. 2440**, SPIE.

4. Levenson, M.D. and Visnawathan, N.S. (1982) Improving resolution in photolithography with phase-shifting mask, *IEEE Trans. Elec. Devices* **ED-29**, 1828-1836.

5. Lin, B.J. (1993) Phase-shifting masks gain an edge, *IEEE Circuits and Devices* **2**, 28-35.

6. Toh, K., Dao, G., Singh, R., and Gaw, H. (1990) A chromeless phase-shifted mask: A new approach to phase-shifting masks, *Proceedings of 10th Annual Symposium on Microlithography* **Vol. 1496**, SPIE, 27-53.

7. Nakajima, M., Yoshioka, N., Miyzaki, J., Kusnose, H., Hosono, K., Morimoto, H., Watakabe, Y., and Tsukamoto, K. (1994) Attenuated phase-shift mask with a single layer absorptive shifter of CrO, CrON, MoSiO, and MoSiON film, *Proceedings of Optical/Laser Microlithography VII* **Vol. 2197**, SPIE, San Jose, CA, 111-121.

8. Wong, A., Pierrat, C., Vasdev, P., and Neureuther, A. (1994) Phase-shifter edge effect on attenuated phase-shifting mask image quality, *Proceedings of Optical/Laser Microlithography VII* **Vol. 2197**, SPIE, San Jose, CA, 122-129.

9. Kido, M., Szabó, G., Cavallaro, J. R., Wilson, W. L., and Tittel, F. K. (1994) A new phase shifting method for high resolution microlithography, *Proceedings of Optical/Laser Microlithography VII* **Vol. 2197**, SPIE, San Jose, CA, 835-843.

10. Tittel, F.K., Cavallaro, J. R., Kido, M., Smayling M. C., Szabó, G. and Wilson, W. L. (1995) Interferometric phase shift technique for high resolution microlithography, *Proceedings of 10th Int'l Symposium on Gas Flow and Chemical Lasers* **Vol. 2502**, SPIE, 617-624.

11. Kido, M., Szabó, G., Cavallaro, J. R., Wilson, W. L., Smayling, M. C., and Tittel, F. K. (1995) Submicron Optical Lithography Based on a New Interferometric Phase Shifting Technique, *Jpn. J. Appl. Phys.* **34**, 43-47.

12. Pack, R., and Bernard, D. (1990) *DEPICT-2 applications for VLSI technology*, Technology Modeling Associates, Palo Alto, CA.

13. Erdélyi M., Sengupta C., Bor Z., Cavallaro J. R., Kido M., Smayling M. C., Tittel F. K., Wilson W. L. and Szabó G. (1995) A new interferometric phase-shifting technique for sub-half-micron laser microlithography, *Proceedings of Optical/Laser Microlithography VIII* **Vol. 2440**, SPIE, 827-837.

14. Luehrmann, P., Oorschot, P.V., Jasper, H., Stalnaker, S., Brainerd, B., Rolfson, R., and Karklin, L. (1993) 0.35 μm lithography using off-axis illumination, *Proceedings of Optical/Laser Microlithography VII* **Vol. 1927**, SPIE, 103-124.

15. Fukuda, H., Imai, A., Terasawa, T., and Okazaki, S. (1991) New approach to resolution limit and advanced image formation techniques in optical lithography, *IEEE Trans. Elec. Devices* **ED-38**, 67-75.

RECENT ASPECTS OF ULTRA SHORT PULSE MID- AND FAR-INFRARED GAS LASERS AND THEIR APPLICATIONS

D. P. SCHERRER, J. KNITTEL, D. B. MOIX, M. BAUMGARTNER
AND F. K. KNEUBUEHL
Institute of Quantum Electronics, Swiss Federal Institute of Technology
(ETH) CH 8093 Zürich, Switzerland
Tel: 1-6332341, Fax: 1-6331077
e-mail: scherrer@iqe.ethz.ch

Abstract

In order to produce ultrashort far-infrared (FIR) laser pulses we have investigated the superradiance and Raman emission of various molecular transitions optically pumped with the pulses of 10 μm-hybrid- or high-pressure single-mode- CO_2 lasers truncated within 10 ps. Furthermore we have truncated directly the FIR laser pulses with the first FIR plasma shutter. In our experiments we have observed new phenomena which include e.g. new interrelations between superradiance, swept-gain superradiance and Raman emission, anticorrelated fluctuations of pump radiation versus FIR emissions, first optical free induction decay (OFID) of 10 μm-CO_2-laser pulses with FIR emitting molecular gases. Finally, we have proved the first spontaneous infrared solitons in stimulated far-infrared Raman scattering on NH_3. We have observed solitons in the depletion region of the infrared pump radiation with corresponding dark solitons in the Stokes pulses and detected the laser phase change connected with the soliton generation.

1. Introduction

The generation of ultrashort coherent pulses at far-infrared wavelengths is of current interest for numerous applications, e.g. real-time spectroscopy of gaseous plasmas, low-energy electron excitations in semiconductors and investigation of the dynamics in semiconductors or in high-T_c superconductors.

Nevertheless, ultrashort FIR pulses cannot be generated by conventional means principally for two reasons. First, the lack of electro-optic materials in the FIR wavelength range precludes the standard production of short pulses with mode-locking or Q-switching techniques. Secondly, the optically pumped FIR gas lasers which represent

273

the most significant class of coherent FIR sources are low-pressure narrow-band systems [1]. The short-pulse performance of these systems is restricted since the conventional lower limit for the duration of an optical pulse generated in an excited medium is the inverse linewidth of the transition.

However, two different approaches have proven to be well suited for short-FIR-pulse generation. They concern direct and indirect methods. Direct methods include FIR-laser systems optically pumped with short CO_2-laser pulses. The resulting FIR pulses can be shorter than the conventional pulsewidth limit for certain FIR transitions due to nonlinear coherent processes such as superradiance and stimulated Raman emission [1,2]. The experimental schemes applied to the direct generation of short FIR pulses by superradiance or stimulated Raman emission comprehend different mode-locked CO_2-laser optical-pumping techniques resulting in FIR pulses of a few 100 ps duration [3,4]. However, the disadvantage of these methods for many applications is the fact that the FIR emission following the pump radiation results in a train of irregular short pulses and not in well defined single pulses.

The indirect methods for the generation of short FIR pulses concern switching of the FIR reflection and transmission of a semiconductor by the creation of an electron-hole plasma with an optical laser [5]. Except for the case where the FIR source is a free electron laser [6], the pulses generated by the indirect methods are generally far longer than those obtained by the direct methods mentionned above. However, they represent well defined single pulses in contrast to a train of irregular short pulses.

2. Ultrashort FIR Single-Pulses Generation

For the generation of ultrashort FIR laser single pulses with a direct method we have investigated the superradiance and Raman emission of various molecular transitions [7] optically pumped by rapidly truncated 10 μm-CO_2 laser pulses produced by an optical-free-induction decay (OFID) 10 μm-CO_2 laser system [8]. In our OFID laser system we have first used a hybrid TEA CO_2 laser which operates only on the discrete vibrational-rotational lines of the CO_2 gas. Secondly, in order to extend the number of available pump lines in FIR gases we have applied a continuously tunable single-mode high-pressure CO_2 laser developed in our laboratory [9]. The experimental arrangements are described in detail elsewhere [7,9]. The main characteristics of our truncated CO_2-pump pulses for our experiments are the following: First, the fast truncation, which occurs within 10 ps [8], produces a sideband frequency structure of the pump pulse. Thus, a broadband CO_2-pumping can therefore induce a broadband FIR emission which yields the short pulses. Secondly, in order to obtain well defined superradiance, it is of prime importance that no overlaping occurs between the pump emission and the resulting FIR emission. This is warranted by the rejection of the trailing edge of the pump pulse by the truncation process.

We have obtained well defined superradiant emission with pulse durations shorter than the conventional pulsewidth limit given by the inverse linewidth of the transition [1]. Thus, we have demonstrated the relevance of the fast truncation which implies a large spectral width of our pump pulse. We have observed the predicted onset of swept-gain superradiance [10] and the appearance of Raman emission above specific pressures in the gas cell. The rise of the Raman emission is related to a decrease of the superradiant emission due to line competition between the two emissions [11,12]. This competition may cause complete suppression of the line-center emission in the case of a long excitation pulse [13]. In our experiments we have found that this competition never results in the complete disappearance of the superradiant emission. On the contrary, we observe for increasing FIR gas pressure in the cell an earlier disappearance of the Raman emission which is due to the typical reshaping of the truncated CO_2-pump pulses in the FIR laser gases which show the development of OFID by increasing pressure in the cell [7]. Therefore, we have a high-pressure regime, without the presence of the irregular Raman emission, which is ideal for the generation of nano- and subnanosecond superradiant single pulses. These FIR single pulses are among the shortest hitherto produced and suited for specific applications such as testing fast FIR detectors [14] or high T_C superconductors [15].

Furthemore, we have systematically investigated the transmitted CO_2-pump pulse. On this occasion we have discovered first that the CO_2-pump radiation and the generated FIR emission strongly interact. This interaction results in anticorrelated fluctuations of the two fields. We have shown that this phenomenon is due to periodic back-and-forth fluctuations of the Λ-like three-level molecular systems [7].

Secondly, our investigations on the evolution of the CO_2-pump pulses in the FIR superradiant cell have revealed, as already mentioned, the mechanism of the development of OFID of the truncated CO_2-pump pulses in the FIR-laser gases. We have demonstrated that OFID is observed for all CO_2-laser-pump lines which induce FIR Raman emission in our system. In addition, we have discovered that OFID can also occur for CO_2-laser lines which do not generate FIR Raman emission in our system depending on the absorption characteristics of the FIR-laser gas. As a result of these studies, we have generated for the first time ps-10 µm-CO_2-OFID pulses with FIR-laser gases instead of the usual hot CO_2 gas as spectral filter [8,16]. We have obtained pulses of 90 ps FWHM. These OFID pulses are therefore longer than the 30 ps CO_2-OFID pulses which can be produced by a similar experimental arrangement yet with hot-CO_2 gas as spectral filter [8]. Nevertheless, in this case a 6 m length cell with about 400 Torr CO_2 gas at 400 °C is required which has to be compared with a glass tube of about 3 m length filled with 2-5 Torr FIR laser gas at room temperature which permits a compacter and simpler experimental design. The main advantages of our new OFID system based on FIR laser gases over the standard ps-CO_2-OFID working with hot CO_2 gas as spectral filter is the possibility of replacing the hybrid TEA CO_2 laser by the single-mode high-pressure CO_2 laser to produce OFID pulses at wavelengths which lie between the discrete vibrational-rotational lines of the CO_2 gas. We have thus generated OFID pulses at 10.55 µm with NH_3 as spectral filter. This wavelength lies 14.6 GHz off the 10P(16) CO_2 laser line and correspond to the sP(1,0) transition of NH_3 at 10.55 µm [17].

3. The First FIR Plasma Shutter

In order to shorten conventional FIR laser pulses or to select a short FIR pulse from a pulse train generated by mode-locked CO_2 optical pumping techniques we have developed for the first time the switching based on a FIR plasma shutter. As already mentioned most switching techniques for FIR radiation developed hitherto rely on the carrier density dependent transmission or reflectivity of semiconductors, e.g. GaAs, Si, irradiated by 337 nm-N_2 or 530 nm-Nd:YAG-frequency doubled laser pulses [5,6]. With these techniques truncation times as short as 160 ps have been reached. However, the need of external lasers for controlling the carrier density make these techniques complex and expensive. On the contrary, our FIR plasma-shutter switching system is easy to construct and requires only one laser, which simultaneously generates the FIR laser radiation and controls the plasma shutter [18]. As a first application we have used our system to truncate 100 ns FIR laser pulses in the wavelength range between 150 and 500 µm. We have thus obtained FIR pulses of variable durations from 5 to 50 ns. We have measured truncation times as short as 700 ps by using an electrically induced plasma as well as a CO_2-laser induced plasma. As a second application we have succesfully applied our system to select a pulse from a FIR pulse train generated by optical pumping with a CO_2 laser in the self-modelocked mode. In this case, the time of the plasma breakdown has been adjusted precisely between the first and second pulse of the FIR pulse train.

4. Infrared Solitons Generated in Far-Infrared Raman Scattering

Spontaneous solitons in stimulated Raman scattering (SRS) were first demonstrated in CO_2-laser pumped para-H_2 [19] and subsequently also in frequency doubled Nd:YAG-laser pumped para-H_2 [20-22]. The origin of such spontaneous solitons is the rapid phase variation of the Stokes field which occurs at a noisy level. It creates essentially a phase mismatch of π between pump and Stokes field. Gain and loss conditions are thus reversed. This leads to the soliton excitation [19-20].

In our experiments we have proved the first spontaneous solitons in SRS experiments with a far-infrared (FIR) molecular laser gas, i. e. rotational SRS in NH_3 at 58 µm and 72.6 µm optically pumped at 10.37 µm and 10.35 µm with pulses of variable shapes and durations produced by our OFID CO_2-laser system. In a first step we demonstrated the presence of solitons in the depletion region of the transmitted CO_2-pump pulses accompanied by dark solitons in the generated FIR radiation. We have measured solitons of about 1 ns duration. Secondly, in a more complicated experiment where part of the pump beam was separated before entering the Raman cell and combined with the transmitted beam to produce interference, we detected the predicted phase changes of π connected with the soliton generation in SRS.

5. Conclusion

In our studies we have investigated new methods to produce short FIR single pulses. On this occasion we have observed new phenomena and demonstrated that optical pumping of various molecular gases by rapidly truncated CO_2 pulses results in well-defined superradiant pulses of a duration considerably shorter than the inverse linewidth of the transition. Thus, we have shown the relevance of the fast truncation of the pump pulses in order to reduce the duration of the FIR superradiant pulses. We have observed the onset of swept-gain superradiance and the rise of Raman emission up specific pressures in the gas cell. The rise of the Raman emission is connected with a decrease of the superradiant emission due to line competition between the two emissions. This is obvious from our pulse measurements. In addition, we have found that this competition results in an earlier disappearance of the Raman emission due to the typical evolution of our truncated pump pulses which exhibit OFID by the transmission through the FIR gas cell. Therefore, we have an emission regime where only superradiant emission is present without the superposition of the irregular Raman emission. This is ideal for the generation of nano- and subnanosecond FIR single pulses. Moreover, we have discovered that FIR gases are efficient spectral filters for the generation of OFID pulses. Thus, we generated for the first time CO_2-OFID pulses of about 100 ps duration with FIR gases as spectral filters. We have also investigated an indirect method for short FIR single-pulse generation based on a FIR plasma shutter. We have measured truncation times of 700 ps and shown that our system permits to produce FIR pulses of variable durations as well as to select an ultrashort pulse from a pulse train generated by mode-locked optical pumping technique in a simple way. This is relevant for applications where single pulses are needed. Finally, we have demonstrated the first spontaneous infrared solitons in stimulated far-infrared Raman scattering on NH_3. These solitons develop in the depletion region of the infrared pump radiation with corresponding dark solitons in the Stokes pulses. Our measurements have also confirmed the predicted laser phase change of π during the soliton generation .

6. References

1.   DeTemple, T.A. (1979) Pulsed Optically Pumped Far Infrared Lasers, in K. J. Button (eds.),
      Infrared and Millimeter Waves, Vol. 1, Academic Press, New York, USA , pp. 129-184.
2.   Drozdowicz, Z., Temkin, R. J. and Lax, B. (1979) Laser Pumped Molecular Lasers-Part I: Theory,
      IEEE J. Quantum Electron. **QE-15**, 170-178.

278

3. Lemley, W. and Nurmikko, A. V. (1980) Generation of Ultrashort Pulses in Synchronous Pumping of Near-Millimeter Wave Lasers, *Int. J. Infrared and mm Waves* **1**, 85-94.

4. Schatz, W., Heusinger, M. A., Nebosis, R. S., Renk, K. F. and Lang, P. T. (1993) 100 ps Far-Infrared Laser Pulses From Optically Pumped D_2O, *Infrared Phys.* **34**, 339-344.

5. Salzmann, H., Vogel, T. and Dodel, G. (1983) Subnanosecond Optical Switching of Far Infrared Radiation, *Opt. Commun.* **47**, 340-342.

6. Burghoorn, J., Kaminski, J. P., Strijbos, R. C., Klaassen, T. O. and Wenckelbach, W. Th. (1992) Generation of Subnanosecond High-Power Far-Infrared Pulses by Using a Passive Resonator Pumped by a Free-Electron Laser, *J. Opt. Soc. Am.* **B9**, 1888-1891.

7. Scherrer, D. P. and Kneubühl, F. K. (1993) New Phenomena Related to Pulsed Far-Infrared Superradiant and Raman Emissions, *Infrared Phys.* **34**, 227-267.

8. Kälin, A. W., Kesselring, R., Hongru, Cao and Kneubühl, F. K. (1992) Optical Free Induction Decay (OFID) 10 μm CO_2 laser systems, *Infrared Phys.* **33**, 73-112.

9. Knittel, J., Scherrer, D. P. and Kneubühl, F. K. (1994) The First Optical-Free-Induction-Decay System With a Tunable Single-Mode High-Pressure CO_2 Laser, *Infrared Phys. Technol.* **35**, 67-71.

10. Ehrlich, J. J., Bowden, C. M., Howgate, D. W., Lehnigk, S. H., Rosenberger, A. T. and DeTemple, T. A. (1978) Swept-Gain Superradiance in CO_2-Pumped CH_3F, in L. Mandel and E. Wolf (eds.), *Coherence and Quantum Optics IV*, Plenum Press, New York, pp. 923-937.

11. Nishi, Y., and Murai, A. (1990) FIR Laser Emissions from Population Inversion Transition by TEA-CO_2 Laser Pumping, *Int. J. Infrared and Millimeter Waves* **11**, 309-322.

12. Wiggins, J. D., Drozdovicz, Z. and Temkin, R. J. (1978) Two-Photon Transitions in Optically Pumped Submillimeter Lasers, *IEEE J. Quantum Electron.* **QE-14**, 23-30 .

13. Dupertuis, M. A., Siegrist, M. R. and Salomaa, R. R. E. (1984) Competition between Raman and Line-Center Oscillations in Optically Pumped Far-Infrared Lasers, *Phys. Rev.* **A. 30**, 2824-2826 .

14. Waldman, J., Moix, D. B., Scherrer, D. P., Kneubühl, F. K., Goodhue, W. D., Mueller, E. R. and Coulombe, M. J. (1992) Multiquantum-Well Detection of Nanosecond Far-Infrared Superradiant Pulses at Temperature above 77 K, *Infrared Phys.* **33**, 487-491.

15. Moix, D. B. (1995) PhD Thesis, ETH Zurich, to be published.

16. Yablonovitch, E. and Goldhar, J. (1974) Short CO_2-Laser Pulse Generation by Optical Free Induction Decay, *Appl. Phys. Lett.* **25**, 580-582.

17. Knittel, J., Scherrer, D. P. and Kneubühl, F. K. (1995) High-Pressure Single-Mode CO_2 Laser with ps-Plasma Shutter, *Optical Engineering,* in press.

18. Knittel, J., Scherrer, D. P. and Kneubühl, F. K. (1994) Plasma-Shutter for Far-Infrared Laser Radiation, *Infrared Phys. Technol.* **35**, 655-659.

19. Drühl, K., Wenzel, R. G. and Carlsten, J. L. (1983) Observation of Solitons in Stimulated Raman Scattering, *Phys. Rev. Letters* **51**, 1171-1174.

20. MacPherson, D. C., Swanson, R. C. and Carlsten, J. L. (1989) Spontaneous Solitons in Stimulated Raman Scattering, *Phys. Rev.* **A 40**, 6745-6747.

21. Akiyama, Y., Midorikawa, K., Obara, M. and Tashiro, H. (1991) Measurement of π Stokes Phase Jump in Spontaneously Initiated Stimulated Raman Scattering, *J. Opt. Soc. Am.* **B 8**, 2459-2465.

22. Gakhovich, D. E., Grabchikov, A. S. and Orlovich, V. A. (1993) Spontaneous Solitons in Short Length Geometry of Stimulated Raman Scattering, *Opt. Commun.* **102**, 485-490.

CO LASERS - STATE OF THE ART AND POTENTIAL OF APPLICATIONS

A. Ionin
P.N. Lebedev Physics Institute of Russian Academy of Sciences
53 Leninsky prospect, 117 924 Moscow, Russia
fax (095) 132 0425

I. Spalding
The Laser Centre, United Kingdom
fax 44(0) 1235 848296

Abstract

The recent progress in the Research and Development (R&D) of CO lasers and their applications in science and technology is discussed.

Key words: CO laser, laser physics, laser R&D, laser applications.

1. Introduction

The CO laser has a few advantages compared to CO_2 lasers: a shorter wavelength ($\lambda \sim 5$-6 μm); a higher specific output power (energy) and laser efficiency; a lower plasma shielding effect; a lower diffraction limit for angular divergency (theoretically better focusability); a higher damage threshold for optical materials; feasibility or using optical fiber transmission; "transparency windows" in the atmosphere; a more effective interaction with structural materials and human tissues; order of magnitude longer pulses for repetitively pulsed lasers. The main disadvantage of the CO laser (apart from using a toxic gas mixture) is that a low temperature mode of operation is needed for getting the higher laser efficiency. Due to that fact the CO laser is inferior to the CO_2 laser both in research worker's attention and the scale of applications. Nevertheless a permanent progress in studying the CO laser

279

W. J. Witteman and V. N. Ochkin (eds.), Gas Lasers - Recent Developments and Future Prospects, 279–289.
© 1996 *Kluwer Academic Publishers.*

physics, the R&D of the lasers themselves and their applications takes place. The main trends in those areas are discussed below.

2. A study of the CO laser physics

Kinetic processes taking place inside the active medium still has to be studied both theoretically and experimentally. The effect of vibrational deexcitation by electron collision and the influence of the phenomenon on the vibrational distribution function of the CO laser is very important [1,2]. The theory predicts 20% laser efficiency for the first overtone of the CO laser [3]. Experimental equipment is developed for measuring the translational temperature and vibrational distribution [4].

Optical pumping of CO molecules is very promising for studying the laser kinetics. Since the first research work in the field [5] a few other papers were published [6-8]. An excitation of high vibrational levels ($v > = 40$) was observed with a CO laser operating on the main ($1 - > 0$) vibrational band [8].

Explosive absorption, predicted in [9], and calculated for CO lasers in [10] and experimentally observed in [7] can take place in high power CO laser systems.

Amplification of CO laser radiation and its peculiarity due to the multiline spectrum [11] should be taken into consideration, especially for high power laser systems.

The optical quality of the beam became recently the object of research for CO lasers. For instance, a lensing effect has been observed in [12] for a supersonic CO laser.

The phase conjugation effect, that could be useful for improving the optical quality, was observed experimentally with an efficiency of 11% (for a closed loop scheme) [13] and 0.2% (for a classical scheme) [14] for degenerate four-wave mixing inside an inverted CO laser medium. Theoretical calculations also have been done [15].

Spectroscopy of the electric discharge can be useful for measurement of the translational temperature of $CO + N_2$ mixtures [16] and, of course, for the determination of the chemical contents [16] when is very important for long run CO lasers.

Isotope separation in an electric discharge is also a very impressive effect [17] though there are not yet positive results.

3. Research and development of CO lasers

3.1. DIFFERENT MODES OF GENERATION

CW, pulsed and repetitively pulsed (RP) modes of operation have been studied since 1964 when the first CO laser was launched by Patel. The results have been published in different reviews ([18,19] for instance). Therefore we are considering here only some specific modes of operation.

Overtone lasing has been studied since 1977 [20]. In developing an overtone CO laser [21] there the following results for a pulsed CO laser were obtained: $Q_{out} = 50$ J, $\eta = 5\%$, q = 10 J/l Amagat [22]. Quite recently $Q_{out} = 4$ J, $\eta = 1\%$, q = 10 J/l Amagat were obtained ($\lambda = 2.8$ - 4.2 µm, v = 14->12 - 40->38) [3] for a pulsed mode, though an efficiency of 20% was predicted [3]. In a CW mode the overtone laser operates with an output power of 0.5 - 1 W (v = 2500 - 3800 cm^{-1}) [23].

Lasing on the 1->0 transition [24] is obtained with optical pumping of solid, liquid and gas phases of CO molecules (see [8]). The output power of the laser on the 1->0 transition is less than 1% of the total output power.

Frequency selected Q-switching was studied in the 70's [25] and the output energy did not exceed 10^{-6} J, though for a nonselected Q-switched mode of operation the laser output reached a few Joules with an efficiency up to 5% [26]. Just a few months ago an output energy of 0.5 J with an efficiency of 0.5% for a short pulse ($\tau \sim 5$ µs, $\lambda \sim 6$ µm, $\Delta v \sim 5$ - 40 cm^{-1}) was obtained at the Lebedev Institute.

Transient processes have been researched for a CO laser switched between two transitions [27].

3.2. LASER GEOMETRY

A transverse flow, transverse excited (TE) geometry mainly is applied to high power CO lasers (see below).

Laser tube, sealed-off and axial flow geometry. Here should be mentioned [28], where a specific output of 29 W/m with 15% efficiency at 300 K was obtained with a sealed-off laser and 4 kW from a fast axial flow laser, developed at Laser Ecosse, UK on the basis of an industrial CO_2 laser [29].

A *waveguide structure* is used for a compact and frequency tuned CO laser. For instance, an output power of 0.5 W is obtained for a laser with an active volume of 1.5 x 1.5 x 180 mm^3 (Δf = 550 MHz, $P_{10-9}(17)$, t = -20°C, p=180Torr) [30].

A *slab geometry* combining large area electrodes with a waveguide gap is very promising for development of compact lasers with relatively high output power. For a multimode regime P_{out}=120W (t = - 30°C), η = 17%, P_{out} = 15.5 kW/m^2; P_{out} = 50 W (T = 300 K). For a laser with an unstable resonator P_{out} = 80 W, η = 12% (t = -30°C) [31]. This specific output power was increased up to 26kW/m^2 recently [32].

3.3. PUMPING METHODS

A *DC discharge* is applied to TE lasers (P_{out} >= 10 kW) and for lasers with a tube geometry. A multiple-beam sealed-off CO laser with P_{out} ~ 100 W was developed in [33].

The RF discharge is widely applied to waveguide, slab, TE and fast axial flow lasers. The latter are characterized by a high output and laser efficiency (1 kW, f = 27 MHz, P = 2.4 kW/m, t = 50 min. in [34] and 4 kW in [29]).

MW discharge is also used for pumping. A fast axial flow CW CO laser with P_{out} = 440 W and η=8% (T = 300 K, f =2.456 GHz, P_{in} = 6 kW, p_{out} = 1.4 kW/m) was created in [35].

3.4. COOLING METHODS

Liquid nitrogen is used for cooling laser mixtures to get a higher efficiency.

Adiabatic expansion of the CO laser mixture through a supersonic nozzle gives a possibility of cooling also. A 1 kW CW closed cycle CO laser [36] and a 7 kW (τ = 10 s) [37], 100 kW (τ = 10^{-3} s) [38,39], 200kW (τ = 1 s) [40] open cycle supersonic CO laser were developed.

Cooling without liquid nitrogen (by ordinary refrigerator methods, by thermoelectric effect) is also used.

3.5. DECREASING THE WORKING TEMPERATURE

It is very important to obtain effective lasing without deep cooling or at room temperature for extensive applications of CO lasers.

CO$_2$ laser installations could be used. An electron - beam controlled discharge (EBCD) CW CO laser (P = 1 kW, η = 5%) operating at room temperature was developed in [41] in such a way.

A room temperature operation mode was also realized in [42] with very high efficiency of 15% (P$_{out}$=1 kW, RF discharge, TE, T = 286 K, v = 33 m/s, without Xe).

3.6. SCALABILITY

According to their output power CO lasers could classified in the following way:
* waveguide lasers (~1-10 W);
* slab, gas flow, diffusion cooled, DC, RF and MW discharge lasers (~10^2 -10^{3}W);
* fast axial flow, TE and EBCD lasers ("10^3 -10^5 W).

CO lasers with output powers much higher than 1 kW and operating times more than 1s are enumerated below:
* EBCD, CW, 10 kW, Russia, the 80's [43];
* EBCD, RP, τ = 10 s, 10 kW, Russia [44];
* TE, DC, CW, 5 - 15 kW, Japan [45];
* supersonic, CW, RF, τ = 10 s, 7 kW, Germany & France [37];
* axial flow, CW, RF, 4 kW, UK [29];
* EBCD, CW, τ = 5 s, 85 kW, Russia [46];
* supersonic, EBCD,CW, τ = 1 s, 200 kW, Russia [40].

4. Delivery of laser radiation

4.1.LASER PROPAGATION THROUGH THE ATMOSPHERE

The propagation strongly depends on the spectrum of the CO laser. The process of propagation itself influences the spectral content and optical quality of the radiation on target, workpiece etc.

Absorption due to water vapor is different by a few orders of magnitude for neighboring spectral lines. Special measures should be taken for matching the CO laser spectrum with atmospheric transparency windows [47].

Thermal blooming was observed quite recently for a high power CO laser [37]: heating of the air spreads and distorts the CO laser beam, changing the near-field intensity distribution and also decreasing the relative maximum intensity in the far-field distribution by a factor of 2 (P_{out} = 2 kW), humidity being changed from 0.35 up to 11.8 g/m^3.

4.2. OPTICAL FIBERS

Using optical fibers for 5 μm radiation creates the possibility of flexible power delivery to a workpiece. Though different kinds of materials were used for fiber optics, quite good results were obtained for chalcogenide glass fibers. The absorption is 0.45 dB/m and 0.3 - 0.45 dB/m for AsS Glass/Teflon and GeAsS Glass/Teflon fibers, according to [48].

5. Applications in science and technology

5.1. SPECTROSCOPY

The CO laser gives an opportunity to be used for spectroscopic problems not only in the main spectral band (~5-6 μm), but also near the wavelength of ~3 μm. An overtone CO laser ($\Delta v=2$: 10->8 - 35->33) operating on 330 lines in the 3800-2500 cm^{-1} spectral band and frequency tuned within $\Delta f=200kHz$ was developed in [23].

5.2. FREQUENCY MEASUREMENTS

The CO laser, frequency locked on the $v=1->0$ transition ($\Delta f<10kHz$) could be used as a new secondary frequency standard in the mid-IR [49].

5.3. LASER MONITORING

Laser monitoring was successfully used with ^{12}CO and ^{13}CO lasers and optothermal detection of NO_2 (60 ppb), H_2O (7 ppb), NH_3 (300 ppb) and $CO(CH_3)_2$ (100 ppb) [50,51] was realized.

5.4. MEDICAL TREATMENT

Due to the different absorption of laser radiation in blood CO lasers can be more promising for surgery than CO_2 lasers. The human tissue cutting rate is the same for a 15W CO and an 80W CO_2 laser. Coagulation of blood vessels is much better with a CO laser [33].

5.5. LASER MATERIAL PROCESSING (more detailed information see in [29,52-54])

The phase-transformation hardening efficiency of a 1kW CO laser is the double at that of a CO_2 laser of comparable power (70^0 incidence) [52].

Cutting has been applied to steel, non-ferrous metals and ceramics. A 3kW CO laser was successfully used for cutting 80mm steel, though a 3kW CO_2 laser was able to cut only a 20mm work-piece [55]. Though these results are under question at the moment (optical quality of the lasers was not the same one), different experiments demonstrated that the cutting of steel with a CO laser is by factor 1.25 better than cutting with a CO_2 laser. For Al this factor is 2.0-2.5, and for ceramics Al_2O_3 : 1.35-2.0 [52]. Copper with a thickness of 10mm was cut by a 5kW CO laser [56]. A 5kW CO laser was also applied for underwater cutting of 10-75mm steel [57].

Welding of steel of thickness of 1-4mm was demonstrated to be a factor 2 better than in case of a CO_2 laser when using Ar shroud gas instead of He[58]. The welding quality of Al is much higher [59]. Experiments are underway now within the framework of Eureka Project EU113.

6. Eureka Project EU113 (CO Eurolaser)

For an investigation on the possible use of CO lasers for medical applications, for materials processing and for the development of commercial CO lasers in Europe with output powers from 50W up to a few kW the Eureka Project EU113 (CO Eurolaser) was launched [29]. There are 18 organizations from France, UK and Russia taking part in the Project. The duration of the current phase of the Project is from 1992 till January 1997.
The Project consists of five Sub Projects:
* the development of a sealed-off CO laser of good beam quality with a laser power up to 50W for medical and low-power industrial applications;
* the development of a 1kW CO laser for laser material processing;

* the development of a 3-4kW CO laser for the same purpose;
* the investigation of the possibility of CO lasers for industrial applications;
* the development of optical fibers for transportation of 5 μm CO laser radiation with an output power up to 1 kW.

7. Conclusions

Steady interest to CO laser physics, R & D of CO lasers and their applications takes place in France, Germany, Japan, USA, UK and Russia. There will be physical problems we face in future, especially for high power CO lasers. For some applications CO lasers do have advantages over CO_2 and Nd-YAG lasers. Though CO lasers do not oust CO_2 lasers from laser market, they do find their niche in science and technology on account of their successful combination of laser spectrum, output power and optical quality.

8. Acknowledgments

The joint French-British-Russian activity on CO lasers has been supported by NATO Linkage Grant 930917, NATO Computer Network Supplement (CN.SUPPL.940970) and by French, British and Russian governmental sources.

9. References

1. Napartovich,A.P. (1995) Physics of high-power CO lasers, Gas Lasers-Recent Developments and Future prospects, *NATO ARW 950443, Moscow, Abstracts*, 20.
2. Igoshin,V.I. et al (1994) Calculation of the electron energy in the discharge plasma of a pulsed EBCD CO laser on the basis of an analysis of the temporal characteristics of its radiation, *Kvant. electron*, **21**,429-432 (in Russian).
3. Belykh,A.D. et al (1995) Pulsed CO laser based on the first vibrational overtone, *Kvant. electron*, **22**, 333-340 (in Russian).
4. Bahir,L.P. (1995) IR radiometr with a system of automatic control "Lotos-FPI" for diagnostic of CO laser, *V-th Int. Conf. Industrial Lasers & Laser Applications'95, Shatura, Russia, Abstracts*, 67.
5. Rich,J.W., Bergman,R.C. et al (1975) Vibrational-vibrational pumping of carbon monoxide initiated by an optical source, *Appl. Phys. Lett*. **27**, 656-658.
6. Cohn,D.B., Parazzoli,C.G. et al (1986) Optical pumping of CO by a convective flow CO laser, *IEEE J. Quant. Electron*, **22**, 723-729.
7. Anan'ev,V.Yu., Ionin,A.A. et al (1987) Nonlinear absorption and transformation of CO laser spectrum by CO molecules excited in EBC discharge, *Kvant. Electron*, **14**, 2018-2020 (in Russian).

8. Urban,W. et al (1989) Treanor pumping of CO initiated by CO laser excitation, *Chem.Phys.* **130**, 389-399.

9. Oraevsky,A.N. et al (1985) Explosive absorption of radiation, *Kvant.Electron*, **12**, 2290-2299 (in Russian).

10. Schmelev,V.M. et al (1985) Optical unstability of molecular gas at nonisothermal condition, *Chem.Phys.*,**4**, 873-879 (in Russian).

11. Anan'ev,V.Yu., Ionin,A.A. et al (1989) Pulsed EBCD CO laser amplifiers, *Kvant.Electron.*, **16**, 9-27 (in Russian).

12. Schellhorn,M. and von Bulov,H. (1995) Imrovement of the beam quality of a gasdynamically cooled CO laser with an unstable resonator using a cylindrical mirror, *Optics Letters* (to be published).

13. Belousov,D.V., Gurashvili,V.A. et al (1991) Phase conjugation of CO laser radiation, *XIV Int.Conf. on Coherence and Nonlinear Optics, Leningrad, Abstracts*, **1**, 177.

14. Afanas'ev,L.A., Ionin,A.A., Kotkov,A.A. et al (1994) Active medium of molecular CO_2 and CO lasers as a nonlinear component of a phase-conjugating mirror, *Kvant.Electron*, **21**, 557-560 (in Russian).

15. Berdyshev,A.V., Napartovich,A.P. et al (1994) Formation of amplitude gratings in the active medium of CO laser in the multifrequency radiation field, *Kvant.Electron.*, **21**, 91-96 (in Russian).

16. Azharonok,V.V., Gurashvili,V.A. et al (1995) Optical and spectroscopic diagnostics of working medium of high power technological electric discharge CO and CO_2 laser, *V-th Int.Conf.Industrial Lasers & Laser Applications'95, Shatura, Russia, Abstracts*, 16 .

17. Rich,J.W. and Bergman,R.C. (1986) *Isotope separation under vibrational-vibrational exchange, in Nonequilibrium vibrational kinetics, edited by M. Capitelli*, Springer-Verlag, Berlin, Heidelberg, 313-338.

18. Rich,J.W. (1982) Relaxation of molecules under the exchange of the vibrational energy, in *Applied Atomic Collision Physics*, **3**, Gas Lasers, Academic Press, New York, 125-176.

19. Ionin,A.A., Kovsh,I.B. et al (1984) *Electric discharge high pressure IR lasers and their applications, Radiotechnics*, v.32, VINITI, Moscow.

20. Bergman,R.C. and Rich,J.W. (1977) Overtone bands lasing at 2.7-3.1μm in electrically excited CO, *Appl.Phys.Lett*, **31**, 597-599.

21. Basov,N.G., Danilychev,V.A., Ionin,A.A. et al (1978) Cooled EBCD CO laser on two-quantum transitions of CO molecule, *Kvant.Electron*, **5**, 1855-1857 (in Russian).

22. Basov,N.G., Ionin,A.A. and Kovsh,I.B. (1985) The electoionization CO laser: a multiwavelength IR oscillator, *Infrared Physics*, **25**, 47-52.

23. Bachem,E., Urban,W. et al (1993) Recent progress with the CO-overtone $\Delta v = 2$ Laser, *Appl.Phys.B* **57**, 185-191.

24. Djeu,N. (1973) CW single line CO laser on $V = 1 -> V = 0$ band, *Appl.Phys.Lett.*, **23**, 309-310.

25. Nurmikko,A.V. (1975) Forced mode locking of a single-line high-pressure CO laser, *J.Appl.Phys*, **46**, 2153.
26. Anan'ev,V.Yu., Basov,N.G., Ionin,A.A. et al (1985) Efficiency increasing for EBCD Q-switched CO laser by lasing of pulse train, *Kvant.Electron*, **12**, 1666-1670 (in Russian).
27. Goltjaev,O.M., Kornilov,S.T., Prozenko,E.D. et al (1991) Research of transient processes at modulation of cavity length of waveguide CO and CO_2 lasers, *J. of Appl.Spectr.*, **54**, 588-593 (in Russian).
28. Peters,P.J.M, Witteman,W.J. et al (1980), Efficient simple sealed-off CO laser at room temperature, *Appl.Phys.Lett.*, **37**, 119.
29. Spalding,I., Clucas,A., Ionin,A. et al (1995) Eurolaser activities and EU113 achievements in particular, *V-th Int.Conf.Industrial Lasers & Laser Applications'95, Shatura, Russia, Abstracts*, 13.
30. Gerasimchuk,A., Kornilov,S. Protzenko,E. et al (1989) Frequency tuned carbon monoxide waveguide laser with RF pumping, *Appl.Phys.B* **48**, 513.
31. Zhao,H., Baker,H.J. and Hall,D.R. Area scaling in slab rf-excited carbon monoxide lasers, *Appl.Phys.Lett*, **59**, 1281-1283.
32. Villarrea,F., Hall,D.R. et al (1995) High pressure CW molecular-gas lasers using narrow-gap slab wavequides, *CLEO'95, Technical Degest*, 44.
33. Aleinikov,V.S. and Masychev,V.I. (1990) *Carbon monoxide lasers*, Radio i Svyaz, Moscow (in Russian).
34. Hall,D.R. et al (1992) New technology for industrial carbon monoxide laser, *Proc. of LAMP'92, Nagaoka, Japan*.
35. Luo,X., Schafer,J.H., Uhlenbush,J. (1994) High power room temperature operating CW CO laser excited by microwave discharge, *Proc. SPIE*, **2502**, 69-74.
36. von Bulov,H. and Zeyfang,E. (1990) Gas dynamically cooled CO laser with RF-excitation: design and perfomance, *Proc. SPIE*, **1397**, 499-502.
37. von Bulov,H.and Schellhorn M.(1994)High power gas dynamically cooled CO laser with unstable resonator, *Proc.SPIE*,**2502**,63-68.
38. Klosterman,E.L. and ByronS.R. (1979) Electrical and laser diagnostics of 80kW supersonic CW CO electric discharge laser, *J. Appl. Phys.*,**50**, 5168-5175.
39. Ionin, A., Kotkov,A. et al (1990) Supersonic EBCD CO laser, *Proc.SPIE*,**1397**, 453-456.
40. Dymshitz,B.,M. et al (1994) CW 200 kW supersonic CO laser, *Proc.SPIE*,**2206**
41. Ionin,A., Mayerhofer,W., Zeyfang,E. et al (1994) Room temperature repetitively pulsed EBCD carbon monoxide laser, *Proc.SPIE*,**2502**, 44-50
42. Uehara,M., Kanazawa,H. and Kasuya,K. (1994) Recent studies of high power CO laser under room temperature operation, *Proc.SPIE*,**2502**, 38-43.
43. Averin,A., Basov,N. et al(1982) CW technological EBCD CO laser with output power of 10 kW,*Kvant. Electron.*, **9**, 2357-2358 (in Russian).
44. Averin,A., Basov,N., Ionin,A. et al (1990) Repetitively pulsed EBCD carbon monoxide laser, *Kvant. Electron.*,**17**, 561-562 (in Russian).

45. Kuribayashi,S., Sato,S. et al (1992) Current status of the high power CO laser program, *Proc.LAMP'92, Nagaoka, Japan*, 51-54.

46. Golovin,A.,Gurashvily,V. et al (1995) E-beam sustained CW cryogenic CO laser with a subsonic gas flow, *8-th Laser Optics Conf., St. Petersburg, Russia, Technical Digest*, **1**, 140-141.

47. Anan'ev,V.Yu., Ionin,A.A. et al (1987) Master-oscillator-amplifier electroionization carbon monoxide laser system and propagation of its radiation through the atmosphere, *Int.J.Infrared and millimeter waves*, **8**, 549-5712.

48. Sato,S. et al. (1992) Multihundredwatt CO laser power delivery through chalcogenide glass fibers, *CLEO'92, Anaheim, USA*.

49. George, T., Urban,W. et al (1991) Saturation stabilisation of the CO fundamental-band laser, *Appl.Phys.,B* **53**, 330-332.

50. Gerasimchuk,A., Kornilov,S., Protsenko,E. et al (1992) Selective optothermal delection of NO_2 and H_2O with a frequency tuned CO waveguide laser, *Appl.Phys.,B* **55**, 503-508.

51. Kornilov,S., Protsenko,E. et al (1994) Investigation of absorption of acetone and ammonia inside lasing band of a waveguide CO laser by optothermal method, *J. of Appl.Spectr.*, **61**, 210-214 (in Russian).

52. Maisenhalder,F. (1992) Materials processing by CO lasers, *Proc.LAMP'92, Nagaoka, Japan*,**1**, 43-50.

53. Kanazawa,H. et al (1992) Comparison of cutting properties between CO laser and CO_2 laser, *Proc.LAMP'92, Nagaoka, Japan*, **1**, 665-670.

54. Spalding,I. et al (1993) Industrial laser developments in the UK, *AIAA Paper* **93-3152**, 1-8.

55. Sato,S., Fujioka,T. et al (1988) Cutting of steels by high power CO laser beam, *Proc. &th ICALEO* **88**, 324-331.

56. Sato,S. et al (1990) Metal cutting with high power 5μm wavelength CO laser, Report to a seminar at Gas Lasers Lab, Lebedev Physics Inst., Moscow.

57. Beppu,S. et al (1992) Local drying underwater cutting of reactor core internals by CO laser, *Proc.LAMP'92, Nagaoka, Japan*, 661-664.

58. Schellhorn,M. and von Bulov,H. (1994) CO laser deep penetration welding: a comparative study to CO_2 laser welding, *Proc.SPIE*, **2502**, 664-669.

59. Schellhorn,M. (1995) Application of high power CO laser in aluminium welding,*V-th Int.Conf.Industrial Lasera & Laser Applications'95, Shatura, Russia, Abstracts*, 26.

RECENT ADVANCES IN HIGH-POWER LASERS IN THE KURCHATOV INSTITUTE

V.Yu.BARANOV

National Research Center-Kurchatov Institute

Introduction

The first results in the field of different types of high power gas lasers creation were obtained in the Troitsk Branch of Kurchatov Institute of Atomic Energy (now TRINITI) more than 25 years ago [1]. A phenomena of the transformation of the arc discharge into diffusion glow discharge by gas flow action [2](opened in 1966) gave a direction of further researches. The researches were carried out simultaneously in two main fields:

-developing of different types of lasers: continuous wave (CW), high repetition rate pulsed lasers (HRR), with self-sustained and non self-sustained discharge, with opened and closed operation cycle etc.;

-investigations of physical processes which are determine the efficiency of lasers.

High results obtained in the field of low-temperature plasma physics (different types of nonstabilities, kinetics); processes, influenced on laser beam divergence (active medium non homogeneity, resonator optics, radiation behavior in nonlinear gain and absorption active medium) allow TRINITI to take leader place among developers of gas discharge lasers and the same other facility.

At present time TRINITY has a large experience in the technological application of their lasers in industry. For example, CW CO_2 lasers with 1-5kW power are widely used for metal processing and hardening. Based on HRR CO_2 lasers the modern technologies were elaborated in TRINITI - cutting of composite, ceramic and other materials; laser isotope separation technology etc. Several types of lidars had been developed in TRINITI - the Doppler wind lidar; the Differential Absorption lidar based on optically pumped ammonia laser; excimer lidar to measure ozone profile in atmosphere and others .

A few other applications of the laser developed in TRINITY are also described and discussed.

W. J. Witteman and V. N. Ochkin (eds.), Gas Lasers - Recent Developments and Future Prospects, 291–300.
© 1996 *Kluwer Academic Publishers.*

1. Carbon Dioxide CW and HRR Lasers with output power up to 100 kW

1.1. CW CO_2 LASER FOR INDUSTRIAL APPLICATIONS

The Troitsk Institute for Innovation and Fusion Research traditionally devotes much effort to scientific and engineering studies of CW gas lasers and their applications. We study the problem, conducting research on the basic physics of gas discharge, optical and quantum effects in lasin media, laser beam interaction with a variety of materials and substances.

Among the lasers designed at the TRINITI, the closed fast flow electric discharge CW CO_2 laser LT-1 produces 0.5-9kW of the output power is a multipurpose tool used in many industrial applications [3]:

- precision welding of steels and alloys up to 5mm thick at velocities up to 2.5 meters per minute;
- metal, wood and ceramics cutting;
- surface treatment (local hardening, laser alloying and cladding) with the rate of up to 120 cm^2/min.

The original technical decisions patented in number of countries (USA, England, France, Japan, etc.) were realized in the installation. Laser pulley has smaller dimensions as compared with known foreign lasers and controlled sectioned electrode system has no analogies in world practice

Broad spectrum of LT-1 functions has defined the success of its application in more than 10 different industrial fields. These lasers are commercially available since 1989. Their most spectacular achievements are:

- cutting of radioactive nuclear fuel elements at nuclear power plants, welding of key parts of nuclear reactor safety systems and heat exchangers (Fig. 1, 2)

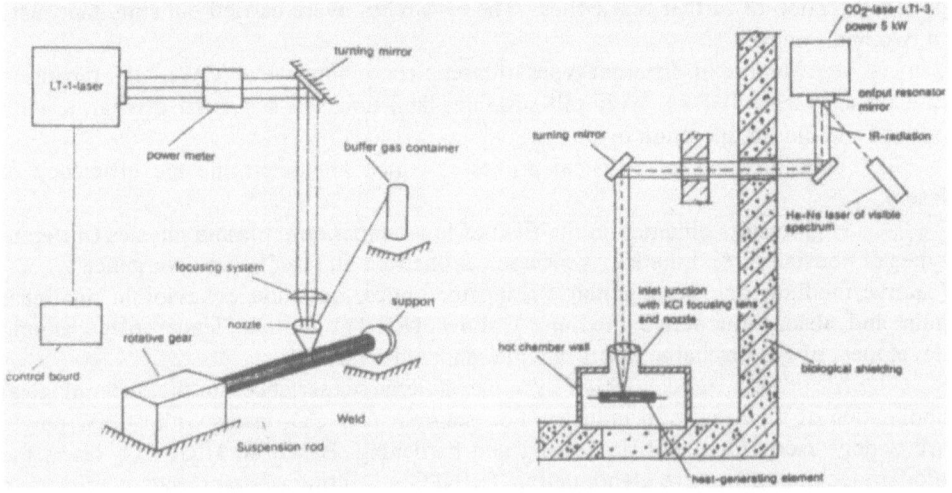

Figure 1. Welding of elements of nuclear reactor   Figure 2. Laser cutting of heat-generating element of nuclear reactor

- surface treatment of blast furnace tuyeres doubling their life time;
- hardening of cutting tools and stamps;
- laser carbonization of low-carbon steels (for example, four - fold improvement of pile driver durability has been achieved).
- increasing of corrosive resistance of carbon and stainless steels by using laser welding, melting, surface alloying which can solve the problem of corrosion untraditional, secure higher level of properties in comparison with standard types of corrosion protection.

In our institute there have been works on the creation of industrial CO_2 lasers with the average radiation power up to 50 kW in recent years. Among them there are CW CO_2 laser pumped by self-sustained discharge and HRR CO_2 laser pumped by e-beam controlled discharge.

1.2. HIGH POWER SELF-SUSTAINED-DISCHARGE CO_2-LASER

Further input power increasing up to 50kW CO_2 lasers exited by a self-sustained discharge became possible after the sectional many electrode cathode system, consisting of laminated electrodes, and has been achieved due to solution of a number of basic challenges [4]:

- discharge glow-mode stabilization at a higher pressure (100Torr) gas flow;
- ensuing of an active medium efficient homogeneous excitation by a glow discharge of a large volume (hundreds of liters);
- ensuring of high power laser radiation extraction at small (15cm) distances along the flow with the outcoming problems f cooled metaloptics creation

A scheme of the CW fast-flow open-cycle CO_2 laser pumped by a self-sustined gas discharge is presented in Fig 3.

A CW direct current discharge in the gas flow is supported by means of electrode system which consists of a flat copper anode and a cathode plate, cathode elements being sectioned both across and along the gas flow.

The main parameters of these lasers are shown in Table 1. The first column represents the laser parameters when using a conventional CO_2 - N_2 - He gas mixture. These results clearly demonstrate the fact of quite a high laser efficiency when no expensive helium is used.

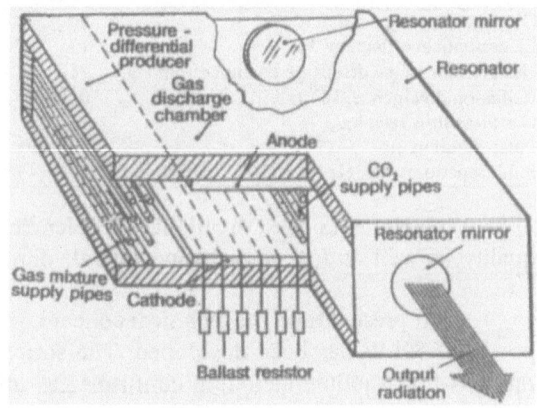

Figure 3. CW CO_2 laser pumped by the self-sustained discharge

TABLE 1. The main parameters of CW CO_2 laser

Parameters	Gas mixture		
	CO_2-N_2-He 0.06:1:0.5		CO_2-air 0.05:1
Output radiation power, kW	50		35
Electrooptical efficiency, %	10-12		6-9
Gas pressure in the discharge chamber, mm Hg	50-120		30-80
Radiation divergence, 10^{-3} rad	0.5-1.0		0.5-1.0
Gas mass flow rate, kg/s	1.5		2.0

1..3 HIGH POWER CO_2 LASER WITH A NON-SELF-SUSTAINED DISCHARGE.

In constructing CO_2 laser with a non-self-sustained discharge the compromise has been found between the necessity of making homogeneous high speed flow of the working gas mixture and that of creating homogeneous electric field in the discharge gap. The problem has also been solved which allowed the creation of periodically pulsed accelerator of electrons with an average electron beam current density up to $100\mu A/cm^2$ and electron energy of about 200keV [5].

The use of a non-self-sustained gas discharge technique allows us to maintain a stable mode of the discharge operation under the gas mixture pressure of about 1 atm. The optical cavity coincides with the discharge zone. The main laser parameters are shown in Table 2.

TABLE 2.   The main parameters of HRR CO_2 laser

Parameters	Gas mixture		
	CO_2-N_2-He 1:6:3		CO_2-air 1:9:0.06
Output radiation power, kW	50		35
Electrooptical efficiency, %	13		6-9
Gas pressure in the discharge chamber, mm Hg	760		30-80
Radiation divergence, 10^{-3} rad	0.5-1.0		0.5-1.0
Gas mass flow rate, kg/s	4		2.0
Pulse duration, μs	50-100		50-100
Pulse repetition rate, Hz	100		100

Such lasers have the highest efficiencies, specific lasing powers, and radiation quality as well as low masses and overall dimensions per an output radiation power unit.

At present time the technical concept of creation a transportable CO_2 laser with power of 50kW has been developed. The structure of such mobile system consists of the gas flow module including confusor, gas mixture, beam chamber, gas blower and diffusor.

On the ground of the above mentioned conception a mobile laser technological complex on the base of 50kW HRR pulsed CO_2 laser has been developed and is being built nowadays. Putting it into experimental operation is planned to take place in 1996.

Fig.4 shows an outward appearance of the mobile laser technological complex MLTC-50 which is based on two vehicles

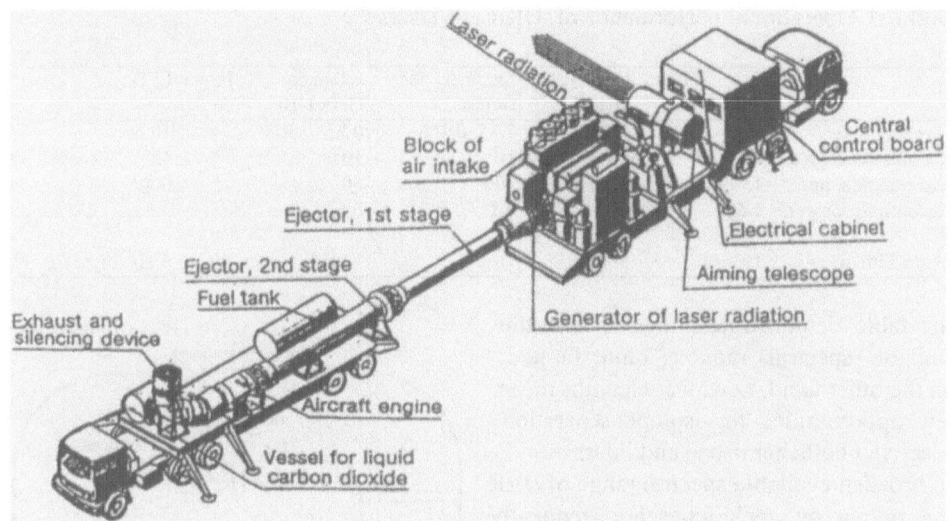

Figure 4. Mobile Laser Technological Complex MLTC-50

Some applications of this complex may be:

- remote cutting of metal and armored concrete constructions in a course of demontage of emergency repair on a nuclear power plant, oil and gas boring wells in a process of emergency work after an earthquake and other natural disasters, while cutting up ships and submarines to metalscrap;
- decontamination of concrete surface by "peeling" at demontage works on nuclear power plant;
- removal of paint layers and other pollution from surfaces;
- fight with locust raids;
- cleaning of coast line from oil products at oil emergency overflow and cleaning of water surface from oil layer including thin iridescent layer that cannot be removed by other methods.

2. High-repetition-rate pulsed gas discharge CO_2 lasers (the output power 1-3 kW) and excimer lasers (the output power up to 1 kW)

The Troitsk Institute for Innovation and Fusion Research is prominent in Russia as a center for design of HRR pulsed gas lasers. This type of lasers sources is known for its unique capability to produce high average power beam (usually in the range from several hundred watts to a few kilowatts) as a train of pulses repeating hundreds times a second instead of a more traditional continuous wave. Typically, a single pulse width is shorter than the pause between successive pulses by the factor of a few thousand, therefor the pulse instant power is thousand of times higher than the average output power.

Numerous experiments on laser applications would be impossible without adequate equipment (Fig.5) . Its key element is lasers developed, designed, and manufactured at the TRINITI. Some of their characteristics are listed in the Table 3.

TABLE 3 Operational performance of HRR gas lasers

	CO$_2$ laser Svertchok	CO$_2$ laser Dyatel	Excimer laser Gefest-2	Excimer laser Gefest-10	LPD
Wavelength, μ	9-11	9-11	0.25, 0.308	0.25, 0.308	10.6
Max.pulse energy, J	2	10	1	10	12
Max. rep. rate, pps	1500	400	100	50	600
Max.output power. kW	1.5	3	0.1	0.5	3.5
Pulse width, ns	10^4	500	70	100	5·10^4

The table demonstrates, however, that the available spectral range is quite limited. On the other hand, new wavelengths mean new opportunities for isotope separation, selective photochemistry, and lidars. To broaden available spectral range of HRR laser output, two techniques are frequently used: Raman scattering and optical pumping. Excimer lasers output is shifted to near ultraviolet or blue - green visual regions by Raman scattering. HRR CO$_2$ lasers are used for optically pump other molecular gas lasers (NH$_3$, C$_2$D$_2$, CF$_4$, and others) and thus to produce far infrared output with wavelengths between 10 and 20 microns. The unique set of tunable HRR infrared lasers with output

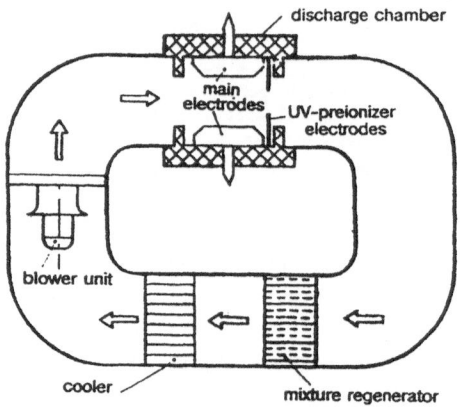

Figure.5. The scheme of HRRP" CO$_2$ laser "Dyatel-3

power up to several hundred Watts and repetition rate up to 150 pulses per second is available now in Troitsk.

The TRINITI has invested many years of research in the development of technologies based on HRR lasers. During these years, scientists and engineers have gained invaluable experience in practical work with HRR machines. Isotope separation is perhaps the most mature HRR technology existing at the Troitsk Institute and ready for industrial launching

Over almost two decades, The Troitsk Institute for Innovation and Fusion Research in collaboration with the Institute of Spectroscopy studies the process of multiphoton dissociation (MPD) of molecules by infrared radiation. Multiphoton dissociation is the process in which a molecule absorbs radiation quanta exiting its vibrations until it breaks apart because of these vibrations. The crucial point is that a molecule efficiently absorbs only the light with definite wavelengths (they are called resonant). The set of resonant wavelengths is a molecule's "bar code", it is unique for every gas. Moreover the values of resonant wavelengths are determined by masses of atoms constituting the molecule. This particular fact is what makes the MPD so attractive for the isotope separation.

Demonstration facilities for sulfur and carbon isotope separation have been created as a result of collaboration between the TRINITI and several other institutes in Russia. These systems are aimed at optimization of MPD process and evaluation of its cost efficiency. Preliminary experiments show that laser produced isotopes are far cheaper as compared to any conventional technique.

Results of basic scientific research, existing technological cooperation with other institutes allow to start building a laser isotope separation plant. Light element's isotopes produced there can answer a wide variety of demands in many technologies. These isotopes can be readily used in medicine, agriculture, and environmental monitoring.

A trend of more extensive laser applications in environmental and ecological studies is now quite evident. The most advanced area of their applications is lidars, systems for remote optical sensing of Earth's atmosphere.

Lidars developed at the TRINITI use two types of HRR beam sources, carbon dioxide and excimer lasers. Therefor probing is performed using either infrared or ultraviolet spectral bands. The lidar based on XeCl excimer laser can measure ozone concentration at the altitudes between 10 and 30 kilometers. This device may help to analyze one of the gravest concerns of the humankind, the damages in the ozone shield of our planet. The same system can measure concentration of dust particles in the air. If the excimer laser is used to pump a dye laser, the spectral range available for probing becomes much wider, extending possible applications of a lidar. A tunable lidar can detect and measure typical pollutants produced by burning of fossil fuels: nitrogen and sulfur oxides. A more delicate method of beam wavelength conversion, Raman scattering, enables us to measure also concentrations of more than ten other components of waste gases.

The lidar, all its auxiliary systems, and a personal computer for data acquisition and processing can be easily installed in an ordinary truck wagon. Mobility

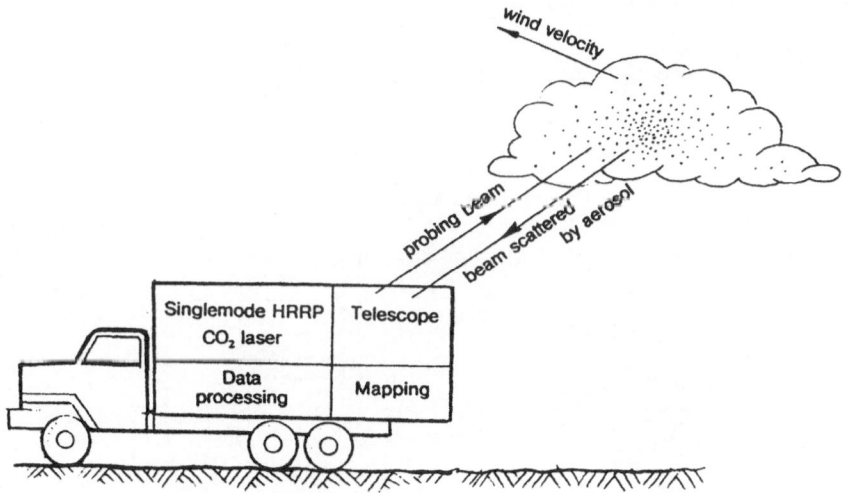

Figure.6. Wind measurements of Doppler lidar

298

of the lidar makes it especially suitable for emergencies, and the first field test for our system has taken place during emergency work following the Chernobyl Nuclear Power Plant accident. The lidar has passed the test excellently, for more than a month measuring aerosol concentrations above the damaged NPP unit.

The wind lidar is being developed at the Troitsk Institute on the basis of a HRR carbon dioxide laser with stabilized frequency output (Fig.6). Its operation is a direct application of Doppler effect: the light scattered by a dust particle carried by the wind changes its frequency, the frequency shift depending on the particle (and therefore wind) velocity. Accuracy of the measurement is between 1 and 3 meters per second that is sufficient for almost any practical purpose. Mobile ground-based wind lidar car within ten minutes gather the data of wind conditions above the area 20 km in diameter. Again nowhere is this information more vital than in emergency when by accident a drifting cloud of dangerous substances has been released. A similar air- or space-borne lidar can be used to obtain global three-dimensional wind map which would significantly improve reliability of weather forecasts.

3. Carbon Monoxide Lasers.

Another very promising type of gas lasers is a CO laser. In this field the TRINITI also holds national leadership. Several unique CO systems are now operational in Troitsk, CW lasers of two types (100W, with self-sustained electrical discharge and 10kW, with an electron beam controlled discharge) are among them (Fig.7). Our HRR discharge pumped CO laser produces 2kW of the average output power.

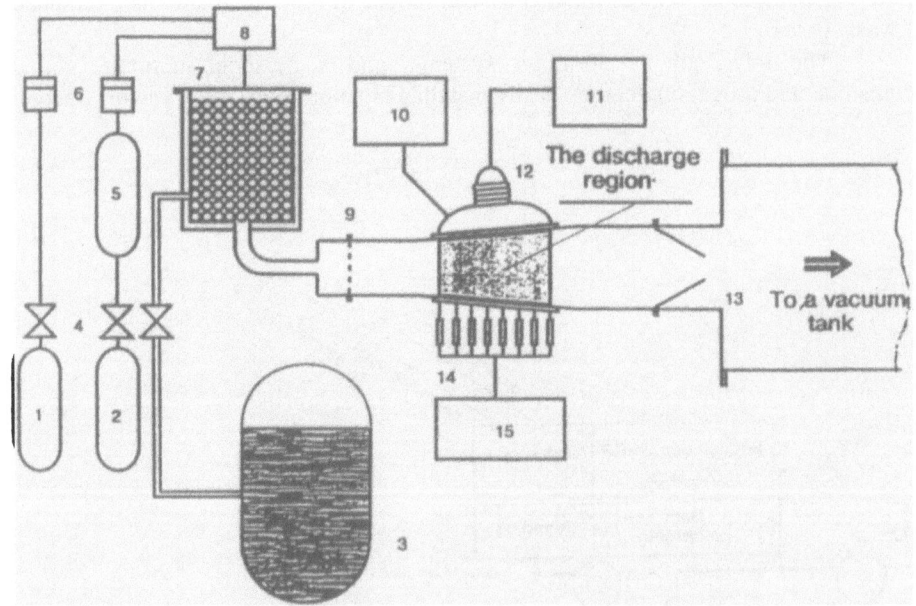

Figure 7. Schematic view of the e-Beam-Systained Cryogenic CW CO-Laser with a Subsonic Gas Flow.
1-bulb with N_2; 2-bulb with CO; 3-liquid nitrogen tank; 4-gates; 5-gas cleaner; 6-cutoff valves; 7-gas cooler; 8-mixer; 9-throttling grid; 10-e-gun vacuum system; 11-e-gun power supply system;

12-e-gun; 13-control gate valve; 14-ballast resistors; 15- discharge power supply system.

CO lasers demonstrate several advantages over CO_2 lasers, mostly due to its shorter (5-6 microns) output wavelength. First their optical components can be made of cheaper materials (CaF_2, for example). Second, optical fibers can be used to deliver the beam to the processed area. Third, absorption of a CO laser beam by metals is usually better as compared to CO_2, therefore less output power is required for the same effect. Besides, CO laser beam heating is more localized than when a CO_2 laser is used.

Wavelength of a CO laser output can be easily converted to near infrared region (2.8-3.2 microns). This is especially important for isotope separation and other wavelength sensitive applications, because new molecules can be involved which do not absorb the wavelengths available for other lasers. For example, separation of rare oxygen-17 and oxygen-18 isotopes has been demonstrated experimentally.

Finally, mobile CO lasers exhibit better operational performance due to their high efficiency (30%) and high specific energy output (100 J/g). A CO laser system producing the same output power as a well designed CO_2 laser system, weighs ten times less.

The future of CO lasers will probably be connected with oxygen isotope separation and biomedical applications.

4. Pulsed Laser Systems With Pulse Width of 1 Ns on CO_2.

The experimental research program at TIR-1 facility was directed to develop CO_2 laser systems with highest characteristics: high (up to 10^{15} W/cm^2) intensity at focal spot on a target, wide pulse duration range (0.3ns - 500ns), laser beam with diffractive divergence. Such laser can be used in the field of:
- interaction of laser radiation in Laser Thermonuclear Fusion research program;
- laser beatwave electron acceleration;
- nonlinear transformation of CO_2 laser radiation frequency;
- heavy elements ion beam generation with charge of $Z \leq +(30-35)$.

To solve these tasks the CO_2 laser was created with different output parameters and technical reliability. Simultaneously there were carried out some researches of:
- self-sustained and non self-sustained discharge in gas mixture CO_2-N_2-He;
- nonlinear processes accompanying short pulse radiation, interaction coherence influence on pulse form changing;
- radiation self-focusing influence in resonant absorption gas cells on distortion of spatial laser beam profile.

During the work in TIR-1 there were created next CO_2 lasers:
- An electrical discharge laser with UV preionization of specific characteristics such as $\tau = (25-100)$ns, $\alpha_o = 4 \times 10^2$cm^{-1}, E=5J.
- Power pulse laser, controlled by electron beam τ=500ns, E=3kJ.
- Laser facility TIR-1 consists of preliminary optical generator, gain system and equipment to form spatial temporal characteristics τ =(0.3-2)ns, E=300J of diffractive quality and high contrast radiation. This system produces up to 10^{15}W/cm^2 on a target

- Power pulse generator based on electroionization module, used in multicharge laser ion source of Pb(+30-+35), τ=30ns, E=120J.
- Developing of HRR laser system consists of preliminary optical generator and electrical discharge gain module. Its spatial temporal characteristics τ =(10-30)ns, E=150J, pulse repetition rate 3Hz are suitable for multicharge ion source of heavy elements in real operating accelerators.

At TIR-1 there was carried out the research of CO_2 laser radiation (10^{12} - 10^{15})W/cm^2-interaction with plasma. Using of X-ray diagnostics, plasma scattered radiation spectrum measuring, diagnostics method used multicharge ion parameters in laser plasma in the experiments allow to make numerical modeling to describe laser plasma state at different moments of its heating and scattering. The obtained results are very useful in the field of laser radiation interaction with hot plasma.

Plasma ion component measurements gave the significant information about ionization and recombination processes in CO_2 laser plasma. Obtained data allow now to go over to the technical developing of multicharge ion source with required output characteristics.

5. Acknowledgment

The materials presented appeared as a result of the activities of a large group of specialists from TRINITI. The author wish to thank all participants of this work particularly Dr.A.V.Rodin, Dr.V.G.Naumov and Dr. A.V.Gurashvili.

6. References

1.Baranov V.Yu. (1966) Some effects observed under the researches of discharge in the ionized and nonionized gas flow. *Teplophisika vysokix temperatur*, v 4, N 5

2. Baranov V.Yu. (1970) *Some questions of the investigations of gas discharge plasma and creation of strong magnetic fields* L.,Nauka,39-74.

3.Kosyrev F.K. et al. (1978) LT1-2 5kW commercial process laser unit. *Avtomaticheskaya Svarka*, N 10.51-52

4. Artamonov A.V., Blochin V.I. (1977) A study of the electric discharge chamber of the fast-flow CO2 laser. *Sov.J. Qvant. Elec.* , v.4, N 4, 581-586.

5. Vostrikov V.G.,Naumov V.G. et al. (1982) On the effect of the specific pump power on the efficiency of the atmospheric -pressure electroionization CO2 laser. *Sov.J. Qvant. Elec*, v.9, N 2, 413-415.

CO$_2$ LASERS WITH FLEXIBLE PARAMETERS AND THEIR USE IN TECHNOLOGICAL APPLICATIONS

O.B.DANILOV, V.V.DANILOV, V.V.LYUBIMOV,
N.N.ROSANOV, A.I.SIDOROV
*Research Institute for Laser Physics, Science Center "S.I.Vavilov
State Optical Institute"*
12, Birzhevaya line, St.Petersburg, 199034, Russia

1. INTRODUCTION

One of the main developments of industrial CO$_2$-lasers is beam controlling with conservation of the energetical laser efficiency. Usually this problem is solved by deflectors, which are in contact with the laser beam. But the high speed and precise beam control necessity favours a system of master-oscillator-amplifier or a laser-injector-regenerative amplifier. In these systems a small power master-oscillator or injector forms the control beam and the final amplifier or regenerative amplifier keeps the laser effficiency at a high level by adjusting the power of the control beam to the necessary level.

The paper considers two problems: the intracavity control of the beam divergence of CO$_2$-lasers (both pulsed and CW) and the regenerative amplification of a given wave front laser-injector signal.

2. TEA-CO$_2$ LASERS WITH INTRACAVITY LC MODULATORS

2.1. LC MATERIAL FOR $\lambda = 10.6\ \mu m$

Liquid crystal modulators (LCM) are promising devices for laser techniques. Application of LCM in lasers provides a way for controlling the laser parameters from spectral [1] to spatial; as well as for phase conjugation for wave front correction [2]. However, the majority of investigations of LCM in lasers are limited to the visible and near IR-regions.

The objective of our work was the design of a LCM which can be used in CO$_2$ lasers and the investigation of their potentialities. The problems of the intracavity interaction of CO$_2$ laser radiation with the LCM were also studied.

The main problem of LCM applications in CO$_2$ lasers is the high absorption coefficient of the most organic materials in the middle IR-region. By selection of the components of the LCM it was possible to decrease the absorption coefficient to a value sufficiently low for intracavity usage of LCM in CO$_2$ lasers. A particular LCM structure consisted of a 4-component nematic LC based on cyanobiphenils,

W. J. Witteman and V. N. Ochkin (eds.), Gas Lasers - Recent Developments and Future Prospects, 301–322.
© 1996 *Kluwer Academic Publishers.*

302

with a special dope to produce chiral properties. The concentration of the dope corresponded to the pitch of cholesteric helix equal 10 μm. The absorption coefficient of the designed LC was 15 cm⁻¹ for λ=10,6 μm. The mechanism of light modulation of such a LC is the cholesteric-nematic transition in the presence of an electric field, which process is attended by the change of light scattering. Fig. 1 (curve 1) shows the transmission of an LCM for λ=10.6 μm as a function of the applied voltage.

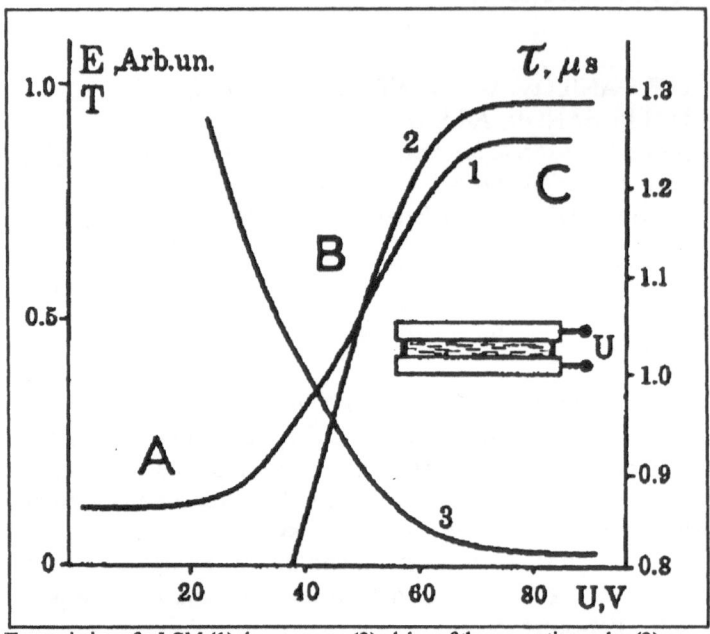

Fig.1. Transmission of a LCM (1), laser energy (2), delay of the generation pulse (3) as a function of the applied voltage (λ = 10.6 μm).

The LCM consisted of two Ge plates with antireflection coatings and a LC-layer 60 μm thick. Curve 1 contains three characteristic regions: A- the region of a focal-conic state of the LC with a high level of light scattering, C- the region of a homeotropic state with a minimum scattering and B- the region of the intermediate state in which the rise of voltage is attended by the rise of the cholesteric helix pitch and the orientation of the LC molecules along the lines of the electric field.

As one can see the described LCM can be used in a laser in two regimes: - the trigger regime in which the LC is located in an A or C state, so the transmission is switched on or off (the switching time is a function of LC thickness and lies in the 2-10 ms region); - the regime of the monotone variation of the transmission (LC is in B -state). It is clear that the most feasible application of the trigger regime is the intracavity modulation of a cw CO_2 laser. This was achieved with a modulation frequency of 100 Hz and an average power of 0,1 W [3]. The second regime provides a monotone variation of the intracavity losses of the laser. This enables us to control the energy (Fig.1, curve 2) and the delay of the generation pulse from the pumping pulse (Fig.1, curve 3). Curves 2,3 were obtained for a TEA-CO_2 laser with plane mirrors and an intracavity LCM like the one shown on Fig.1.

303

2.2. BEAM DIRECTION CONTROLLING IN THE TRIGGER REGIME (A-C REGIME).

In the pulsed TEA-CO$_2$ laser trigger regime an LCM can be used for beam direction control. The experimental arrangement is shown in Fig.2. The compact TEA-CO$_2$ laser (discharge dimensions: 10x10x150 mm, pulse duration: 1-1.5 µs) contained a conjugate resonator which consisted of lenses L1,L2, output mirror M, diaphragm D1 and a LCM as controllable mirror. The LCM consisted of a dielectric layer with a (mxn)-matrix of high reflective electrodes, a LC-layer and Ge window with a nonreflective cover as common electrode. The number of reflective electrodes (*mxn*)=(*10x5*) is 50 and the diameter is 0.35 mm. In a conjugate resonator such an LCM works like a spatial filter applying the voltage to one of the electrodes so that one spatial mode or one beam position can be achieved. Thus, in the described laser, 50 beam directions can be choosen. In [4] we have shown that a one lens conjugate resonator and a special construction of the LCM enables us to produce (mxn) beam directions using only (m+n) electrodes. A scanning TEA-CO$_2$ laser described in that work, produced 1024 beam directions.

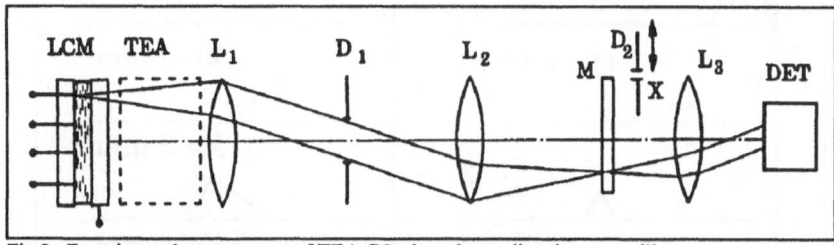

Fig.2. Experimental arrangement of TEA-CO$_2$ laser beam direction controlling.

The scanning speed is limited by the switching (on/off) times of the reflecting elements of the LCM and the relaxation time of the LC after the action of the radiation pulse on it. By decreasing the radiation energy density on the LC the scanning speed can be raised up to 100-200 beam directions per second. It is obvious that in the trigger regime only one beam direction per generated pulse can be achieved.

The energy of the generated pulse for one beam direction was 0,5 mJ. The increasing of energy is limited by the energy density where the destruction of the LC begins (W~ 1,5 J/cm^2).

2.3. BEAM DIRECTION CONTROLLING IN THE B-REGIME

The switching time of the spatial modes can be substantially decreased using the modulation B-regime. As it is shown in Fig.1 (curve 3) the temporal position of the generated pulse is defined by the applied voltage in the B-region. The reason for this is the influence of the level of the intracavity losses on the onset of the generation. So if different voltages from the B-region are applied to some electrodes of the LCM a sequence of spatial modes can be obtained while population inversion in the active medium exists. An initial mode appears on the electrode with minimum losses (maximum voltage), the next modes develop on electrodes with lower voltages consequently.

In the experiment the laser shown in Fig.2 has been used. The voltages from the B-region (U$_1$-U$_3$) were applied to three electrodes. Fig.3a show the oscillograms of the generated pulses for one beam direction (1) (U_2, $U_3=0$), two (2-4) ($U_1 > U_p$, $U_3=0$) and three beam directions (5) ($U_1 > U_2 > U_3$) during the existence of inversion. The duration of the pulses was shortened because of the high intensity of the radiation. This causes a strong interaction between the radiation and the LC which "switches off" the generation on a particular mode. The minimum time between two beam directions was 200 ns.

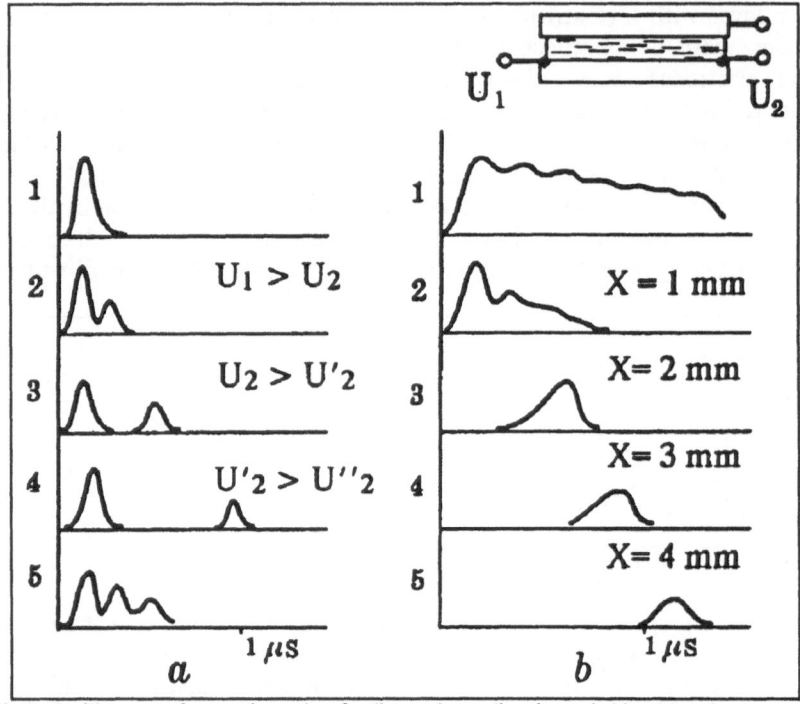

Fig.3. Oscillograms of generation pulses for discrete beam direction switching (a) and continuous shift of beam (b) in B-regime.

By the use of the B-regime a continuous change of the beam direction can be obtained. The LCM for that case is shown in the upper part of Fig.3b. It consisted of a Ge window (common electrode) and a high-resistant high-reflective layer on the second plate of the LC-cell. This construction permitted us to produce a permanent distribution of the electric field in the LCM and if $U_1 \neq U_2$ are located in B-region, a permanent distribution of intracavity losses exists in the cross-section of the resonator.

In this experiment, for the spatial temporal measurements of the laser dynamics, a diaphragm D_2 (Fig.2) have been used, placed near the output mirror outside the resonator.

Fig.3b shows oscillograms of laser generation without D_2 (1) and for different positions of D_2 (2-5). Just as in the previous case, the generation arises in the spatial region of minimum losses and spreads to the regions of higher losses. The obtained angle velocity of the beam direction was $1,5 \cdot 10^4$ rad/s. It must be noted that the dynamics of the mode development in this case is essentially unstable.

2.4. BEAM PROFILE CONTROLLING WITH AN APODISING LC-DIAPHRAGM (B-REGIME)

The foregoing method of shaping the intracavity losses distribution have been used with an LC-diaphragm (Fig.4, upper part). To obtain losses with a central symmetric distribution one electrode has been located in the center of the plate and the second one along the perimeter, in the shape of a ring. The "softness" of the diaphragm and its diameter could be controlled by the voltage values (U_1, U_2) and their difference.

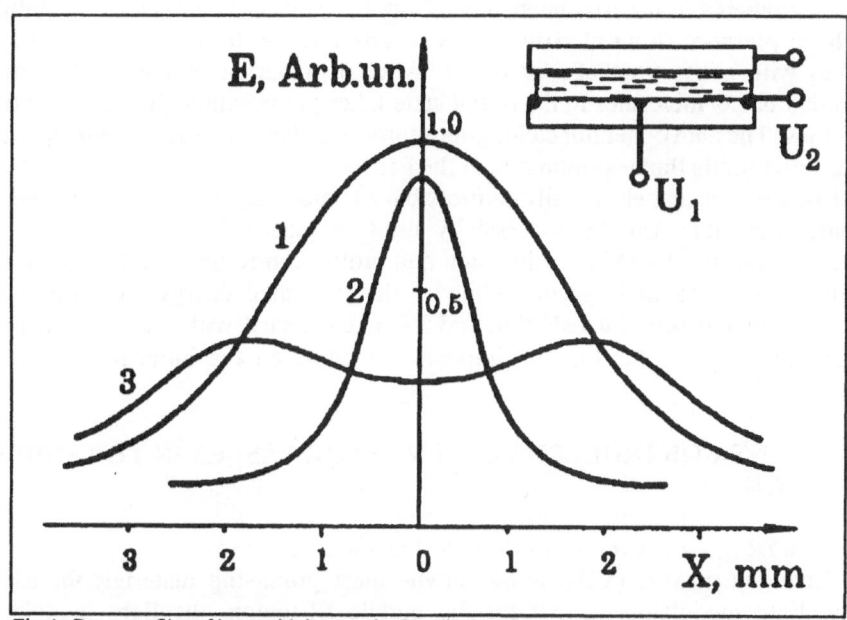

Fig.4. Beam profiles of laser with intracavity LC-diaphragm.

Investigations of the beam profile controlling were carried out in a TEA-CO_2 laser with a resonator which consisted of a plane output mirror and a LC-diaphragm as a second mirror. Fig.4 shows beam profiles in the near zone for $U_1 = U_2 > 60$ V (1) a "full-opened" diaphragm, $U_2 < U_1 < 60$ V (2) : "ordinary" diaphragm and $U_1 < U_2 < 60$ V (3) : a "negative" diaphragm. The energy of this laser was 20-30 mJ. The increase of energy caused an increase of the interaction between the radiation and the LC and was attended by the beam profile deformation.

2.5. INTRACAVITY INTERACTION OF RADIATION WITH LC

High intensity radiation in intracavity conditions causes changes of the optical parameters of a LC. As it was shown in our experiments [5] the processes which occur in a LC under the action of radiation can be divided to short-term and long-term processes and manifests themselves in disorientation of LC molecules. This in turn, causes the rise of light scattering in the LC.

Among the short-term processes which take place in a LC during the radiation pulse, thermal induced processes are dominant in a particular LC, because of its relatively high absorption coefficient. The heating of the LC by radiation results in a pressure rise and in the appearance of an acoustic wave which spreads from the center of the irradiated zone with sound velocity (1.5 km/s): the opto-acoustic effect. Disorientation of LC molecules produced by this effect causes the dynamic rise of intracavity losses. This influences the dynamics of the pulse generation and leads to pulse shortening, mode deformation and for high energy densities (W > 200 mJ/cm^2) to a suppression of the main mode (TEM$_{00}$). In the last case the generation transfers to higher order modes. For W < 50 mJ/cm^2 these effects are negligibly small.

Long-term effects in a LC are associated with two processes: 1. The large thermal gradients in the irradiated zone of the LC produces convective instabilities which, in presence of an electric field, converts into electrohydrodynamic (EHD) vortices with a life time of ~20 ms. 2. After degradation of the EHD vortices relaxation of LC molecules to the initial state takes place with a characteristic time of ~30 ms. The above-mentioned long-term processes became essential for W > 100 mJ/cm^2 and limits the repetition rate of the generated pulses.

It was shown that electrically controllable LC modulators based on cholesteric-nematic transitions can be successfully used in pulsed TEA-CO$_2$ lasers. By intracavity use of a LCM beam direction and profile controlling can be obtained in several regimes, as well as controlling of the generated energy and laser pulse delay. Interaction between radiation and LC is connected with thermal and opto-acoustic processes in the LC and imposes the limits on the intracavity radiation energy density.

3. CURRENT-CONTROLLED VO$_2$-SLM'S FOR LASERS IN THE MIDDLE IR-REGION

3.1. SPECTRAL CHARACTERISTICS OF VO$_2$-SLM'S.

Vanadium dioxide (VO$_2$) is one of the most promising materials for use in spatial light modulators (SLM) for the middle IR-region. the light modulation mechanism in VO$_2$ is based on a semiconductor-metal phase transition which occurs during the heating of VO$_2$ up to T = 50-75°C and is attended by a significant variation of its optical characteristics. To form a spatial profile of transparency or reflectance in VO$_2$-SLM different techniques can be used: heating by laser beam [6], by electron beam [7], or by passing current through film heaters with the appropiate configuration. Electron beam heating is the most-used method for VO$_2$-SLM controlling. However, in spite of the high spatial resolution and high switching speed [8], the use of this method is limited because an electron beam with a current density in the focus of 2-5 A/cm^2 and an electron energy of 15-20 KeV provokes a failure of the VO$_2$ film and the degradation of the optical characteristics of the SLM. In addition, the presence of an electron beam limits the amount and constitution of the optical coating of the SLM, which sets the spectral region of its application and the values of its reflection in switched-on and -off conditions.

In this paper the construction of high-speed VO$_2$-SLM's for the middle IR-region, controlled by film heaters, designed for the intracavity control of the laser generation is described. The optical characteristics of SLM's in steady state and dynamic regimes are presented.

The study of the peculiarities of the variation of the optical constants during a phase transition in VO$_2$ thin films has shown that it is possible to create controllable thin-film structures not only for $\lambda = 10.6$ μm [7], but also for the spectral region of 3-6 μm which is of significant practical interest for chemical and CO-lasers controlling. Samples of VO$_2$-SLM's were fabricated on SiO$_2$, Al$_2$O$_3$ and Ge substrates. Dielectric interference films were formed from standard pelleted materials (Y$_2$O$_3$, Al$_2$O$_3$, BaF$_2$, ZnS, ZnSe etc) by electron beam evaporation; metal films by direct current magnetron sputtering. The coating technique of VO$_2$ films was identical to the one described in [9], but had its own specific peculiarities connected with the parameters of the used equipment (Z-400, "LH"). Spectral characteristics of some achieved thin-film structures with VO$_2$ are shown in Fig.5a. Curves 1,1' and 3,3' are consistent with 8-layered structures, 2,2' - VO$_2$ film without additional optical coatings.

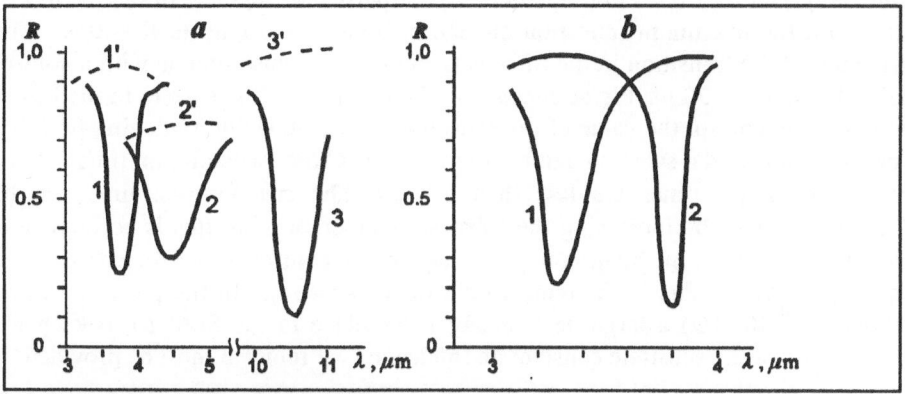

Fig.5. Spectral characteristics of VO$_2$-SLM's.
   a) - experimental structures with direct change of reflection, 1, 2, 3 - T = 25°C, 1', 2', 3' - T = 75°C;
   b) - calculated structures with inverse change of reflection :1 - T = 75°C, 2 - T = 25°C.

Calculations, performed for several promising constructions for the spectral region for 3-5 μm have shown, that in relation to the thickness of films it is possible to achieve SLM's not only of a common type in which the transition of VO$_2$ to the metallic phase is accompanied by a strong increase of reflection, but also of a radically new type with an inversion of the sign of the reflection. The optical characteristics of VO$_2$-films used in the calculations are in agreement with data from [10, 11, 12]. The results for one calculated model are shown in Fig.5b. As one can see, the 10-layered construction which can be achieved by common materials at $\lambda = 3.8$ μm, during the phase transition in VO$_2$ has a reflection change from R = 0.16 up to R = 0.96. In the same system at $\lambda = 3.3$ μm the inverse process takes place from R = 0.98 in the semiconductor phase down to R = 0.21 in the metallic phase. Such systems can be used for laser generation switching-off, for laser pulse shape control (in combination with a VO$_2$ modulator of the common type) and for the laser spectrum control. It should be noted that the comparatively narrow spectral regions of the structures with VO$_2$ produces some difficulties in their fabrication as it needs the precise measurement of the film thickness. However, experimental results for $\lambda = 10.6$ μm show that fabrication with intermediate measurement of the spectral characteristics makes it possible, in some cases, to eliminate errors which appear as a result of the discrepancy between the calculated and the real film optical parameters by adjusting of the film thickness.

3.2. CONSTRUCTION AND PARAMETERS OF MATRIX VO_2-SLM'S.

The results, presented below, are achieved for a VO_2-SLM designed for use in a CO_2-laser (λ = 9-11 μm). In Fig.6 the construction of a one-dimension VO_2-SLM (a) and the configuration of the heaters (b) are shown. The SLM consists of a dielectric substrate which is coated by a matrix of metal heaters (2 ,Cr or Al). Then follows a dielectric film (3), a VO_2-film (4) and a system of λ/4 films (5) which sets the values for minimum and maximum reflection coefficient of the SLM. The heaters used in the experiments were 0.3 mm wide and up to 45 mm long, with 0.1 mm clearance between them.

The control of the SLM is performed by passing a current through one or several heaters. Therewith, the heat released in them is transmitted into the VO_2-film and produces a reflection modulation of a particular segment of the SLM. λ = 10.64 μm the maximum reflection change is from R = 0.1 up to R = 0.98. The dynamics of the transition in the SLM was studied in a pulse regime with a control pulse duration τ = 20 μs. Fig.6c shows that the variation of the current in the heater permits us to control the value of the reflection of the SLM and switching-on time. For I = 75 mA and a substrate temperature T = 30°C the switching-on time is 8-10 μs. Switching-off times are less then 5-7 μs. The switching-on time can be decreased to 3 μs by increasing the substrate temperature but this is accompanied by a decrease of the modulation depth. The specific control power for τ = 10 μs is equal 6-10 W/cm^2 due to the temperature of the substrate. In the pulse-repetition regime (10^4-10^5 Hz) a large heat release takes place in the SLM so to keep the temperature of the substrate constant an intensive heat removal must be provided.

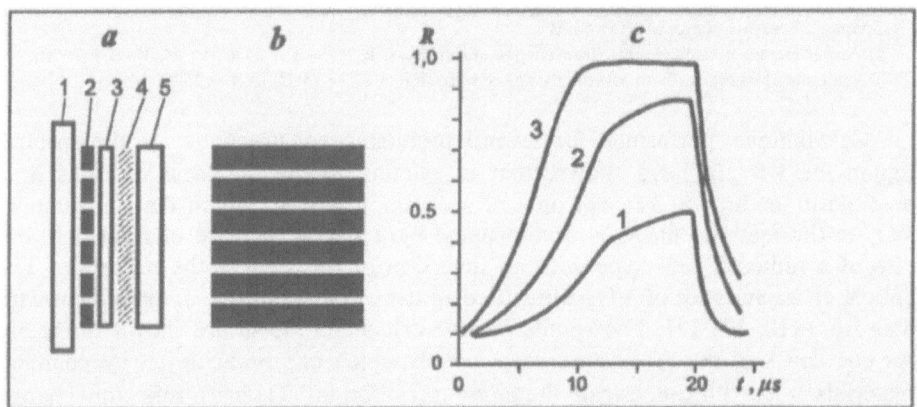

Fig.6. The construction of one dimension SLM (a); configuration of heaters (b); time dependence of SLM's reflection for pulse control (c), 1- I = 25 mA, 2 - 50, 3 - 75, τ = 20 μs.

The use of the two described one-dimensional SLM's in a laser with a conjugate resonator makes two-dimensional beam scanning with SLM's positioned as in [13] possible.

A two-dimensional SLM can be achieved by combining on one substrate two matrix heaters perpendicular to each other (Fig.7,b). This type of a SLM (Fig.7a) consists of a substrate (1), the first group of heaters (2), a dielectric film (3, Al_2O_3), the second group of heaters (4), a dielectric film (5, Al_2O_3), a VO_2-film (6) and λ/4-films (7). The values for the control currents of each group of electrodes are set so that the control power $P_{1,2}$ = 0.5 P_0, where P_0 is the power needed for the transition of VO_2 into the metallic phase. In this case, in the intersection of switched-on heaters the reflection of the SLM will be R = 0.98 (Fig.7c) and at exterior regions 0.5 or 0.1 (at switched-off heaters).

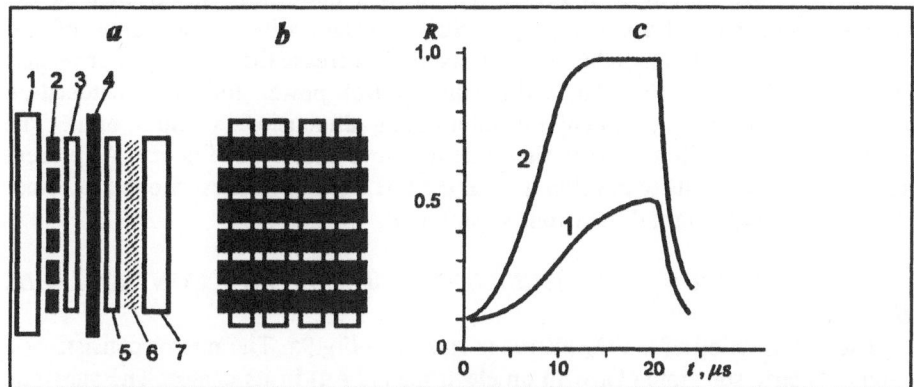

Fig.7. Construction of two-dimensinal SLM (a); configuration of hearters (b); Time dependence of SLM's reflection (c): 1 - in intersection of switched-on heaters, 2 - off intersection, $I_{1,2}$ = 20 mA, τ = 20 μs.

Thus the maximum modulation factor of a two-dimensional SLM for optimum relation of control currents $I_{1,2}$ is approximately 2 times less than the one-dimensional one. But this does not exclude the possibility for its application in compact lasers with a relatively low amplification.

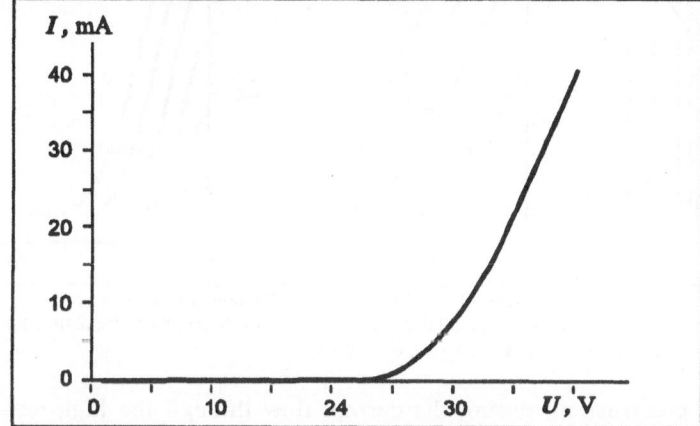

Fig.8. Voltage-dependent control current in film heater of two-dimensional SLM.

The modulation factor of a two-dimensional SLM can be maximised with the following construction. It differs from the above-described (Fig.7a,b) in that it contains two groups of low-resistant electrodes 2, 4 (Fig.7a) instead of high-resistant heaters and a high-resistant film heater 3 placed between them, instead of the dielectric film. In this construction the control current flows through film 3 in the intersection of the switched-on electrodes. This affords a maximum SLM reflection change: from R = 0.1 up to R = 0.98. However, if the film heater has a linear resistance, the current which flows through the switched-on intersection, will be attended by currents through other, "switched-off" intersections. The power released in the last ones is equal to 1/9 of the switching-on power so they do not decrease the contrast but for a SLM with a large amount of intersections this produces a considerable rise of the consumed power and an additional heating of the substrate. We developed a film heater with a non-linear dependence of the resistance via a control voltage (Fig.8). Such a heater which is a system of two films (dielectric and semiconductor) permits us to decrease the excess power release with a factor of 10^4-10^5. Therewith, the control power for one switched-on intersection comprises ~50 mW for intersection dimensions 0.3x0.3 mm^2. The switching time of such a SLM is the same as above-described (Fig.6c). At present work is underway by us to develop a similar SLM with 10^4 intersections (100 lines x 100 columns) and light dimensions 40x40 mm^2.

3.3. CONSTRUCTION AND PARAMETERS OF A VO_2 CONTROLLABLE MIRROR.

The construction of a VO_2-mirror is shown in Fig.9a. The mirror consists of a round dielectric substrate (1), with an electrode (7, Ag) in its center. The substrate is coated by a high-resistant film heater (3, Ge). Along the perimeter of this film is a round electrode mounted (2, Ag). Then follows a dielectric film (4, ZnS), a VO_2-film (5) and $\lambda/4$ films (6). The light diameter of the mirror is equal to 20 mm.

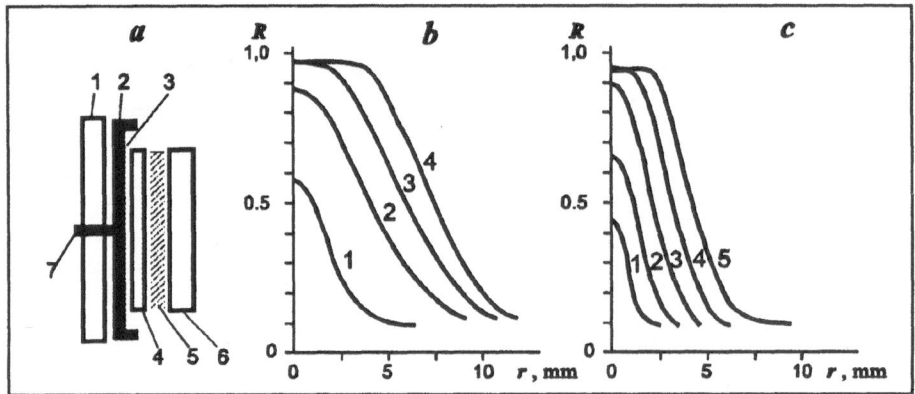

Fig.9. Construction of VO_2-mirror (a); spatial dependence of reflection in steady-state regime (b): 1 - U = 30V, 2 - 50, 3 - 60, 4 - 70; spatial dependence of the reflection in the pulse regime for $\tau = 20$ μs (c): 1 - 2 μs, 2 - 4, 3 - 6, 4 - 8, 5 - 10, U = 200 V.

In this construction, during the current flow through the high-resistant film heater, the spatial distribution of the heat release is defined by the distribution of the current density which increases from the perimeter to the center. The distribution of the specific power released in such construction, can be represented as:

$$P(r) = \frac{U^2 h}{\rho \ln^2(r_1/r_0)} \cdot \frac{1}{r^2}$$

where: U: control voltage, h: thickness of film heater, ρ: its specific resistance, r_1: radius of exterior electrode and r_0 - radius of central electrode.

If a semiconductor film heater is used (to get a more precise dependence P(r)) the temperature dependence of the specific resistance $\rho(T)$ must be taken into account, which for most semiconductors can be represented as:

$$\rho(T) \sim exp(\Delta W / 2kT)$$

The spatial distributions of the VO_2-mirror reflection for different direct currents (steady-state regime) are shown at Fig.9b. The shape and the width of the reflection profile depends on the following factors: the control voltage, the thermal conductivity of the substrate, the thermal dependence of the specific resistance of the film heater and on the steepness of the thermal dependence of the VO_2 reflection. With a proper choise of this factors the spatial distribution of the reflection can be approached by Gaussian and super-Gaussian shapes. In Fig.9c the dynamics of the VO_2-mirror reflection for pulse control is represented. In such regime the width of the profile is less than in a steady-state regime. The reason for this is the smaller influence of the thermal conductivity of the substrate for short control pulses.

The damage threshold of VO_2 matrix SLM's and mirrors is defined by the absorption coefficient of VO_2 and the maximum reflection coefficient. The described VO_2-devices were successfully operated under intracavity conditions with light power densities of 500-600 W/cm^2 for continuous operation and energy densities of 1.5 J/cm^2 in the pulsed regime for generated pulse durations of 1-1.5 µs.

It is evident from the presented results that the control of VO_2-SLM's by thin-film heaters instead of electron-beam control makes it possible to keep the advantages of the second method as high-speed switching and high spatial resolution, and to have a benefit for other parameters: reliability, simplicity of construction and compactness. Also, the removal of the restriction of the number and composition of optical coatings makes an expansion of the spectral region of SLM's application possible. The increase of the SLM's reflection in the metallic phase also increases the damage threshold.

4. REGENERATIVE AMPLIFICATION OF A LASER-INJECTOR SIGNAL WITH A GIVEN WAVE FRONT

4.1. THE GENERAL IDEA
The reflection regenerative ring amplifier scheme is presented in fig.10 [14]. For the first time this scheme has been reported in 1972 [17].

For classical injection locking ("conditionally stable regime") one can write:

$$G_0 \cdot R > 1 \qquad (1)$$

where　　　　G_0 - the small signal (power) gain per one pass;
　　　　　　　　R - the power reflection coefficient of the input-output mirror.

In this case the laser can generate power without a controlling signal. But with the presence of a controlling signal and with the right turning:

$$G \cdot R < 1 \tag{2}$$

where G is the saturated gain per one pass.

Let's quote the relation for the ratio of the output power (I_{out}) to the laser-injection power (I_i) [14]:

$$\frac{I_{out}}{I_i} = \frac{[g(1-q)-r]^2 + 4rg(1-q)\sin^2(\Theta/2)}{(1-rg)^2 + 4rg\sin^2(\Theta/2)} \tag{3}$$

where $g^2 = G$; $r^2 = R$; q is the ratio of absorbed power by input-output mirror to falling power; θ is the phase shift per one pass along the ring resonator (or the angle of the resonator turning);

$$\Theta = kL = 2\pi L/\lambda \tag{4}$$

where L is the amplifier perimeter. Taking (3) as a basis it can be made the energetic value of the regenerative regime of CW lasers which had the most intensive development. These are the CO_2-laser, the COIL and the HF-overtone laser with unsaturated specific active medium gains for CO_2 of $\sim 10^{-2}$ cm^{-1} [14], for I - O_2 (Δ) $\sim 5 \cdot 10^{-3}$ cm^{-1} [15] and for the HF-overtone laser $\leq 10^{-3}$ cm^{-1}[16] respectively.

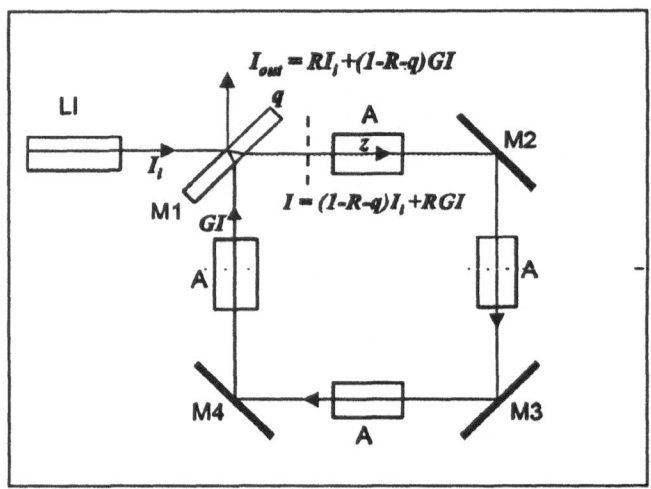

Fig.10. Scheme of a reflective regenerative ring amplifier.
LI - laser-injector; M_1 - input-output mirror; M_2, M_3, and M_4 - 100% mirror;
A - gain medium.

First of all the value shows that the conditionally stable regenerative regime allows to extract from the active medium almost so much energy (power) as in the free lasing regime.

The reflective regenerative amplifier theory [14] does not give a solution for the energetic scaling problem, besides the only one that is the resonance tuning of many series injector-regenerative amplifier systems. This question we'll consider below. Formula (3) shows the possibility to realise a ratio (I_{out}/ I_i) ~ 100 by the right combination of parameters for all our interesting cases.

In fig.11 a LOS is presented, which consists of an injector and ring reflective amplifier. But different from fig.10 here we make the following step: the amplifier ring resonator has been made self conjugative by an intracavity system of one by one (1:1) telescopes (T) which re-images plane (1) (in fig.11) on the itself after a pass along the resonator perimeter. The given phase distribution injector signal formation in plane (1) can be realised by different schemes. In fig.11 two from these schemes are shown. In the first case (a) a self conjugative resonator with intracavity STLM which is combined with a 100% mirror (M_5) is employed as laser-injector. The objective (O_3) together with the intracavity objective (O_2), realizes the reprojection in plane (2), where an angle controlled plane wave is formed in plane (1) on the input-output mirror (M_1) of the regenerative amplifier. In the second case (b) the given phase distortions are input for the laser-injector signal by the matrix of phase modulators (PM), that is reprojected in plane (1) by the objectives (O_2) and (O_3). Both methods may be combined in a single adaptive LOS that makes the formation and controlling of the laser beam and also the compensation of the optical system distortion after analysing of a wave front (the system of AWF is absent in fig.11).

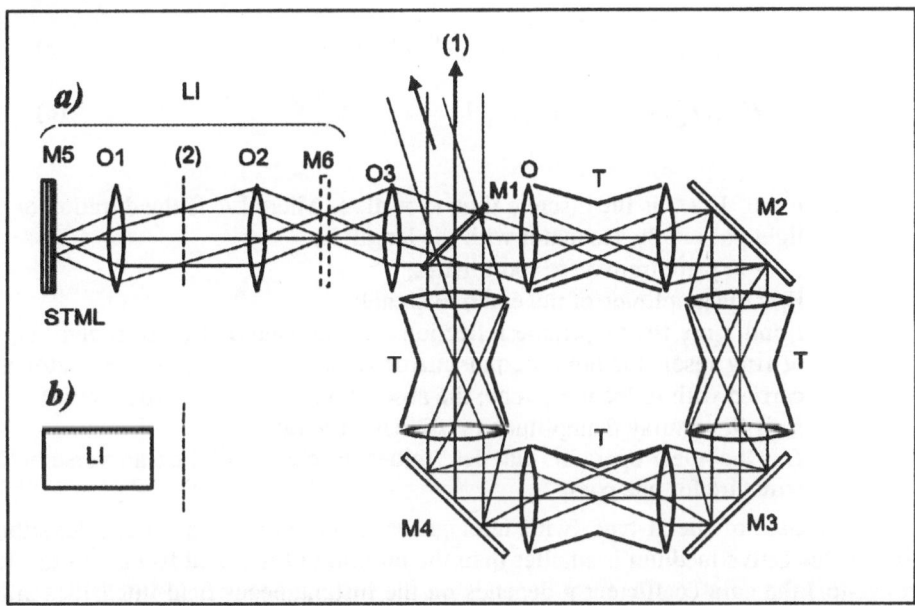

Fig.11. The scheme with a ring self-conjugative regenerative amplifier. T - intracavity 1x1-telescope; M_i- mirrors. a) laser-injector (LI) with intracavity beam scanning; M_5- 100%-mirror with STLM; M_6- output mirror; O_1 , O_2- intracavity objectives; O_3- re-imaging objective. b) laser-injector with outcavity matrix of phase modulators (PM).

One can say that the efficiency of the laser energy (power) extraction from the active medium of the ring reflective amplifier in fig.11 will be almost the same one as from the system in fig.10. This is because the telescope objectives are the laser windows simultaneously and the amplifiers active medium is placed between the mirrors and the telescope objectives and there is vacuum between the objectives (the intra telescopes). In this case the systems in both figures are almost identically with respect to the additional losses.

Below we will consider two important questions: (1) the influence of the frequency shift between the laser-injector and the ring regenerative amplifier on the efficiency of the energy extraction in the controlled regime and (2) the influence of the ring selfconjugative resonator optical deformations on the amplification process of the given wave front signal.

First of all let's determ the initial equations:

4.2. THE INITIAL EQUATIONS

In general the description of the field dynamics in a regenerative amplifier is a very difficult task because of the many laser modes (longitudinal and transverse) under the gain profile. The external signal can phase a part of these modes (with nearby frequencies) and excitate the pulse regimes with the participance of other modes. In this paper we shall use a simplified model of a regenerative amplifier [14] where two clear separated groups of longitudinal modes exists: phased modes (with a complex amplitude E) and unphased modes (with a complex amplitude F). The transverse structure of the field is not fixed. In this case the spatial-time field structure into the regenerative amplifier is described by the following equations:

$$E_{n+1}(\vec{r}_\perp) = \tau \cdot E_{i,n}(\vec{r}_\perp) + rg_n e^{i\Theta} \cdot \hat{L} \cdot E_n(\vec{r}_\perp) \tag{5}$$

$$F_{n+1}(\vec{r}_\perp) = r \cdot g_n \cdot \hat{L} F_n(\vec{r}_\perp) \tag{6}$$

Here   n = 0, 1, 2... is the discrete time t_n = nL/c, where L/c is the duration of a light pass in the resonator with the length L;
$r_\perp$ = x, y are transverse coordinates;
E_i is the amplitude of the external signal;
r and τ are the amplitude reflection and transmission (correspondingly) of ring resonator input-output mirror (other mirrors of the resonator is carried with reflection of one; the absorbtion is not took into account);
g_n is the saturated amplitude coefficient of gain;
L is the linear operator, that determines the change of the transverse field structure for one pass.

For a laser of the A-type (with homogeneous line broadening; the relaxation time of the active medium is smaller than the duration of the field formation in the resonator) the gain coefficient g depends on the instantaneous field intensities and in general on the transverse coordinates. Usually this dependence is prompted by the transverse inhomogeneity of the pumping and losses and also by the inhomogeneity of gain saturation. In this case the gain coefficient is determined by the sum of the radiation intensities:

$$I_\Sigma = |E|^2 + |F|^2 \qquad \text{that is} \tag{7}$$

$$g_n = g(I_{\Sigma,n}) \tag{8}$$

This dependence is determined by the follow relation (here z is the longitudinal co-ordinate):

$$\frac{I_\Sigma(z=0)}{I_s} = \frac{\alpha_0 L - \ln(g^2)}{g^2 - 1} \tag{9}$$

that is the integral of follow equation for plane waves:

$$\frac{dI}{dz} = \frac{\alpha_0 I}{1 + (I/I_s)}, \tag{10}$$

where α_0 is the unsaturated specific gain coefficient (in cm^{-1}), and I_s is the saturation intensity.

4.3. ABOUT THE FREQUENCY MATCH OF THE LASER-INJECTOR AND THE RING REGENERATIVE AMPLIFIER

In this part we use the plane wave approximation (the transverse field structure is not taken into account).

1. In the case of a stationary regime the time dependence is absent. Because $I = |E|^2$ and $I^{(0)} = |F|^2$, that is $(1 - r \cdot g \cdot e^{ikL}) \cdot E = \tau \cdot E_i$ or:

$$\left|1 - rg \cdot e^{ikL}\right|^2 I = |\tau|^2 \cdot I_i \tag{11}$$

$$(1 - |rg|^2) I^{(0)} = 0 \tag{12}$$

Now let's consider two distinctive cases:

1) $I^{(0)} = 0$; the lasing of the ring laser has been supressed completely which is named "the net regenerative regime". From (12) one can see that for the stability of the locking regime it is needed that

$$|rg|^2 < 1 \tag{13}$$

Here equation (11) determines the $I(I_i)$ dependence for various shifts of the injector and the regenerative amplifier and this dependence is nonlinear.

2) $I^{(0)} \neq 0$, that is the regime when the ring lasing and the injector controlled lasing are co-existing ("the regime of co-existence"). Conform with (12) in this case $|rg|^2 = 1$, that is $g = (1/r)$ and with (9):

$$\frac{I_\Sigma}{I_s} = \frac{\alpha_0 L - \ln(1/r^2)}{(1/r^2) - 1} \tag{14}$$

Here in the denominator the losses take place, and in the numenator the overstepping over threshold. The condition rg=1 in equation (11) leads to the following relation:

$$4 \cdot \sin^2(kL/2) \cdot I = |\tau|^2 \cdot I_i \qquad (15)$$

And thus, in the regime of co-existence the dependence $I(I_i)$ is linear.
Let's determine the value of I_i for which the regime of co-existence is possible:

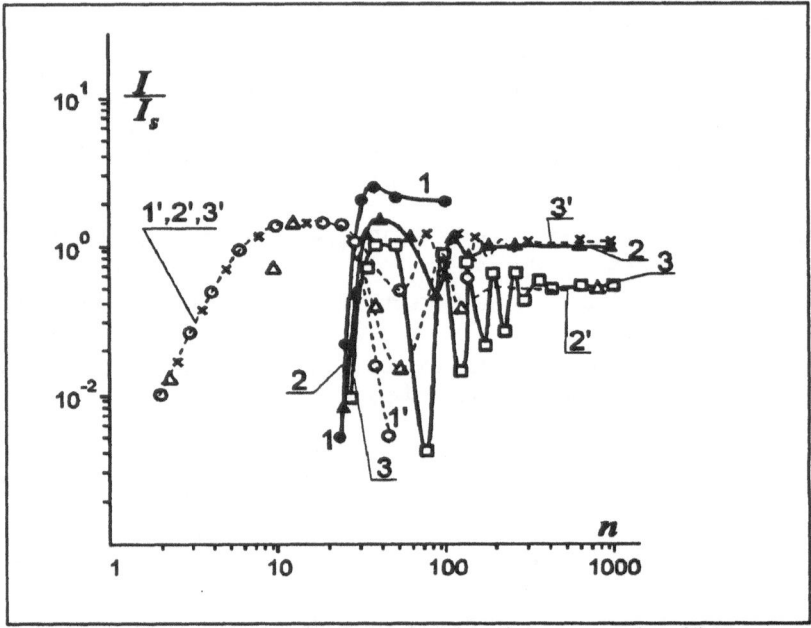

Fig.12. Intra ring amplifier intensity (in units of saturation's intensity) versus number of light wave passes along the ring resonator. The injector signal is given in the time of 25-th pass of own lasing of the ring amplifier. 1, 2, 3 - the intensity locked by injector ($|E|^2$); 1', 2', 3' - the intensity non-locked by injector ($|F|^2$); $G_0 \cdot L = 2$; $r^2 = \tau^2 = 0.5$; $I_{i,cr} = 6.4 \cdot 10^{-2} \cdot I_s$; $I_{\Sigma} = 1.3 \cdot I_s$; 1, 1' - $I_i = 0.1 \cdot I_s > I_{i,cr}$; 2, 2' - $I_i = 4 \cdot 10^{-2} \cdot I_s < I_{i,cr}$; 3, 3' - $I_i = 2 \cdot 10^{-2} \cdot I_s \ll I_{i,cr}$

Because $I < I_{\Sigma}$, for realising this regime it is needed that:

$$I < I_{i,cr} = \frac{4 \sin(kL/2)}{|\tau|^2} \frac{L - \ln(1/r^2)}{(1/r^2) - 1} I_s \qquad (16)$$

From (14) it is clear that $I_{\Sigma} = I^{(0)} + I$ is constant during the change of the external signal I_i from 0 to $I_{i,cr}$ It means that in this range the power redistributes itself easily from fully unphased radiation (I=0) to fully phased radiation by the injector ($I^{(0)} = 0$). In the case of precise tuning $I_{i,cr} = 0$ the stationary regime of locking is possible only. With appearance of a frequency shift a critical value I_i (that is $I_{i,cr}$) also appears, which is determined by (16). It is necessary to overcome the value $I_{i,cr}$ for realising of the "net regenerative regime" and this value increases with the increase of the frequency shift.

2. The case of the non-stationary regime (equations (5), (6) and (8)) is presented in fig.12, 13, and 14, where the dependencies of the intracavity radiation's intensity (the own lasing and injector controlled lasing) as a function of passes number along the ring resonator of the amplifier are presented. Fig.12 corresponds to the case where the own lasing has been achieved itself and then the injector signal is given. Fig.13 shows the picture of the field establishment in the ring resonator when the injector signal is present at once before the development of the own lasing. Both regimes correspond to a shift between the injector and the ring resonator of $kL/2=0.157$ rad (which is $\Delta\nu = 5$ MHz for $\lambda = 10.6$ μm and $L = 5$ m).

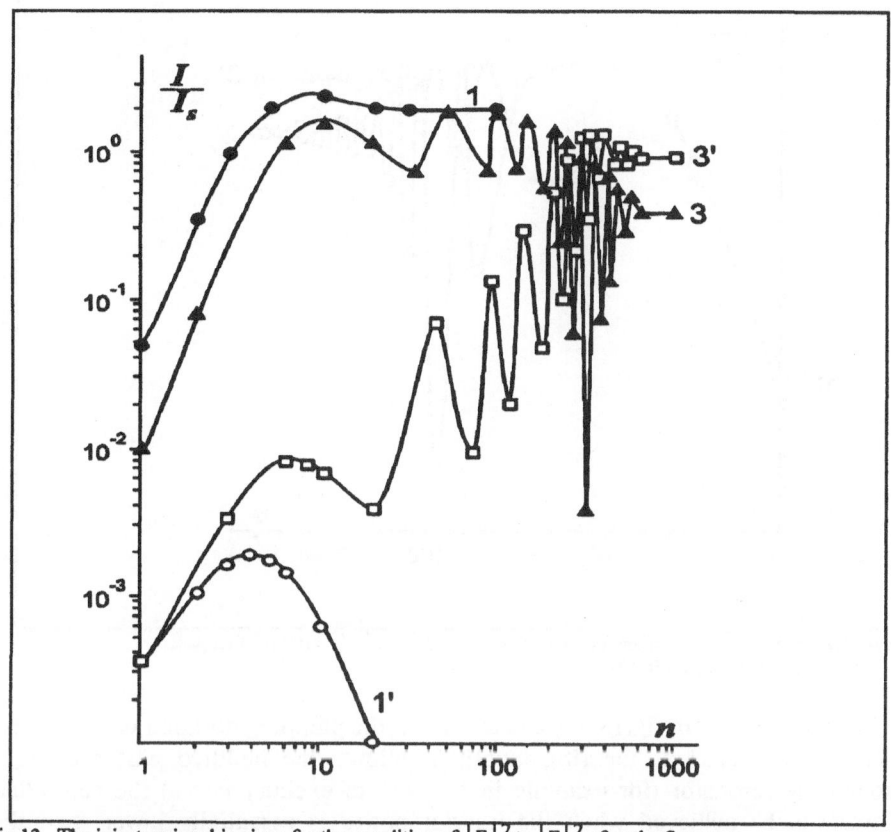

Fig.13. The injector signal is given for the condition of $|E|^2 > |F|^2$ after the first pass (other signs as in Fig.12).

First of all let's notice that independent from the start conditions the system comes to predictable stationary states. In this case if the injector power is higher than the critical value, the own ring amplifier lasing is completely suppressed and the injector controlled field build up takes place practically without oscillations (curves 1 and 1' in fig.12 and 13).

A totally different picture is found when the injector power is close to the critical one. In this case the process in the stationary regime, when both fields are co-existing, has an oscillating nature and depends on the start conditions essentially. In fig.13 one can see the case (curves 3 and 3') when after the first pass in the ring resonator the controlled power (by the injector) is one order of magnitude higher than the non-controlled power. In this case although the injector power is three times lower than the critical power, during one hundred passes the controlled power is one order higher than the non-controlled power and during only five passes the controlled power achieves the level of the saturation power (curve 3 in fig.13). Let's mark (see fig.14) that if after the first pass the injector controlled power is less than the non-controlled power the achievement of the stationary regime takes place under the condition that one the of fields is absent.

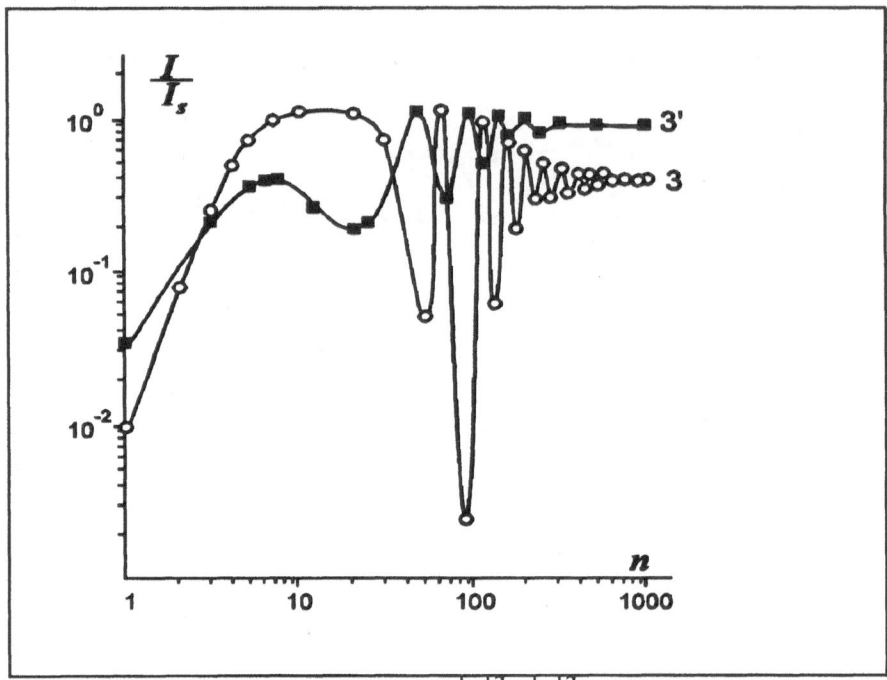

Fig.14. The injector signal is given for the condition of $|E|^2 < |F|^2$ after the first pass (other signs as in Fig.12).

From the results in fig.13 one can see that if the pumping duration of the A-type laser which is used as the ring amplifier, allows one hundred passes of light through the resonator (for example in the case of excimer lasers) the controlled regime can be achieved when the injector power is essentially lower than the critical value. This result coincides with the one of [19, 20], where the regenerative regime was investigated for an excimer laser with an unstable telescope resonator and explains the [21] result where the injector locking regime of the excimer ring amplifier has been realised for a large frequency shift. As stated above one can think to make a series injector regenerative LOS for excimer lasers: the value of the saturation energy $W_s \sim (h\nu/\sigma_t) \sim (10^{-18}/2\cdot10^{-16}) \sim 5\cdot10^{-3}$ J/cm^2 (where σ_t is the cross-section of the lasing transition). Suppose the frequency shift is maximum which means that $(kL/2)=(\pi/2)$ (injector frequencies are in the centre between the longitudinal modes of the regenerative amplifier) in correlation with (16) $I_{i,cr} \sim 2\cdot I_s$, and the critical pulse energy $W_{i,cr} \sim 10^{-2}$ J/cm^2. Because it is clear that an excimer laser system in which in spite of the frequency shift the previous cascade locks the frequency of the following one is real.

4.4. ABOUT THE SELF-CONJUGATIVE RING RESONATOR OPTICAL DEFORMATION INFLUENCE ON THE PROCESS OF THE GIVEN WAVE FRONT REGENERATIVE AMPLIFICATION

The following approximations will be made:

1) The stationary regime: the injector power is higher than the critical one and the frequency shift is very small;

2) Only the phase deformation is considered;

3) The active medium gain is homogeneous along the cross-section.

For these conditions the intensity of the non-phased modes $|F_n|^2 = 0$ and instead of equations (5), (6) one can employ the stationary equation for the phased radiation only.

The equation for the amplitude of the wave which is propagating in the ring amplifier is:

$$E = \tau \cdot E_i + rg \cdot \hat{L}E \qquad (17)$$

In this case the output wave (from the amplifier) has the following amplitude distribution:

$$E_{out} = \tau \cdot g \cdot \hat{L}E - rE_i \qquad (18)$$

Here the "minus" sign is determined by the phase shift in the time of the injector field reflection from the input-output mirror. For analysis of the wave change during its propagation in the amplifier it is easy to use the solutions of the ring resonator modes [22]:

$$\varphi_k = exp(i\chi_k) \cdot L\varphi_k \qquad (19)$$

where φ_k is the solution of equation (19) (resonator mode) and $exp(i\chi_k)$ is the eigen value.

If E can be determined as a solution of equation (19) solution:

$$E_i = a \cdot \varphi_k \qquad (20)$$

then the solution of equation (17) is the following:

$$E = b \cdot \varphi_k \qquad (21)$$

where $b = \dfrac{a\tau}{1 - rg \cdot exp(-i\chi_i)}$ \qquad (22)

In this case E_{out} is determined by the relation:

$$E_{out} = \tau \cdot g \cdot \hat{L}E - r \cdot E_i = \frac{g \cdot exp(-i \cdot \chi_k) - r}{1 - g \cdot r \cdot exp(-i \cdot \chi_k)} E_i \qquad (23)$$

or for $1 - rg \ll 1$ and $|\chi_k| \ll 1$

$$E_{out} \sim \frac{(1/r) - r}{1 - g \cdot r \cdot exp(-i \cdot \chi_k)} E_i \qquad (24)$$

Let's consider the amplification of the wave which does not confirm with any mode of the ring resonator. That means:

$$E_i = \sum a_k \cdot \varphi_k \qquad (25)$$

In this case the solution of equation (17) is the following:

$$\varphi = \sum b_k \cdot \varphi_k \qquad (26)$$

where $b_k = \dfrac{a_k \tau}{1 - rg \cdot exp(-i\chi_k)}$ \qquad (27)

In (27) it is seen that considerable deformations of the gain signal which are determined by non-equation of the eigen values $exp(i\chi_k)$ for different modes, prevent the achievement of an extremelyl high amplification of the regenerative amplifier. Because it is clear that a self-conjugative resonator, which has the highest degree of mode degeneracy [17], is most favourable for realising a high amplification with conservation of the phase-amplitude distribution in the front of the gain signal wave.

Let's consider more in detail the process of the gain wave's deformation. In this case if all modes would have eigen values equal the one of the highest quality mode $(exp(i\chi_0))$, phasing of the wave would take place in the amplifier and the coefficients in the series (26) are determined by the relation:

$$b_{0k} = \frac{a_k \tau}{1 - rg \cdot exp(-i\chi_0)} \qquad (28)$$

Let's determine the wave deformations by relative additions for these coefficients:

$$\frac{b_{0k} - b_k}{b_{0k}} = \frac{rg \cdot [exp(-i\chi_0) - exp(-i\chi_k)]}{1 - rg \cdot exp(-i\chi_k)} = \frac{i\chi_k - i\chi_0}{(1 - rg) + i\chi_k} \qquad (29)$$

In a self-conjugative resonator, as well in a concave one [13], the phase deformation which appears because of the low machining precision and the adjustment of the elements, leads to remission of the mode degeneracy:

$$<(\chi_0 - \chi_k)^2> = 4\pi^2 <\left(\frac{\Delta L}{\lambda}\right)^2> \qquad (30)$$

where ΔL is a local deviation of the resonator length from the average value (the wave aberration). This leads to the following value for $<\left|\chi_k\right|^2><(1-rg)^2$:

$$< \left|\frac{(b_{0k}-b_k)}{b_{0k}}\right|^2 >= \frac{<\chi_k^2>}{(1-rg)^2}\left(1-\frac{<\chi_k^2>}{|rg|^2}\right) \tag{31}$$

It follows that the maximum power gain for which the front deformation of the gain wave is still not large, is:

$$\frac{I_{out}}{I_i}=\left(\frac{E_{out}}{E_i}\right)^2 < \frac{1}{4\pi^2<(\Delta L/\lambda)^2>} \tag{32}$$

From the strength of (32) it is seen that if $(I_{out}/I_i) = 100$, the optical quality of the ring self-conjugative resonator must be $\sim (\lambda/60)$ which is a rather difficult condition. Because it is interesting to understand the possibility of the optical deformation compensation by the leading into input signal the conjugative phase deformation. Let's present the regenerative amplifier output signal as a series with an eigen function of the ring self-conjugative resonator:

$$E_{out}=\sum c_k\cdot\varphi_k \tag{33}$$

After some calculations which is basically similar as above, the following relation is obtained for the series coefficients:

$$c_k \sim \frac{(1/r)-r}{(1-gr)+i\chi_k}a_k \tag{34}$$

It is clear that from of the given output field E_{out} with the coefficients c_n, the input signal (which is the injector signal) has to have a series coefficient (in 25), which is determined by the following relation:

$$a_k = c_k\frac{(1-gr)+i\chi_k}{(1/r)-r}=c_k\frac{(1-gr)}{(1/r)-r}(1+i\frac{\chi_k}{1-gr})$$

$$=\left[c_k\frac{(1-gr)}{(1/r)-r}\right]exp(+i\frac{\chi_k}{1-gr}) \tag{35}$$

According to equations (35) and (34) one can see that for the necessary wave front (WF) in order to obtain the conditions for the regenerative ring amplifier optical deformation, the injector wave front must be equal to the summation of two fronts: A) WF-1 which we have to obtain (in the series coefficients of equation (25) the values in the square brackets of equation (35) will determine WF-1) and B) WF-2 which is conjugative to the wave front determined by the amplifier deformation (the sign (+) in the exponent of the relation (32)). Let's mark that the correlation of χ_k with the ring amplifier optical deformation (30) leads to the condition that the conjugative deformation of WF-2 in the input signal has to be more than the own amplifier deformation in the factor $(1-gr)^{-1}$.

5. CONCLUSIONS

1. Liquid crystal spatial-time light modulators for intracavity CO_2-laser operation have been elaborated and can be successfully used in pulsed TEA-CO_2 lasers.

2. Compact current-controlled VO_2-spatial-time light modulators have been elaborated for the middle IR-region; the switch time is ~ 5 µs; R_{max} is $\sim 98\%$.

3. The principal possibility of a given wave front laser-injector signal regenerative amplification has been shown.

6. REFERENCES

1. Berenberg V.A., Danilov V.V., Reznikov Yu.A. (1993), "Liquid crystals in laser optics", *J. Opt. Technol.* (Sov.),**60**, 487-512.
2. Richard I., Maurin J., Nuigenard J.B. (1986), "Phase conjugation with gain at CO_2 laser line 10,6 µm from thermally induced gratings in nematic liquid crystals" *Optics Communications*, **57**, 365-370.
3. Danilov V.V., Saveliev D.A. (1986), "The same peculiarities of CO_2 laser modulation by liquid crystal modulator" *J.Techn.Phys.* (Sov.), **31**, 741-744.
4. Danilov V.V., Danilov O.B., Sidorov A.I. (1991),"Peculiarities of TEA-CO_2 scan laser with LC SLM", *J.Techn.Phys.* (Sov.), **36**, 1402-1407.
5. Danilov V.V., Sidorov A.I. (1993), "The dynamic of cholesteric liquid crystal under the influence of IR-radiation pulse", *Molecular Materials*, **2**, 97-101.
6. Roach W. (1971), "Light-induced gratings on vanadium dioxide films", *Appl.Phys.Lett.*, **19**, 453-458.
7. Chivian J.S., Scott M.W., Case W.E.,Krasutsky N.J. (1985), "An improved scan laser with a VO_2 programmable mirror", *IEEE J. of Quant.Electr.*, **QE-21**, 387-395.
8. Welch A., Burzlaff B., Cunningham W. (1981), "Electronically scanned CO_2 laser radar techniques", *SPIE*, **300**, 153-161.
9. Duchene J.,Terrailon M., Polly M. (1972), "Preparation of VO_2 thin films and their optical characteristics", *Thin Solid Films*, **2**, 231-237.
10. Parker A., Verleur H., Guggenheim H. (1966), "Determination of optical constants of VO_2 thin films", *Phys.Rev.Lett.*, **17**, 1286-1293.
11. Chain E. (1991), "Optical properties of vanadium dioxide and vanadium pentoxide thin films", *Appl.Opt.*, **30**, 2782-2790.
12. Case F. (1991), "Improved VO_2 films for infrared switching", *Appl.Opt.*, **30**, 4119-4126.
13. Daks M., Powell C. (1968), "A fast digitalized scan laser", *IEEE J. of Quant. Electr.*, **QE-4**, 648-654.
14. Smith K., Tomson R.M. (1978), "Computer modelling of gas lasers", Plenum Press, N.Y..
15. Kalinovsky V.V., Kirillov G.A., Nicolaev V.D. (1992), Proceeding SPIE, 1980, 138-147.
16. Averianov V.P., Danilov O.B., Zhevlakov A.P., Ivanovskaya M.I., Leschenko D.O., Tulsky S.A., Beljaev A.A., Karelsky V.G., Lyashedko P.G., Maksimov Yu.P. (1994), "CW laser with overtone of HF-molecule in the laser with out cavity resonator", *Optika i spectrosc.* (Sov.), **78**, 1-3.
17. Spenser M.B., Lamb W.E. (1972), "Laser with a transmitting window", *Phys.Rev. A*, **5(2)**, 884-892.
18. Danilov V.V., Danilov O.B., Sidorov A.I. (1991), "About the possibility of a beam scanning speed increasing in TEA-CO_2 laser with LC SLM", *Pisma v JTF* (Sov.), **17**, 58-61.
19. Bigio I., Slatkine M. (1983), "Injection-loking unstable resonator eximer laser", *IEEE J. of Quant.Electr.*, **QE-19**, 1426-1436.
20. Baranov V.Yu., Borisov V.M., Stepanov Yu.Yu. (1988), "Electrical discharge eximer lasers with galogenides of noble gases" (Sov.), Energoatom, Moscow.
21. Paiola T.I., Mc Dermod J.S., Laudenslager J.B. (1982), "A wavelengh scannable XeCl oscillating amplifier laser system", *Appl.Phys.Lett.*, **40**, 1-3.
22. Kalinina A.A., Lyubimov V.V., Nosova L.V. (1991), "Scanning laser with ring conjugate resonator", *Optica i spectroscop.* (Sov.), **70**, 182-187.
23. Belousova J.M., Danilov O.B., Lyubimov V.V. (1967), "The question of concave resonator CW laser spectrum", *Journ. of Techn. Physics* (Sov.), **3**, 1134-1139.

HIGH-POWER INDUSTRIAL CO$_2$ LASERS EXCITED BY A NONSELF-SUSTAINED GLOW DISCHARGE

N.A.GENERALOV, M.I.GORBULENKO, N.G.SOLOV'YOV,
M.YU.YAKIMOV, V.P.ZIMAKOV
Institute for Problems in Mechanics,
Russian Academy of Sciences
101, Prospect Vernadskogo, Moscow, Russia, 117526

1. Introduction

Nonself-sustained glow discharges assisted by periodic-pulsed discharges (PSD - pulser sustained discharge) are used to produce the active medium in fast-transverse-flow electric-discharge industrial CO$_2$ lasers (power range 1-40 kW CW or time average). This method was proposed by Reilly [1] and Hill [2], and important practical results were reported by Shashkov et al.[3], Seguin et al.[4-6] and Generalov et al.[7-9].

Two different schemes have been used to create pulse assisted DC glow discharges (see figure 1).

The characteristic feature of the discharge schemes used in [1-6] is that both the assisting pulsed avalanching voltage and the steady (or also pulsed) main discharge voltage are applied to the same electrode as shown at the top picture of figure 1. The spatial uniformity of the discharge is provided by multielement electrodes and by additional photoionization.

In the other scheme developed by Generalov et al.[7-9] (bottom picture in figure 1) an avalanching periodic-pulsed voltage is applied to an additional pair of large area plane electrodes isolated from the plasma by dielectric sheets. This type of assisting discharge is called electrodeless (capacitively coupled) periodic-pulsed discharge (EPD). EPD is characterized by a high peak power required to produce an uniform ionization in the discharge gap between the dielectric sheets and a relatively low time-average power. Vibrational excitation of the uniformly ionized medium is produced by a steady-state nonself-sustained discharge, called electrodeless pulse sustained discharge (EPSD) or main discharge. The DC voltage of the EPSD is applied to tubular metal electrodes (cathode and anode) placed at the edges of the discharge chamber. A fast gas flow is directed from cathode to the anode perpendicular to the axis of the optical resonator. The other two walls of the chamber have holes to let the laser radiation pass to cavity mirrors placed outside. The optical axis of the resonator is folded in a Z-shape so that the laser beam completely fills the active volume. This scheme was studied in the Institute for Problems in Mechanics, Moscow. On the base

W. J. Witteman and V. N. Ochkin (eds.), Gas Lasers - Recent Developments and Future Prospects, 323–341.
© 1996 *Kluwer Academic Publishers.*

324

of the study a series of industrial CO_2 lasers (named : "Lantan") with output powers from 1 to 5 kW were designed [10]. An experimental model of a 10 kW CW CO_2 laser ("Tsyklon") [7-8] was also developed. These lasers demonstrate a high efficiency, enhanced beam quality [11], wide possibilities for power control, low gas consumption and a high reliability.

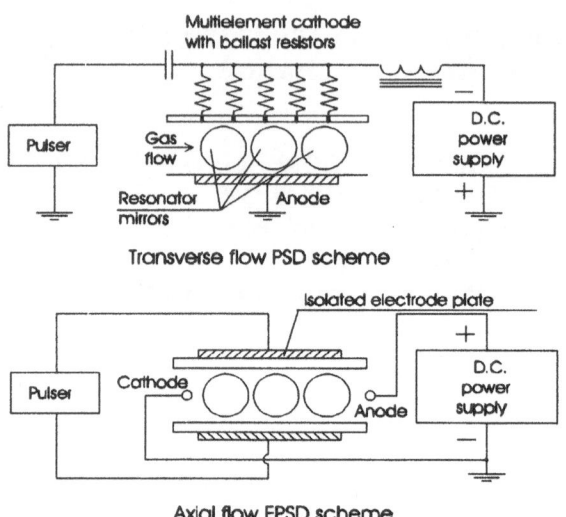

Figure 1. Two schemes of a nonself-sustained glow discharge assisted by a periodic-pulsed discharge .

The high performance characteristics of these lasers are achieved due to the advantages of the excitation scheme used : discharge optical homogeneity, simple electrode system, low discharge chemical activity and the original power control scheme.

This paper is dedicated to the detailed description of the discharge physics and the applications of the EPSD method in high power industrial CO_2 lasers. The achieved results and prospects for further applications of this method are considered.

1. Electrodeless capacitive periodic-pulsed discharge

The electrodeless capacitively coupled periodic-pulsed discharge (EPD) is classified among the relatively well studied volume pulsed discharges applied particularly to pulsed gas lasers. The characteristic feature of EPD is that the electrodes to which the potential pulse is applied are isolated from the discharge gap by dielectric sheets (see figure 2). This unusual kind of pulsed discharges has been investigated theoretically and experimentally together with CO_2 laser applications [7-9,18,19].

In order to understand better the processes in the discharge, let us consider a simple model: The voltage pulse applied is assumed to be a step function with a peak voltage value U_0 and relatively short rise time. The initial free electron density N_{e0} is taken to be uniform over the discharge volume. In a periodic-pulsed discharge a considerable electron density remains after the preceding pulse. The potential distribution in the discharge gap is taken to be uniform; sheaths of spatial charge near the dielectric sheets are assumed to be so thin that the potential drop is negligible. With these assumptions the equivalent circuit for EPD may be defined as shown in figure 2. R_g is the plasma column electric resistance depending on the free electron density N_e, C_g is the discharge gap capacity and C_d is the dielectric sheet capacity.

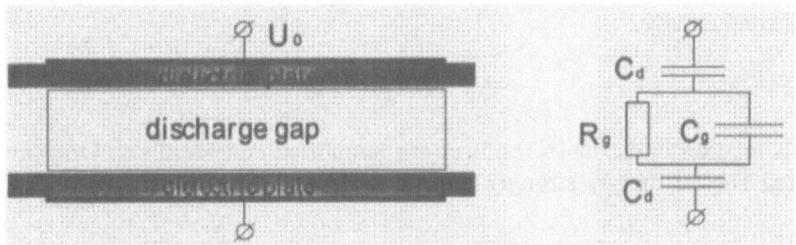

Figure 2. Electrode arrangement and equivalent scheme of the EPD.

This model is very similar to that analyzed in [7] and called the electrotechnical model. Within this model the evolution of the discharge electric field, current and electron densities are described by a system of non-linear ordinary differential equations which may be numerically integrated to obtain the plasma electric field and the current and electron density waveforms as shown in figure 3. It is useful to understand the general relations and make evaluations on the base of simple physical considerations.

The voltage value U_0 is assumed to be high enough to start an avalanche ionization in the discharge gap. One can define the characteristic ionization time T_i as

$$T_i = (v_i - v_a)^{-1} \qquad (1.1)$$

where v_i is the ionization rate constant (sec^{-1}) and v_a the dissociation attachment rate constant. Equation (1.1) describes that free electrons are produced in electron shock processes and disappear by dissociative attachment processes. On the base of the calculated values of v_i and v_a taken from [20,21] we have obtained values of T_i at different E/p (E is the electric field strength and p is gas pressure). A gas mixture with ratio $CO_2/N_2/He : 1/6/12$ at pressure $p = 30$ Torr was taken as example. The results are presented in the Table 1.

If the rise time of the applied voltage pulse is short with respect to characteristic ionisation time correspondent to the voltage peak value U_{0P} the electron density begins to rise with a characteristic time $T_i(U_{0P})$.

When the electron density achieve a considerable value the electric current begins to flow in the plasma column. Positive and negative charges are then separated at the surfaces of the dielectric sheets and screen the electric field in the plasma column.

The ionization process in the plasma is practically terminated as the potential drop in the plasma column falls below a certain value due to the plasma polarization process

326

described in the previous paragraph. The characteristic time of polarization within the electrotechnical model is $R_g C_d/2$ where R_g is the electric resistance of the plasma column and C_d is the electric capacity of the dielectric sheet. The condition for the end of the ionization then may be defined as :

$$T_i \approx R_g C \qquad (1.2)$$

The free electron density N_e may be roughly estimated as proportional to the plasma conductivity :

$$N_e \sim 1/R_g \sim C/T_i \qquad (1.3)$$

This relation is the main result of our simplified consideration. One can see from (1.3) and Table 1 that N_e strongly depends on $U_0 P$.

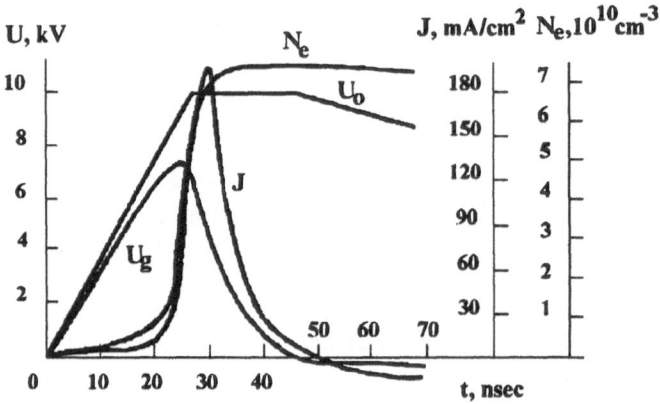

Figure 3. Applied voltage U_0, voltage on discharge gap U_g, as well as current density J and electron density N_e waveforms of EPD obtained from electrotechnical model for discharge chamber similar to that of "Lantan" type laser. Gas mixture $CO_2/N_2/He$ 1/6/12 p = 30 Torr [19].

TABLE 1. Character ionization time at different E/p in a $CO_2/N_2/He$ mixture with ratio 1/6/12.

E/p [V/cm/Torr]	T_i [nsec]
73.8	1
29.3	10
16.8	100
11.7	1000
8.7	steady state

In practice the condition that the voltage front should be short in relation to ionization time generally does not hold, because the peak voltage U_{OP} is usually higher then required (figure 3). In this case the electric field in the plasma starts to decrease before the applied voltage achieves the maximum value. The overvoltage obtained depends on the voltage rate of rise. Hence T_i, the current pulse width and the current amplitude in practice are determined by the voltage rate of rise. To obtain more an effective ionization one should apply a more sharp voltage pulse. There output power of the impulse generator must be sufficiently high to ensure the required current amplitude. Restrictions associated with the output characteristics of the impulse generator lead to an additional decrease of the obtained free electron density.

Pulse repetition rates required to create a quasi-stationary electron density depend on free electrons lifetime. For CW CO_2 lasers free electrons disappear in electron-ion recombination and attachment (negative ion formation) processes. The correspondent lifetimes are within 10-100 μsec and repetition rates are 10-100 kHz.

In figure 4 one can see the bright and dark layers near the dielectric sheets. These layers are similar to those of a common glow discharge and are specific for the cathode. The top and bottom layers correspond respectively to positive and negative current half waves obtained from the trapezoid shape of the applied voltage pulse.

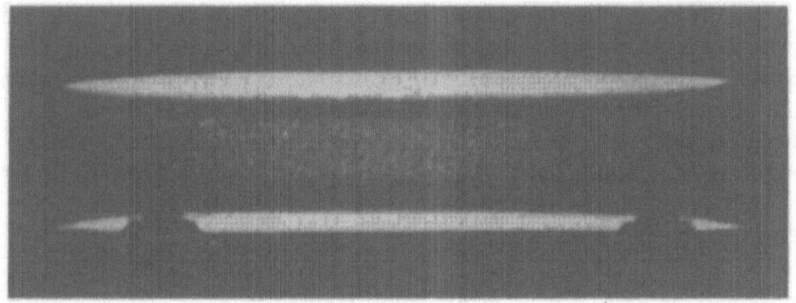

Figure 4. The appearance of a time averaged glow of an EPD in an experimental discharge chamber with disc electrodes.

The difference between these layers and those of the glow discharge is that in the case of a pulsed discharge the layers are not stationary, but only quasi-stationary if the discharge is periodic-pulsed.

The nature of the layers was investigated in [19]. It was shown that the layers formation during not extremely short pulses also is under strong influence of secondary emission processes on the cathode: the formation of a cathode potential drop sheath, secondary avalanches and a high energy electron beam with increased penetration ability. The dark space with an increased electron density near the electrode is the result of these processes: the observed thickness of about 10-15 *cm Torr* is close to that calculated in [19].

If a steady-state electric field is applied to the plasma, such as in a nonself-sustained discharge the layers thickness will decrease. The free electron lifetime between the pulses is determined by recombination processes which in turn depend on the electron kinetic energy. If there is no external electric field applied the electron energy is low but the recombination coefficient is high. For this reason the initial electron density remaining to the beginning of the next pulse (repetition rate 10 kHz) does not exceed 10^9 cm^{-3}, as was stated in [19].

On the other hand, in the presence of a steady-state electric field the recombination coefficient decreases and the initial electron density becomes higher (up to 10^{10} cm^{-3}). As a result the electrode layers thickness decreases, the potential drop becomes smaller and it resembles more a stationary glow discharge cathode sheath. Thus under certain conditions we are able to consider the near electrode potential drop as negligible, such as in the electrotechnical model considered.

When the discharge gap value is comparable to the near electrode layers thickness, near electrode effects lead to nonuniform distributions of the electric field and electron density in the discharge gap [19]. In this case the electrotechnical model fails.

Volume homogeneity of the discharge in the direction perpendicular to the pulsed current may be strongly affected by processes similar to arcing of a conventional glow discharge. These processes are suppressed with reduction of the discharge gap and gas pressure, a decrease in dielectric sheets electric capacity or current pulse width. Fast gas flow in the discharge region also suppresses these instabilities.

A distributed ballast capacity of the dielectric sheets primarily prevents contraction of cathode sheath at the late stage of the breakdown.

The high homogeneity of the EPD in a large volume and a considerable average free electron density attainable at relatively low average discharge power, determines its successful application as ionization source for nonself-sustained discharges. Typical parameters are presented in Table 2.

2. Nonself-sustained DC discharge assisted by an EPD

Application of the EPD to produce a uniformly conductive plasma in a large volume successfully solves two problems encountered in the design of high power CO_2 lasers.

First the EPD technique is able to control the DC discharge parameters and suppresses instabilities which are the obstacle to increase the discharge power. Secondly the EPD ionization ensures a high optical homogeneity of the discharge in a large volume which is difficult to attain in self-sustained DC discharges.

Let us consider the physical properties of the EPSD applied to the high power CO_2 laser "Lantan". The main parameters of the discharge are presented in Table 2.

The arrangement of the electrodes in the discharge chamber of "Lantan" type lasers is shown in figure 5 together with the pulse generation circuit. theeEPD is excited between two plane electrodes spaced 5.5 cm apart, consisting of water cooled metal plates covered by dielectric coatings. The plasma is formed in a plane channel with a rectangular cross section 5.5 x 90 cm^2 with a fast gas flow assigned by the arrow. Two copper tubular main discharge electrodes (cathode and anode) are placed correspondently at the inlet and outlet cross sections of the channel. The electrodes are mounted parallel to each other spaced 26 cm apart.

TABLE 2. Main parameters of EPSD used in a "Lantan-5" industrial CO_2 laser.

Parameter	Value [unit]
Discharge cell:	
Discharge volume	$5.5 \times 26 \times 90 cm^3$
EPD gap	5.5 cm
EPSD gap	26 cm
Gas pressure	45 Torr
Gas mixture $CO_2/N_2/He$	1/15/10
Gas flow velocity	100 m/sec
Current pulse:	
width	100 nsec
current amplitude	130 A
current density amplitude	$60 mA/cm^2$
Main discharge current:	
current amplitude	12.5 A
current density amplitude	$25 mA/cm^2$
time average current at 9 kHz pulse	
discharge repetition rate	5.5 A
Main discharge electric field	$225 V/cm^2$
Main discharge power:	
power density amplitude	$5.7 W/cm^2$
Time average power density at 9 kHz	
pulse repetition rate	$2.5 W/cm^2$
Pulse discharge power:	
peak power density	$32 W/cm^2$
Time average at 9 kHz power density	$0.05 W/cm^2$
Free electron density:	
peak electron density	$6 \times 10^{10} cm^{-3}$
Time average electron density at 9 kHz	$2.5 \times 10^{10} cm^{-3}$

The constant main discharge voltage V is applied directly to the tubular electrodes, without ballast resistors. The voltage value is chosen optimal for vibrational excitation and the spatial average ionization rate of the main discharge electric field is much lower than the time average ionization rate of the EPD. Thus the EPSD is entirely nonself-sustained and have its characteristic properties: a monotonously increasing voltage-current characteristic (see figure 6), current termination when the EPD is switched off and so on.

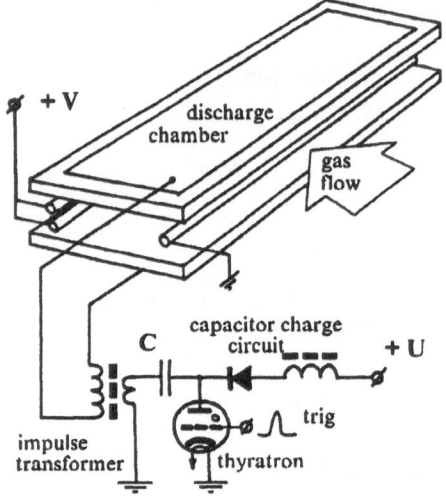

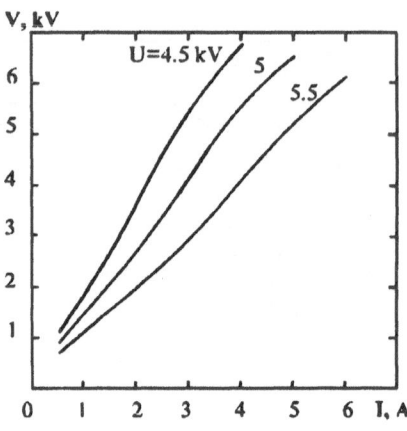

Figure 5. Arrangement of electrodes in the EPSD used in a "Lantan" fast-transverse-flow CO_2 laser together with a typical high-voltage pulse generation circuit.

Figure 6. Voltage- current characteristics of EPSD at different voltages of the EPD power supply. V and U are those of figure 5. I : time averaged EPSD current.

Being created during the discharge pulse the free electrons disappear by mainly dissociative electron-ion recombination and three-body attachment processes or are simply blown away with the gas flow. Following the electron density and gas conductivity the main discharge current also flows in a periodic pulsed manner (figure 7).

The time average current thus depends on the peak electron density (or ionization pulse amplitude) and pulse repetition rate. The discharge power may be controlled either by regulating the high-voltage supply of the EPD as shown in figure 6, or by varying the pulse repetition rate.

The output power thus produced is modulated with the EPD frequency but the kinetics of laser levels under the pressures considered leads to smoothing off the light modulations related to the current modulations at appropriate pulse repetition rates as shown in figure 7.

The mutual arrangement of electrodes is determined by the conditions of discharge homogeneity and convenience for the laser beam extraction. The EPSD perpendicular with respect to optical axes and longitudinal with respect to gas flow.

On the surface of the tubular electrodes cathode and anode potential drop sheaths are formed. The sheaths are equivalent to those of a conventional glow discharge. The phenomenon of normal current density also takes place.

Figure 8 shows the potential distribution in the discharge chamber measured by an electric probe. The electric field is distributed very uniformly over the whole volume excluding the near electrode zones.

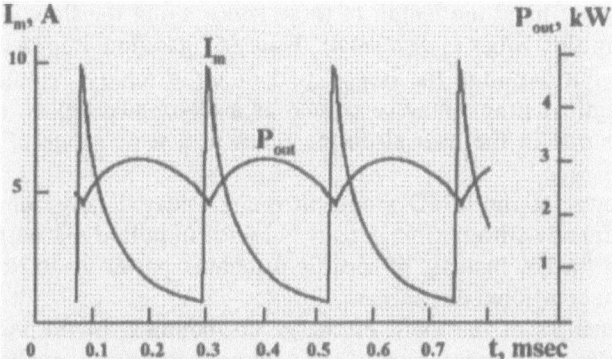

Figure 7. Typical main current I_m and output power P_{out} oscillograms obtained for a "Lantan-5" laser at a 4.5 kHz pulsed discharge repetition rate.

Near (in the region of about 1-2 cm radius) tubular electrodes the current density and electric field are higher simply because of the electrode geometry. The electrode surface area is much smaller than the channel cross section area. This geometry is

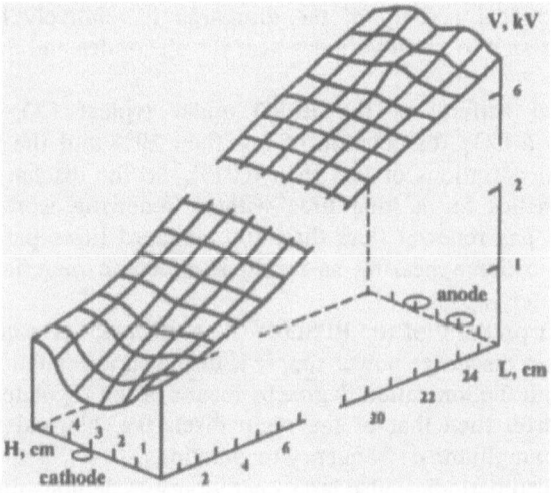

Figure 8. Potential distribution in the near electrode zones of the EPSD.

suitable to fix the positions of current spots on the electrodes without the multielement electrode structure. The increase of the electric strength near electrodes is followed by formation zones, similar to that of dissociative attachment controlled self-sustained discharge with the characteristic electric field strength, electron density, and other properties. The stability of these zones determines the stability of the discharge as a whole, because arcing in this zone leads to discharge failure.

The stability of the discharge is an ultimate problem for successful laser operation. Instabilities occur in near-electrode zones, where both electric field and current density are high. On the other hand the length of these zones along the flow is small, so the gas resident time in this zones is also small. Thus fast gas flow improves the stability to a large extent. Furthermore, the largest part of the discharge volume, where the electric field strength is lower, may be treated as a distributed ballast resistor which stabilises the discharge in the near electrode zones in a wide range of the deposited power.

The critical power of the EPSD arcing in the considered range of parameters is increased almost linearly with plasma density obtained in pulsed discharge. It provides the possibility of a further increase of specific discharge power up to that obtained in steady-state e-beam controlled discharges.

The inhomogeneities of the main discharge are bounded in the vicinities of the metal tube electrodes. The largest part of the volume is uniformly excited by the non-avalanching electric field and non-self sustained current. The near-electrode zones are relatively small, so the efficiency of active volume excitation and optical homogeneity of the active medium are high. The active medium thus has almost a perfect parallelepiped shape, which is favourable to avoid phase distortions of the extracted beam. Thus it is possible to produce perfect laser beams with advanced optical resonators [11].

Primarily due to the non-avalanching value of the main discharge electric field strength the chemical activity of the discharge is relatively low. Plasmachemical processes are initiated in small volumes near the electrodes and during short periods of the EPD pulses.

The chemical activity of the EPSD under typical CO_2 laser conditions is characterized by a CO_2 dissociation of less then 20% and the formation of relative nitric oxides concentrations of less then 0.01%. So the discharge can keep its high power characteristics for a long time without renewing working gas mixture. In practice a small gas renewal (less than 100 standard litres per hour) is required to remove the impurities appearing as result of dielectric materials outgasing (mainly hydrogen and water vapour).

An important property of the EPSD is the perfect power control possibilities. For example, the main discharge power supply is designed unregulated. The laser power is controlled through the ionization degree by means of the regulated EPD power supply, much less powerful then that of the main discharge. Normal laser pulses are also realized by EPD amplitude or frequency modulations.

Super-enhanced periodic-pulse laser operation was demonstrated in the "Lantan-1" prototype industrial laser model designed in 1980 [9]. The main discharge in this model was designed also self-sustained, switched by a thyratron. To prevent the main discharge from arcing, EPD was used as a very uniform preionization. Several tens of kW peak power and 1.5 kW average power at pulse repetition rates up to 200 Hz were achieved.

We can summarize the main advantages of the discharge under consideration as follows: a high optical homogeneity, high efficiency for vibrational excitation, simple electrode system, low chemical activity and a perfect power control.

The following chapters describe the design and characteristic features of the 5 kW fast-transverse-flow industrial laser "Lantan-5" developed on the base of the EPSD.

3. High power industrial CO_2 laser "Lantan-5": design and operational characteristics

The industrial CO_2 laser "Lantan-5" with 5 kW CW output power based on the EPSD technique was designed for metal cutting and welding applications.

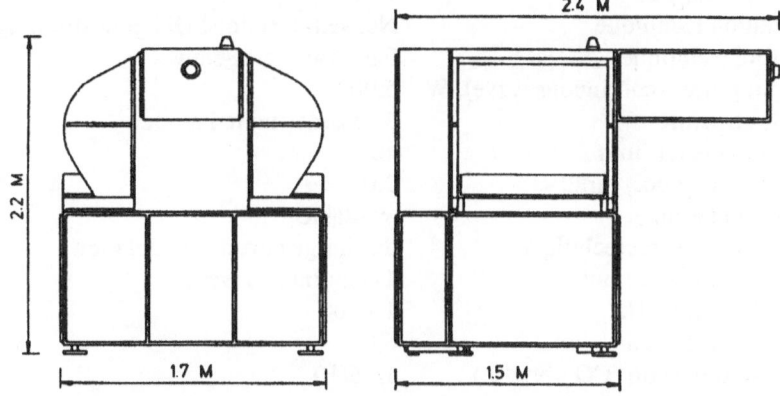

Figure 9. Overall view and dimensions of "Lantan-5": a 5-kW CW CO_2 laser.

"Lantan-5" is a fast-transverse-flow (FTF) industrial CO_2 laser. It is a modification (up to 5 kW) of the previous model "Lantan-3M" described in [10]. These two models are based on the same principles and differ from each other in discharge, pulser and blower power. The general specifications of the Lantan-5 model are represented in the Table 3 and an overall view in figure 9.

The original discharge technique permits "Lantan" lasers to demonstrate a high beam quality, fast power control capability, high efficiency and low gas consumption in a reliable and simple transverse flow discharge geometry.

The output power can be modulated in periodic rectangular pulses or in a trapezoid single pulse. All power regimes are automatically controlled so that this model may be used in flexible processing systems.

A discharge technique used for excitation is described in the preceding chapter. A nonself-sustained discharge is assisted by an electrodeless capacitive periodic-pulsed discharge. The main discharge power supply is designed operating at a constant nonavalanching voltage is optimised for vibrational excitation. The time averaged discharge power is controlled by varying the EPD pulse repetition rate. The time average output power in CW or pulse operation is stabilized by means of a fast backfeed circuit.

Because of the FTF scheme, two parallel axial fans driven by built in water cooled electro-shafts operating at 9,000 rpm are enough for effective gas recirculation. The low operating speed and relative pressure drop less then 1.1 at the absolute pressure of 6 kPa makes the fans very reliable. They ensure a uniform gas flow of 100 m/s in the discharge area.

TABLE 3. General specifications of the "Lantan-5 " industrial CO_2 laser.

Parameter [unit]	Value
Wavelength, μm	10.6
Excitation technique	Nonself-sustained DC glow discharge
Cooling technique	Fast-transverse-flow
Output power (continuous wave), W	5,000
Mode structure	3-rd order mixed mode
Beam diameter, mm	40
Beam divergence, mrad	3.0
Pulse operation	available
Pulse modulation technique	discharge current modulation
Pulse width range, msec	1- continuous wave
Repetition rate, Hz	1 - 300
Gas pressure, kPa	6.0
Gas mixture ratio ($CO_2/N_2/He$)	1/15/10
Gas consumption (St. litres/ hour)	100
Weight, kg	2500
Overall sizes (L x W x H), mm	2500 x 1700 x 2100

The small signal gain of the active medium produced is easily varied from very small values to 0.5 m^{-1} at maximum power.

The "Lantan-5" model was primarily designed with a folded stable resonator, schematically shown in figure 10. The resonator optical length is 7.5 m. The optical axis is folded to obtain 5 passes through the active medium, each of 0.9 m. Thus the resonator optical scheme includes 4 plane folding mirrors a concave rear mirror and a plane ZnSe output coupler with 30% reflectivity.

With a rear mirror of 15 m radius of curvature (semiconfocal configuration) the resonator is operating in a multimode regime with an output beam divergence of about 4.5-5 mrad. Figure 11 represents the high efficiency of beam extraction obtained from "Lantan" with a semiconfocal stable resonator. To improve the beam quality the radius of curvature was increased up to 30 m. The mixed mode order was reduced to approximately 3 so that the divergence was decreased to 2.5-3 mrad. The extraction efficiency was also reduced to approximately 17% at nominal power. Further increase of the rear mirror radius makes the resonator too sensitive to the thermal distortions of the optical elements and resonator structure and consequently the laser efficiency drops substantially.

Multimode operation advantages are a high beam extraction efficiency (up to 20%) and relatively uniform near- and far-field intensity distributions. The beam divergence in this case is 5-10 times higher then the diffraction limited value. This beam quality meets the requirements of some welding and heat treatment applications, but is not good enough for cutting and welding of non-ferrous metals.

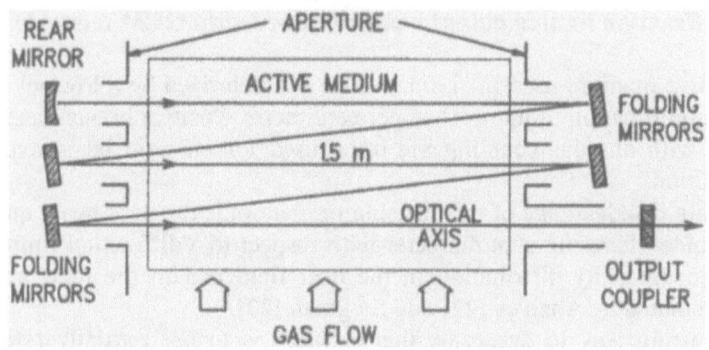

Figure 10. Resonator structure of "Lantan-5".

To obtain TEM_{00} operation with the same resonator structure one can decrease the intercavity aperture diaphragm to approximately 20 mm. The restricted mode volume leads to a decrease of the nominal power of less then 2 kW in this case. The quality of this beam usually does not exceed $M^2 = 1.3$ because of distortions generated by the diaphragm or the higher order modes.

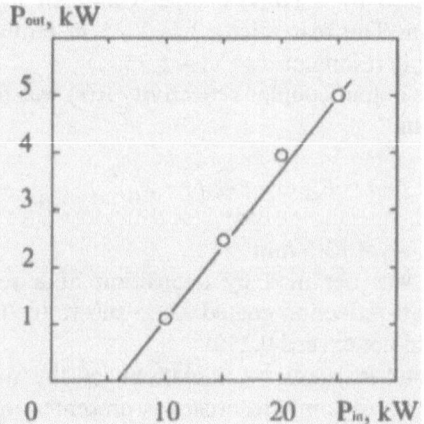

Figure 11. Dependence of the output power P_{out} vs discharge power P_{in} obtained from "Lantan-5" fitted with a semiconfocal stable resonator operating in the multimode regime.

The high homogeneity and optical quality of the active medium obtained with the EPSD technique enables us to achieve a high beam quality together with a high efficiency just by designing an appropriate resonator scheme. The next chapter describes the experiments aimed at extraction of a 5 kW diffraction limited output beam from a "Lantan-5" laser with the help of an unstable resonator with a GRM output coupler [11].

4. Near diffraction limited output beam obtained with GRM unstable resonator

The active medium used in "Lantan-5" is characterized by a Fresnel number $N_F = 8$ and an amplification of up to 0.5% per centimetre. Positive branch confocal unstable resonators with annular coupling are often used for efficient beam extraction under these conditions.

The main disadvantage of the unstable resonators is the poor beam quality revealed in an expanded focusing spot diameter with respect to Vdiffraction limit, as well as a non-uniform intensity distribution in the near field and in the focal spot area. This effect was studied by Anan'ev [22] and Siegman [23].

A promising way to overcome this problem is to use partially reflecting output couplers in conventional type unstable resonators. Successful experimental tests of resonators with graded reflectivity mirrors (GRM) [24], phase-unifying mirrors [17], and simple partially reflecting mirrors [25] were carried out last years.

We have decided to examine the GRM approach at a power level of 5 kW. A super-Gaussian profiled output coupler was produced in the Scientific Research Centre of Technological Lasers, Shatura, Moscow Region.

The positive branch unstable resonator for "Lantan-5" was designed on the base of a stable resonator scheme shown in figure 10. The plane partial output reflector of the stable resonator was replaced by a convex super-Gaussian partial reflector with a radius of curvature of 20 m. The rear mirror has a 35 m radius of curvature so that magnification of this unstable resonator was M = 1.75.

The radial profile of the output coupler reflectivity R(r) was designed and produced to be close to super-Gaussian

$$R(r) = R_a \exp\left(-2\left(r/w_a\right)^n\right) \tag{4.1}$$

where $R_a = 0.68$, $n = 4$, $w_a = 13.5$ mm.

A variable reflectivity was obtained by deposition of a reflecting layer with a variable thickness on an anti-reflection coated ZnSe substrate. Total absorption of the mirror at nominal power did not exceed 0.2%.

The result of the geometrical-optics calculations of the near-field output beam intensity profile in the case of an empty resonator is presented in figure 12 by the solid line together with the experimental values (rectangles) obtained at the output power of 4 kW. An equivalent Gaussian near field intensity profile is shown by dashed line. The ideal Gaussian beam and the considered real beam are treated equivalent if they have equal cross section areas carrying 86% of the total power. The ideal beam is presented as the reference point of beam quality.

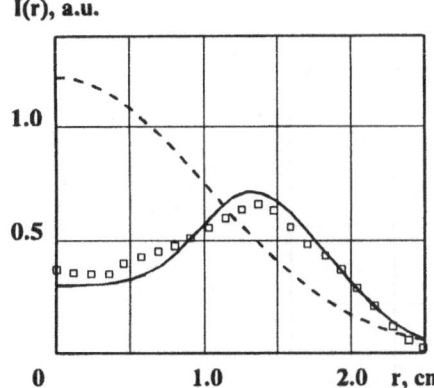

Figure 12. Near-field radial intensity distribution obtained for unstable resonator with a GRM output coupler: rectangles - experimental values, solid line - theoretical calculations, dashed line - equivalent Gaussian beam.

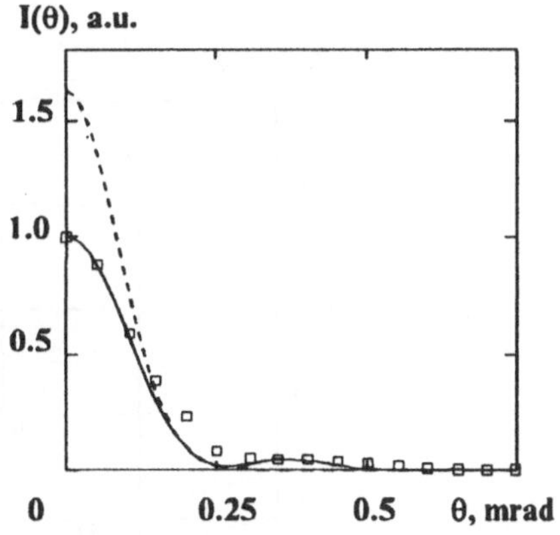

Figure 13. Far-field angular intensity distributions obtained for unstable resonator with GRM output coupler: rectangles - experimental values, solid line - theoretical calculation, dashed line - equivalent Gaussian beam.

One can see that the near-field intensity profile obtained from a GRM resonator appears to be quite smooth and uniform. The difference between the theoretical and experimental profiles lays in the limits of the used theoretical approximation. The beam diameter was slightly spread when the output power was increased which may be connected to variations of the active medium characteristics.

The solid line in figure 13 represents the results of a theoretical calculation of the far-field angular intensity distribution from a given near-field intensity profile with regard to phase distortion in the semi-transparent layer of variable thickness on the output coupler. The rectangles show the far field angular intensity distribution obtained in the focal point of a 14 m lens. These two profiles are normalized to compare the ratio of intensities at the central lobe and at the first ring (or wings). If they were normalized to equal power the intensity of the experimental one would be lower because the experimental curve demonstrates a wider central lobe than the theoretical one. Nevertheless, far-field measurements with calibrated diaphragms revealed that about 80% of the total power is radiated within 0.3 mrad (half angle). It corresponds to that of the theoretically predicted profile.

Taking into account the experimental error the intensity in the central lobe observed for a GRM resonator may be estimated as 1.7-1.9 times lower than that of an equivalent Gaussian beam (dashed line in figures 12,13).

Figure 14 shows the experimental dependence of the output power against discharge power. The efficiency of 13% is achieved at nominal power. Differential efficiency is about 17%, threshold discharge power is 7.5 kW. The last value reveals (in comparison with results presented in figure 11) that the output coupling in this case is about 0.8 and the backfeed ratio 0.2. The last value is close to that obtained from the geometrical-optics approximation.

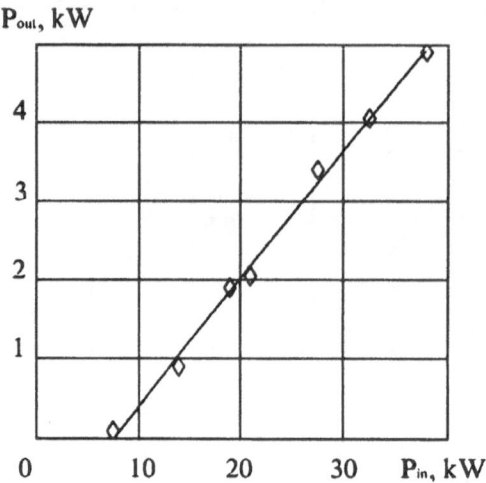

Figure 14. Dependence of output power P_{out} vs discharge power P_{in} obtained from "Lantan-5" fitted with a confocal unstable resonator with a GRM output coupler.

The mirror tilting sensitivity of the resonator was found to be moderate and the intensity profile remains smooth without hot-spot formation. Heat wedge effects in the FTF active medium were negligible in these experiments.

As one can see the extraction efficiency obtained with a GRM unstable resonator is relatively low, particularly with respect to that for a semiconfocal stable multimode resonator, obtained under similar conditions (figure 11). It could be explained by the low order of a super-Gaussian mode excited in the resonator. A difference in extraction efficiency between high and low order super-Gaussian modes was experimentally observed by Serri et al. [24].

In spite of the high extraction efficiency, high order super-Gaussian output couplers seems to obtain a worser beam quality than the low order ones. If one compares the present results for $n = 4$ with results for $n = 8$ from [24] for a far field intensity pattern, one can see that the last has more intensive wings that the other.

Experiments with GRM's reported in literature (for example see [24,26]) as well as the present one, show that the measured far field intensity pattern exhibits more intensive wings than obtained from geometrical optics calculations regardless of phase distortion effects. The discrepancy is obviously caused by phase distortion of the near-field wavefront and the question is what is the source of these phase distortions.

In our case the geometrical optics approach was used to obtain an intercavity intensity distribution with a plane wavefront at the output coupler. The output beam intensity profile was obtained from the intercavity beam intensity profile and output coupler reflectivity profile. The output beam phase profile was determined as a phase shift in the variable thickness layer responsible for the variable reflectivity. As one can see from figure 13 the far field intensity profile calculated on the base of this approach is close to experimentally observed: both exhibit an equal ratio between intensities of central lobe and of the wings. This demonstrates the feasibility of the used theoretical approach.

The present experiments allow us to make the conclusion that the distortion in the variable thickness layer is the main limitation for the beam quality improvement with the help of GRMs.

The experiment demonstrates the possibility of near diffraction limit operation of a 5-kW transverse-flow industrial laser with a super-Gaussian output coupler. The results obtained are close to that obtained by Takenaka et al. [17] with a phase-unifying output coupler and the active medium excited by a medium frequency electrodeless AC discharge and are in good agreement with theoretical predictions on the base of simple geometrical-optics calculations with regard to phase distortion in a variable thickness layer. Good agreement between theory and experiment demonstrates the high optical quality of the EPSD based active medium used in the experiments.

5. Conclusion

Experimental results and developments presented in this review demonstrate the feasibility of the EPSD technique in high power and high beam quality industrial lasers.

In conclusion, let us compare some different excitation methods used in powerful fast-transverse-flow CO_2 lasers to which the EPSD technique may be considered as an alternative:

1. The electron beam sustained discharge [12]. In spite of evident advantages of this method, it has not received wide acceptance for industrial use because of the problems related to e-beam injection in gas and X-rays shielding.

2. Electrodeless (capacitively coupled) AC discharge (10 kHz) has been applied in ML-105,-108 5 and 9 kW lasers produced by MLI Lasers Ltd., Israel [13]. Now these models are not produced anymore because of problems with discharge arcing. Instead of these models the ML-5000 5 kW laser with RF-excitation is produced [14].

3. The RF-discharge (1-100 MHz) characteristics are close to that of an e-beam sustained discharge [15]. Now there are experimental installations (for example, [15,16]) as well as some small-lot production (such as ML-5000) of multikilowatt CO_2 lasers exited by RF-discharges in fast-transverse-flow.

A wide application of this method in transverse flow lasers appears to be limited by technical and economical problems related to the powerful RF- technique. From this standpoint more acceptable is a capacitively-coupled medium-frequency AC discharge (100-200 kHz) [17] to be used in up to 10 kW lasers. This method can be a compromise between AC and RF capacitively coupled discharges.

4. In the case of EPSD most of the problems pointed out above do not take place: the main power is deposited by a safe, reliable and not expensive DC discharge, the time average load on dielectric plates is low and biological shielding is not required. Unfortunately in practice the specific deposited power achieved in an EPSD is lower

than that of e-beam sustained and RF-discharges. Nevertheless the EPSD parameters may be considerably enhanced by application of pulse generators of higher peak power. Very significant is also the optimal choice of dielectric plates characteristics (see relationship (1.3)).

It is not excluded that EPSD may be successfully applied to planar waveguide (or slab) diffusively cooled high power lasers instead of RF-discharge. The geometry of the discharge may be similar to that of presented on figure 5, where the dielectric sheets should be moved together to form a narrow gap required for diffusive cooling.

There are some problems encountered in this way. One of these problems is the substantially increased electron and current densities required for effective laser excitation in a narrow gap. Under typical conditions of a 1 kW planar diffusively cooled laser [27] the specific input power is up to 60 W/cm^2 - ten times higher than in the case of a fast-flow laser (compare to Table 2). To achieve that power density in a DC discharge a current density of up to 250 mA/cm^2 is required. From comparison with data from Table 2 one can conclude that a 600 mA/cm^2 pulse discharge current can produce the required ionization degree. The total pulse current may achieve up to 500 A through an area of 9.5 x 77 cm^2, typical for a 1 kW laser [27].

Nevertheless, a pulsed discharge ionization in a narrow gap may be more effective because of near cathode effects responsible for the Faraday dark space formation and so on. A layer thickness of about 10-15 cm $Torr$, typical for EPD (see Chapter 1), almost equals to the gap width in slab geometry (0.2 x 70 = 14 cm $Torr$ [27]). The effect that prevents the application of a γ-form of RF-discharge in planar waveguide lasers may be feasible for EPSD. In any case, the possibilities of this interesting application require additional studies.

6. References

1   Reily, J.P. (1972) Pulser-sustainer electric-discharge laser, *J. Applied Physics* **43**, 3411-3416.

2. Hill, A.E. (1973) Continuous uniform excitation of medium pressure CO_2 laser plasmas by means o controlled avalanche ionization, *Applied Physics Letters* **22**, 670-673.

3. Artamonov, A.V., Naumov, V.G., Shachkin, L.V., and Shashkov, V.M. (1979) A study of the activ medium of fast-flow CO_2 laser with a nonself-maintained discharge, *Soviet J. Quantum Electronics* **9**, 845 849.

4. Seguin, H.J.J., Nam, A.K., and Tulip, J. (1978) The photoinitiated impulse-enhanced electrically excite (PIE) discharge for high-power cw laser applications, *Applied Physics Letters* **32**, 418-420.

5. Nam, A.K., Seguin, H.J.J., and Tulip, J. (1979) Operational characteristics of a PIE CO_2 laser, *IEEE Quantum Electronics* **15**, 44-50.

6. Nath, A.K., Seguin, H.J.J., and Seguin, V.A. (1986) Optimization studies of a multikilowatt PIE CO_2 lase *IEEE J. Quantum Electronics* **22**, 268-274.

7. Generalov, N.A., Zimakov, V.P., Kosynkin, V.D., Raizer, Yu.P., and Roitenburg, D.I. (1977) Stead externally sustained discharge with electrodeless pulsed ionization in a closed-loop laser. Part I-II, *Soviet Plasma Physics* **3**, 354-364.

8. Generalov, N.A., Zimakov, V.P., Kosynkin, V.D., Raizer, Yu.P., and Roitenburg, D.I. (1980) Stead externally sustained discharge with electrodeless pulsed ionization in a closed-loop laser. Part III, *Soviet Plasma Physics* **6**, 633-638.

9. Generalov, N.A., Zimakov, V.P., Kosynkin, V.D., Raizer, Yu.P., and Solov'yov, N.G. (1982) Rapid-flo combined action industrial CO_2 laser, *Soviet J. Quantum Electronics* **12**, 993-998.

10. Generalov, N.A., Solov'yov, N.G., Yakimov, M.Yu., and Zimakov, V.P. (1991) Application of th electrodeless pulser-sustained glow discharge for development of high-power CO_2 lasers, in *Conference o Lasers and Electro-Optics, 1991*, Optical Society of America, Washington, pp. 322-324.

11. Generalov, N.A., Solov'yov, N.G., Yakimov, M.Yu., and Zimakov, V.P. (1994) High power industrial CO laser "Lantan-5" with graded reflectivity mirror resonator, *J. Pure and Applied Optics* **3**, 533-539.

12. Velikhov, E.P., Pismenny, V.D., Rakhimov, A.T. (1977) Nonself-sustained gas discharge excitin continuous wave CO_2 lasers, *Uspekhi Fizicheskih Nauk* **122**, 419-447 (in Russian).

13. Agm Agmon, P., Hoch, E., Katz, D., Shachrai, A., Zinman, Y., and Fishman, D. (1987) MultiMultikilowatt industrial CO_2 laser, in S.Rosenwaks (ed.), *Gas flow and chemical lasersr ProceProceedings of the 6th international symposium*, Springer-Verlag, Berlin, pp. 275-278.

14. (1991) MLI expands product and activities, *Industrial Laser Review* **6**(2), 18.

15. Hugel, H. (1987) RF excited CO_2 flow lasers, in S.Rosenwaks (ed.), *Gas flow and chemical laser Proceedings of the 6th international symposium*, Springer-Verlag, Berlin, pp. 258-264.

16. Wildermuth, E., Walz, B., Wessel, K., and Schock, W. (1990) Characteristics of a compact 12 k transverse flow CO_2-laser with rf-excitation, in J.M. Orza and C.Domingo (eds.), *Eighth Internationa Symposium on Gas Flow and Chemical Lasers*, SPIE Vol.1397, Madrid, pp. 367-371.

17. Takenaka, Y., Kuzumoto, M., Yasui, K., Yagi, S., and Tagashira, M. (1991) High power and high focusin CW CO_2 laser using an unstable resonator with a phase unifying output coupler, *IEEE J. Quantu electronics* **27**, 2482-2487.

18. Christensen, C.P. (1979) Pulsed transverse electrodeless discharge excitation of a CO_2 laser, *Applie Physics Letters* **34**, 211-213.

19. Raizer, Yu.P., and Shneider, M.N. (1989) Electrodeless capacitive discharge sustained by repetitive high voltage pulses, *Teplofizika Vysokih Temperatur* **27**, 431-438 (in Russian).

20. Lowke, J.J., Phelps, A.V., and Irwin, B.W. (1973) Predicted electron transport coefficients and operatin characteristics of CO_2-N_2-He laser mixtures, *J.Applied Physics* **44**, 4664-4671.

21. Nighan, W.L., and Wiegand, W.J. (1974) Influence of negative-ion processes on steady state properties an striations in molecular gas discharges, *Physical Review A: General Physics* **10**, 922-945.

22. Anan'ev, Yu.A. (1990) *Optical Resonators and Laser Beams*, Nauka Publishers, Moscow (in Russian).

23. Siegman, A.E. (1986) *Lasers*, University Science, Mill Valley, CA.

24. Serri, L., Maggi, C., Garifo, L., De Silvestri, S., Magni, V., and Svelto, O. (1991) Diffraction limited c transverse flow CO_2 laser of high power, in *Conference on Lasers and Electro-Optics, 1991* , Optia Society of America, Washington, pp. 409-411.

25. Mikheev, P.A., Nikolaev, V.D., and Shepelenko A.A. (1992) Unstable resonator with semi-transparen output coupler for fast-flow CO_2 laser, *Soviet J. Quantum Electronics* **22**, 415-418.

26. De Silvestri, S., Magni, V., Svelto, O., and Valentini, G. (1990) Lasers with super-Gaussian mirrors, *IEE J. Quantum Electronics* **26**, 1500-1509.

27. Colley, A.D., Baker, H.J., and Hall, D.R. (1992) Planar waveguide, 1 kW cw, carbon dioxide laser excite by a single transverse rf discharge, *Applied Physics Letters* **61**, 136-138.

HIGH POWER GAS LASER RESEARCH IN GERMANY

W.L. BOHN
DLR Institut für Technische Physik
Pfaffenwaldring 38-40
D-70569 Stuttgart

1. Introduction

Before starting a comprehensive description of high power gas laser research in Germany it seems more than appropriate to address the upcoming competing situation between gas lasers on one hand and solid state and semiconductor lasers on the other. This also reflects the decision of the German Ministry of Research and Education (BMBF) to initiate a new laser research program labeled "laser 2000" which clearly emphasizes high priorities for all areas of semiconductor and solid state lasers. In this program gas laser research is aimed at new concepts and designs for very high power and compact devices.

Although all solid state coherent light sources show undoubtfully many attractive features, gas lasers still dominate when it comes to areas such as very high powers, high beam quality, high coherence properties, and short wavelengths. This is true for today's state of the art and in essence describes the future potential of high power gas lasers.

In a general attempt to describe the future trend of high power lasers, as perceived by the author of this paper, Fig.1 shows rather qualitatively the time evolution of four classes of laser sources:

(a)   high power gas lasers (candidates are CO_2, CO and O_2/I Oxygen-Iodine)

(b)   diode-pumped solid state lasers (DPSSL)

W. J. Witteman and V. N. Ochkin (eds.), Gas Lasers - Recent Developments and Future Prospects, 343–357.
© *1996 Kluwer Academic Publishers.*

344

(c)   diode lasers and
(d)   other laser sources that are not covered by the previous definitions.

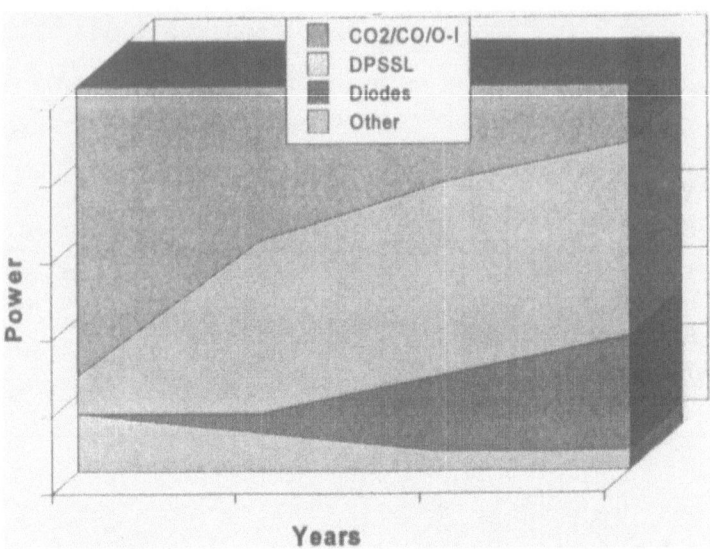

Figure 1. Future trends of high power lasers

Over the years high power gas lasers will successfully confirm their role for all applications where very high powers and especially brightness are needed. Fast progress in diode-pumping has given to solid state lasers the opportunity to emerge as a reliable and high efficient medium power laser source. This trend will certainly increase for the years to come. The diode lasers, on the other hand, will most probably expand into the low power application areas. The pace at which diode lasers, either for pumping solid state sources or in a stand alone unit or array, will conquer the market of low and medium power applications strongly depends on the cost evolution of the diode manufacturing and, hence, on their world-wide production number. Other gas lasers may still survive in the future and cover a small sector of very specific applications. (Some limited numbers of ion lasers may be an example.) Excimer lasers should have been included in this diagram, in particular within the class of high power gas lasers. They surely represent a unique source for the UV and deep UV as needed for many applications in the semiconductor industry. For Germany, however, they will not play any major role since the BMBF decided to stop any research effort on excimer lasers and to leave any related European research and development programs in the last years.

Thus, not denying the tremendous potential of solid state and diode lasers especially for mean and low power applications, this paper advocates the important role of high power gas lasers in particular for high brightness applications.

2. High Power Infrared Gas Lasers

This chapter briefly reviews the almost classical candidates of high power infrared (IR) gas lasers and concentrates on new developments aimed at more compact laser sources or at new approaches for high repetition rate/high average power operation.

2.1 CO_2 AND CO LASERS

Since several years Germany has emphasized radio frequency (RF) as the most efficient excitation technique for CW operated CO and CO_2 lasers. It allows:
- to capacitively couple the electric energy into the discharge without need of metallic electrodes;
- to obtain large volume stable discharges;
- to achieve high power densities ($< 30W/cm^3$) and
- easy modulation of the discharge.

Those advantages are only challenged by higher investment costs as compared to DC excitation.

Fast axial flow RF excited CW CO_2 lasers are nowadays well established in terms of technology up to power levels of nearly 20kW. Their technology has been well implemented into German industry as can be seen by the various laser products on the German market-place.

RF excitation has also been successfully used for the development of CW CO lasers. The RF discharge is applied to a high pressure gas mixture which is subsequently gasdynamically expanded by a nozzle configuration down to temperatures between 150k and 250k. This gasdynamic cooling allows efficient operation of the CO laser. For the first time, a 800W CW output power CO laser with closed gas loop has been reported [1]. More recently, a supersonic CO laser with RF excitation has been developed and is being scaled up to output powers of 8kW in a joint investigation of the French German Research Institute ISL and DLR [2].

2.2 DIFFUSION COOLED CO_2 LASERS

A new development towards a more compact laser unit first reported in [3] is schematically shown in Fig.2. The laser discharge volume in slab geometry is confined between two metallic electrodes which are only

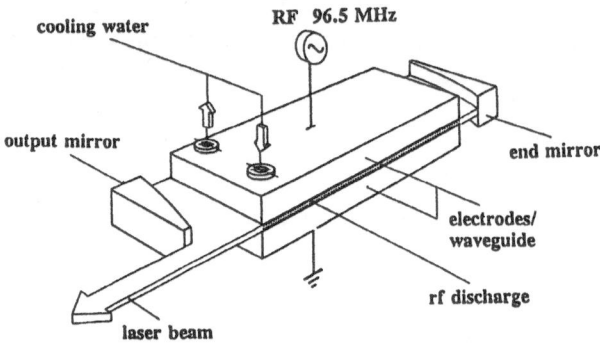

Figure 2. Schematics of diffusion cooled waveguide laser

Figure 3. Diffusion cooled CO_2 wavguide laser

between 1mm and 2mm apart. A 96.5Mhz RF discharge excites this near waveguide laser configuration with an output power of up to 1.5kW and an efficiency of up to 17%. Due to the forced diffusion dominated cooling to the water-cooled electrodes, this laser configuration does not rely on convective gas cooling. Thus, the diffusion cooled waveguide laser operates without any cumbersome gas pumping system. The corresponding hardware of the experimental device is displayed in Fig.3. The system technology has already been transferred to industry. A waveguide RF excited CO_2 laser has simultaneously been developed by the Fraunhofer Institute for Laser Technology (ILT) in Aachen. In this case, however, the slab geometry was extended to a coaxial geometry which leads to a more sophisticated configuration for the optical resonator and still represents an unresolved problem.

2.3 PULSED CO_2 LASERS

Recently, TEA CO_2 pulsed lasers have been developed by industry for power levels up to 2kW. This chapter concentrates on two different technical approaches to achieve high repetition rate, high average power CO_2 laser radiation. In addition, alternative operation with CO laser gas is also reported.

The first scheme is shown in Fig.4 where basically the radiation field produced by a transverse flow RF modulated CW laser is mechanically chopped within the optical cavity of an unstable resonator in a double pass configuration [4]. Fig.5 exhibits the peak pulse power and the energy per pulse as a function of the length of the pumping pulse. The pressure of the

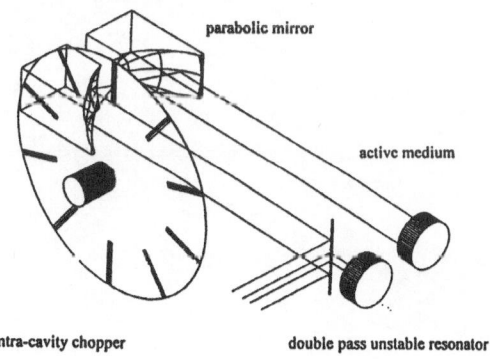

Figure 4. Optically pulsed CO2 laser

348

gas mixture is 90mbar and the open symbols indicate an addition of Xenon

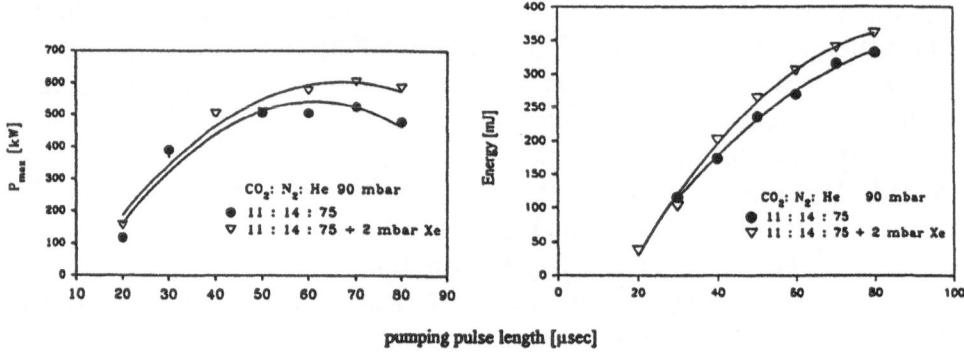

Figure 5. Preliminary results of optically pumped CO_2 laser

gas by 2mbar. Peak powers above 500kW are obtained while the average power is 1.2kW. This approach shows a great potential for high average power, high repetition rate applications: Repetition rates of up to 5kHz can be obtained.

Figure 6. Electron beam system CO_2 laser

The electron beam sustained pulsed discharge [5] represents a different approach for high average power pulsed lasers. A 12 liter discharge volume device which is used as a testbed, is shown in Fig.6. The most important feature of this device operated with a closed gas loop is the recent implementation of advanced pulsed power technology. In particular, IGBT based fast charging devices have been used in a parallel configuration in order to control the applied high voltage during each pulse to a pre-set value with an accuracy of less than a per cent. Newly developed high power switches are also under development. The upcoming technology will lead to a very compact and reliable pulsed power system for the next generation of pulsed CO_2 lasers. The experimentally obtained specific output energies are shown in Fig.7 as a function of specific energy deposited into the gas discharge. The set of CO_2 measurements (1) has been obtained with an unstable cavity with magnification of M = 1.5. The electrical efficiency is approaching 10%.

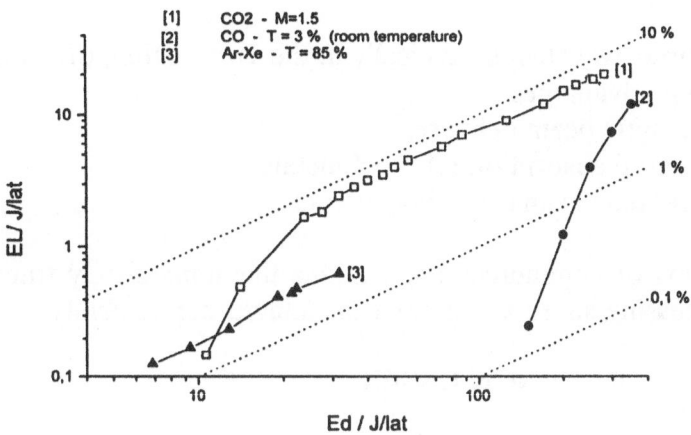

Figure 7. Experimentally obtained specific laser energies

As a testbed the same device has been operated with CO gas at room temperature [6] using a stable optical cavity with a transmission of T = 3%. In this case the efficiency is limited to approximately 3%. A third set of measurements (3) related to the operation with Argon-Xenon will be described in the following chapter.

3. High Power Short Wavelengths Gas Lasers

As already briefly mentioned in the introductory remarks one of the most promising advantages of gas lasers is the combination of high power short wavelength and good beam quality. This is most conveniently expressed in terms of the brightness of the laser radiation given by the formula below

$$B \equiv \frac{PD^2}{\beta^2 \lambda^2 z^2} \tag{1}$$

where P is the laser power, D the beam aperture, λ the wavelength, β the beam quality factor, and z the distance to the target. It can be easily shown that the intensity, I, at the target or at the focus of a lens with focal length, f, is related to the brightness by $B \sim Iz^2$ and $B \sim If^2$, respectively. Those relations clearly demonstrate how beam quality and wavelength may affect the final laser application.

The shorter wavelength, typically in the $1\mu m$ region, offers in addition the following advantages:
-   fiber optic beam delivery;
-   increased absorption for most metals;
-   decreased plasma shielding effects.

In terms of commercial applications this immediately translates into higher processing speed and greater manufacturing flexibility.

3.1 THE ARGON-XENON LASER

A high pressure (2 atmospheres) Argon-Xenon gas mixture has been used in a four liter electron beam driven discharge [7]. The experimental device has not been specifically designed for that task. For practical reasons a slightly modified module of the previously described CO_2 testbed has been accommodated for the operation with Argon-Xenon (see Fig.8). Preliminary results obtained with the stable resonator and a mirror transmission of 85% have been plotted in Fig.7 as a function of the specific energy deposited into the discharge. Substantial improvement is expected after optimization of the device and in particular the pulsed power modulator. Finally, it should be mentioned that the $1.7\mu m$ radiation is eyesafe and exhibits good transmission through the atmosphere.

Figure 8. E-beam pulsed Argon-Xenon laser

3.2 OXYGEN-IODINE LASER (O_2/I)

The O_2/I laser is the only chemically pumped high power laser in operation which exhibits an electronic transition ($\lambda = 1,3\mu m$). The general scheme of the laser system [8] is shown in Fig.9. Chlorine gas is introduced into a

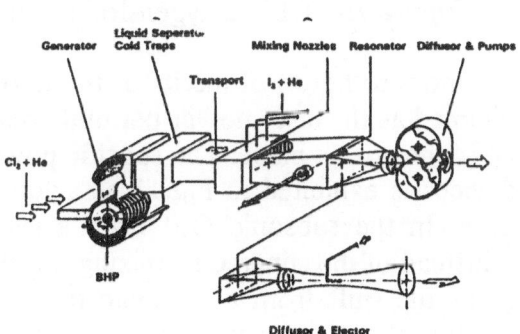

Figure 9. Schematics of O_2/I laser

chemical generator which operates with a mixture of peroxide and KOH in an aqueous solution and transforms the chlorine most effectively into $O_2(^1\Delta)$ excited oxygen molecules. The flow of excited oxygen is conveyed to a supersonic mixing nozzle where molecular iodine is first dissociated and subsequently excited and gasdynamically expanded into the optical resonator. A recompression of the gas is performed by either a diffusor and a pumping system or a diffusor and an ejector. The corresponding hardware of a multikilowatt device that is currently operated by DLR [9] is shown in Fig.10.

Figure 10. DLR Oxygen-Iodine laser

One of the important figure of merits of the O_2/I laser is the specific output power defined as the laser power per unit area of the flow's cross-section at the exit of the nozzle plane. The specific power achieved so far in various laser devices is exhibited in Fig.11 as a function of the reduced chlorine flow rate. In the subsonic O_2/I laser a simple iodine injection manifold is used instead of the supersonic mixing nozzle. As can be clearly seen from Fig. 11 the shift from a subsonic to a supersonic O_2/I laser scheme substantially enhances the specific laser power. The US-Air Force Phillips Laboratory has achieved the highest figure ever published in the literature [10]. For comparison DLR values from preliminary measurements are also shown. Recent results obtained in supersonic O_2/I lasers in China and in Japan are not shown in Fig.11 because the corresponding data are not yet fully available.

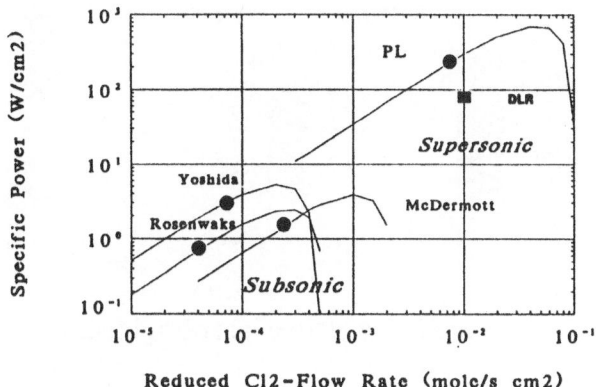

Figure 11. Overview of specific powers for various O_2/I lasers

In Germany, the main research goal is to improve the $O_2(^1\Delta)$ generation, to optimize the I_2 injection and mixing technique and to enhance the laser power extraction efficiency. In addition, dual use aspects of the O_2/I laser are considered in detail. Study work has been performed with regard to the potential of the O_2/I laser as a candidate for flexible, fiber optic supported processing of advanced materials in the automotive and aeronautics industry, in particular aluminum and aluminum alloys [11].

4. Applications of High Power Gas Lasers

As already mentioned in the previous chapter the brightness rather than the absolute output power of the laser is the figure of merit that drives the application. Fig.12 compares the brightness of the three major high power laser candidates including the solid state YAG laser to the classical workhorse CO_2 laser. Above each bar the output power and, in parenthesis, the beam quality factor of the corresponding laser are given. The data have been collected either from available devices (plain bars) or from not yet confirmed (dark hatched) or anticipated (light hatched) specifications.

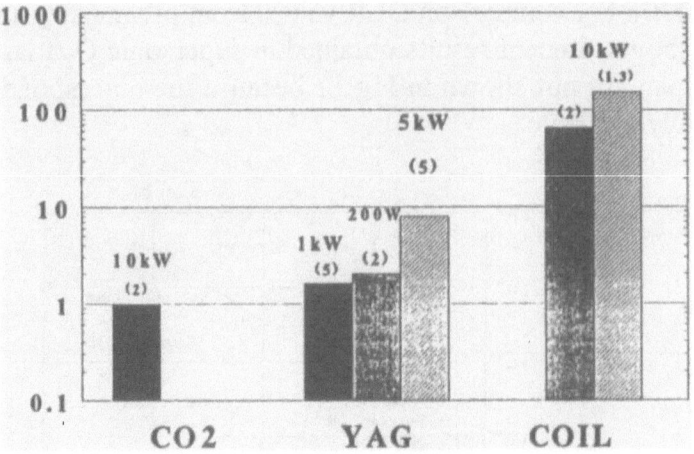

Figure12. Relative laser brightness

Major fields of application for high power gas lasers are in the defence and space sector as well as in the civilian/commercial area. In the latter part, laser cutting and welding has become more and more established in

Figure 13. Large area laser coating

industry. Surface treatment, in particular anti-wear coating, anti-corrosion coating, and transformation hardening arises as an increasingly important application domain. Fig.13 demonstrates a large area coating process performed with a 20kW CW CO_2 laser. More recently, paint removal and rust stripping by high average power high repetition rate gas lasers has been demonstrated. Stereo-stacking is another new area of interest for laser application. Fig.14 exhibits, for example, the rapid growth of a sphere realized by controlled layering with a high power CO_2 laser.

Figure 14. Laser supported rapid prototyping

5. Conclusions

A review of the main German research activities related to high power gas lasers has been given. RF excited diffusion cooled CO_2 lasers have been realized as very compact devices without the usual gas recirculating pumping system. Optically and electrically pulsed CO_2 lasers have been demonstrated and exhibit a promising potential for further power and energy scaling. Their development will be driven by a new range of applications like paint removal and rust stripping on large areas. High brightness has been emphasized as one of the most important figure of merits for high power lasers. Therefore, Argon/Xenon and Oxygen/Iodine lasers which are by far less developed than the CO_2 laser deserve more attention for future exploratory research. In addition, short wavelength lasers offer the advantage of fiber optic beam delivery which translates into greater manufacturing flexibility.

The future potential of gas lasers is clearly identified beyond the power level that may not easily be obtained by solid state or semiconductor lasers especially when it comes to high brightness applications and high coherence requirements.

6. References

1. von Bülow, H. and Zeyfang, E. (1993) Supersonic CO laser with RF excitation, *Review of Scientific Instrument*, Vol. 64, No 7, 1764-1769.
2. Schellhorn, M. and von Bülow, H. (1994) High power gas dynamically cooled CO laser with unstable resonator, *SPIE Proceedings* Vol. 2502, 63-68.
3. Nowack, R., Opower, H., Schäfer, U., Wessel, K., Hall, Th., Krüger, H. and Weber, H. (1990), High power CO_2 waveguide laser of the kW category, *SPIE* Vol. 1276, 18-28.
4. Jung, M., Walz, B., Wessel, K. and Schock, W. (1992) Voruntersuchungen zu einem optisch gepulsten CO_2-Hochleistungslaser, *DLR-IB* 441003/92, 1-19.
5. Beth, M.U., Hall, Th., Mayerhofer, W., (1990) Optimization of discharge parameters of an e-beam sustained repetitively pulsed CO_2 laser, *SPIE* Vol. 1397, 577-580.

6. Ionin, A.A., Mayerhofer, W., Walther, S. and Zeyfang, E. (1994) Room temperature repetitively pulsed e-beam sustained carbon monoxide laser, *SPIE* Vol. 2502, 44-50.

7. Mayerhofer, W. and Beth, M.U. (1992) Design of an e-beam controlled 1kW Ar:Xe laser, *SPIE* Vol. 1810, 439-442.

8. Bohn, W.L. (1992) Chemical oxygen-iodine laser: achievements, problems and future perspectives, *SPIE* Vol. 1810, 465-475.

9. Handke, J., Werner, A., Bohn, W.L., Schall, W.O. (1994) Multi-kilowatt supersonic chemical oxygen-iodine laser (COIL) device, *SPIE* Vol. 2502, 266-271.

10. Avizonis, P.V., Hasen, G.A., Truesdell, K.A. (1990) Chemically pumped oxygen-iodine laser, *SPIE* Vol. 1225, 448-476.

11. von Bülow, H. (1995) Oxygen-iodine laser for industrial applications, *DLR Forschungsbericht* 95-09, 1-92.

12. Wolf, S. and Volz, R. (1995) Use of the laser beam cladding process in plastics processing machine design, *Laser und Optoelektronik* 27 (2), 47-53.

RECENT PROGRESS IN FRANCE ON EXCIMER LASERS AND APPLICATIONS

B.L. FONTAINE

IRPHE-UMR 138 CNRS and GDR 1125 CNRS "LASERMAT", case 918, 13288 Marseille cedex 09, France

Abstract

An overview of recent progress in France on high power excimer lasers and applications is made with emphasis on results obtained through Eureka EU 205 Program and GDR 919 and 1125 CNRS. Focus is made on advances in excimer laser sources for industry and progress in physics of active medium. Researches on excimer laser applications are briefly presented. A new CNRS laboratory, The GDR 1125 CNRS « Lasermat » devoted to studies on high power laser applications is described.

1. Introduction

Research on excimer lasers and applications is very active in Europe. In France, most of scientific community in this field is now associated in cooperative work together with industrial partners. Aim of this paper is to present an overview of recent results obtained in France, strucure of research and projects.

2. Development of high average power excimer lasers

The EU 205 Eureka Program "Excimer Lasers", which started in 1987, associates laboratories and industrials from 5 european contries (France, Greece, RFA, Sueden and UK) for research and development on high average power discharge excited excimer lasers and specific applications with the goal to develop a 1-3 KW average power UV industrial excimer laser. This programm as very recently allowed the achievement in France of two 1 KW average power XeCl lasers (λ=308 nm) with very different laser characteristics. At SOPRA[1] , excitation of a large volume active medium by an X ray preionised C-L-C discharge at allowed 10 Joules laser pulses at 100 Hz prf (VEL demonstrator). At Laserdot/Aerospatiale)[2] 1,1 Joule per pulse at 950 Hz have been obtained by use of a phototriggered fast discharge. Fig. 1 shows laser average power obtained recently on the SOPRA laser while Fig. 2 represents laser efficiency v.s. charge voltage for the Laserdot 635-2 demonstrator.

W. J. Witteman and V. N. Ochkin (eds.), Gas Lasers - Recent Developments and Future Prospects, 359–365.
© 1996 *Kluwer Academic Publishers.*

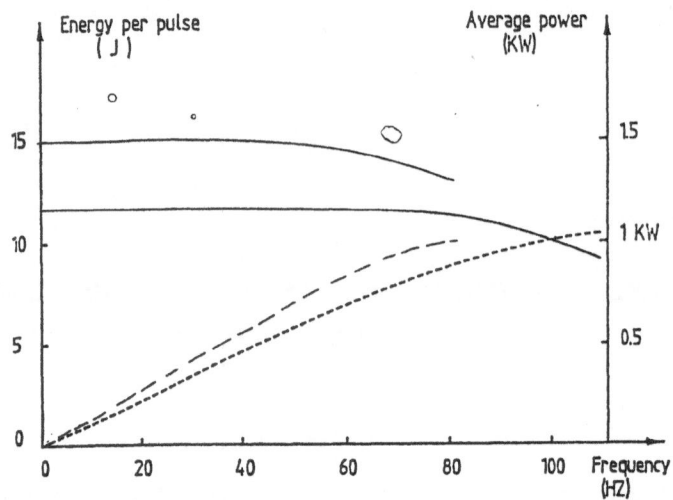

Fig. 1. Energy per pulse and average power as a function of pulse rate frequency for VEL Sopra laser. XeCl, 308 nm (from Ref. 1)

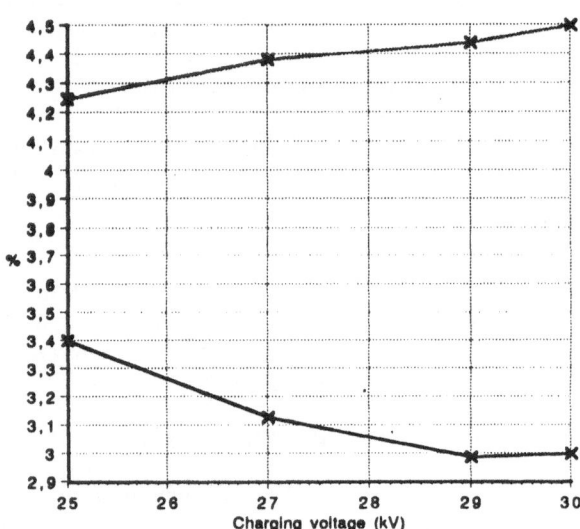

Fig 2. 635-2 Laserdot laser wall-plug and intrinsic efficiency v.s. charging voltage.XeCl, 308 nm, gas laser mixture: 4,5 atm. (from Ref. 2)

3. Physics of XeCl laser

During the last two years progress has been achieved within Eureka EU 205 Program in the frame of federative CNRS laboratories GDR 919 and 1125 CNRS, on excimer laser physics. Particular efforts were made on plasma dynamics and discharge stability, atomic and molecular kinetics, laser modelling, optics, fluid-dynamics and coupled phenomena as well as system technical characteristics and performances.

Results have been obtained, in particular, on:

(a) investigations of an XeCl phototriggered laser. A specific output energy of 7.6 J^{-1} and an efficiency of 3.5 % have been achieved. A self-consistant zero dimensional model have been developped and validated by experiments including emission and laser absorption spectroscopy on various xenon and neon states[3]. On fig 3 is shown a typical recording of current, laser power and neon 3 p($2p_{10}$) state v.s. time.

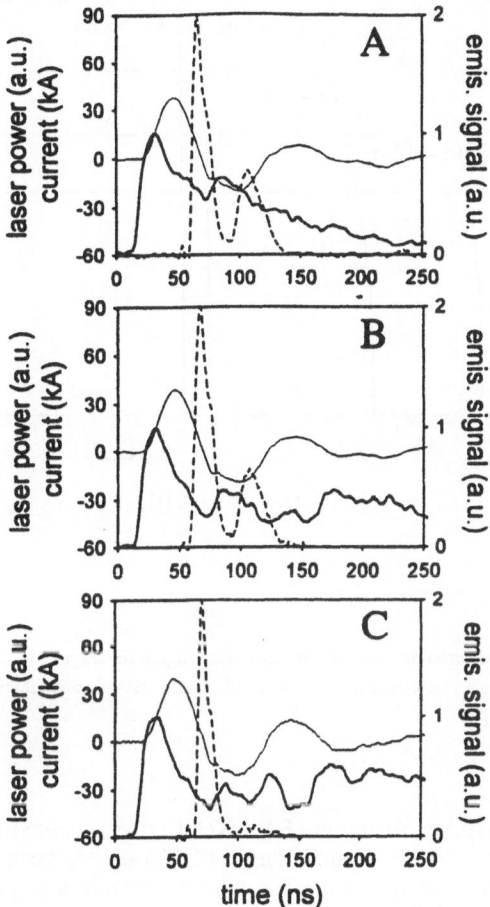

Fig. 3 Measured time variations of current (—), laser power (- -) and neon $3p(2p_{10})$ (—) state emission for a deacreasing electron preionisation density: (A), $6,5 \times 10^8 \, cm^{-3}$; (B), $10^8 \, cm^{-3}$; and (C), $5 \times 10^7 \, cm^{-3}$ (from Ref. 3)

362

(b) investigation of an X-ray preionized large volume XeCl laser. A self consistant modelling of this type of laser has been achieved and experimentaly valided.[4]

(c) theoretical investigations of excimer laser discharge stability. A dimensional modelling has been achieved as well as a study of unstabilities. Cathode sheath formation process in a discharge-sustained laser has been simulated[5].Typical calculated cathode sheath electron density spatial profile v.s. time are shown on Fig. 4.

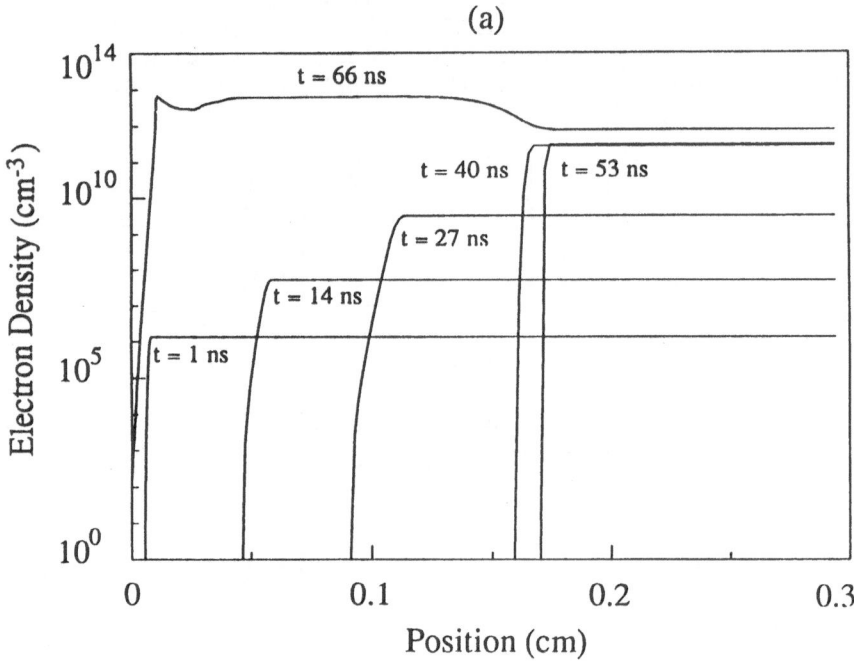

Fig. 4 Spatial distribution of electron density v.s. time in cathod region for a typical excimer laser. Preionization density is assumed to be zero in a region of extension 40 μm from the cathode and equal to 10^7 cm^{-3} in the rest of the discharge gap (from Ref. 5)

(d) high repetition rate spiker-sustainer (S.S.) XeCl Laser. A maximum average power of 220 watts at 1 Khz prf for laser pulse length of 165 ns has been achieved at IRPHE on LUX test-bed[6]. Fig. 5 represents laser average power v.s. pulse rate frequency obtained very recently on LUX with S.S. and C-L-C circuits.

(e) high optical quality, high repetition rate XeCl laser with unstable cavity. A laser power density up to 10^{11} W.cm^2 has been obtained at 800 Hz prf[7].

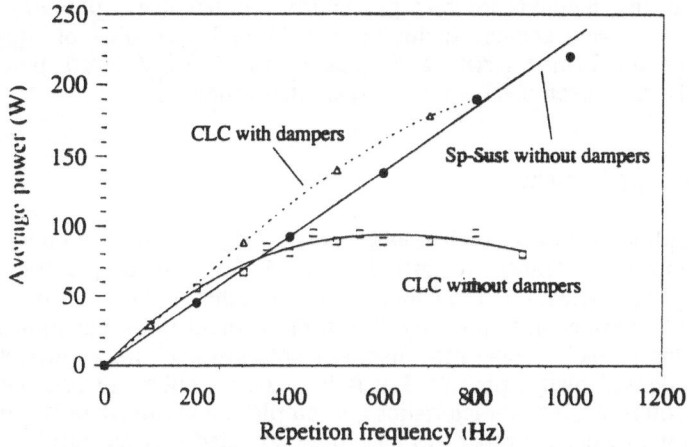

Fig. 5   Laser average power v.s. p.r.f. for S.S. and C-L-C circuits on LUX Test-Bed (from Ref. 6)

f) efficient damping of induced acoustic waves in high prf discharge excited excimer lasers and comparison between S.S. and C-L-C excitation. It has been shown that discharge is much more stable and induced acoustic waves much less intense with S.S. mode[8]. Fig. 6 shows residual pr(essure variations v.s. prf on Lux Test-Bed.

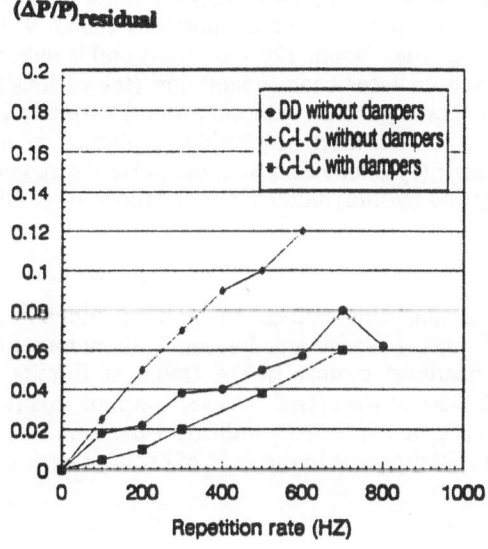

Fig. 6   Residual pressure fluctuations v.s. prf for S.S. anc C-L-C circuits (from Ref. 8)

(g) Researchs on new methods for rare gas halide excimer lasers pumping and new promising excimer lasers shemes including excitation by uncoherent light from a sliding discharge on formed ferrite and ionic excimers VUV lasers potentialities. Interesting results have been obtained recently on these shemes[9,10].

4. Excimer laser applications

Researchs on applications using high average power, high prf or high energy excimer lasers are developped in France in particular in the frame of cooperative researchs (GDR 919 and 1125 CNRS), often in cooperation with other laboratories in Europe.

In situ growth of complex stoechiometry thin films without post-annealing is studied using excimer lasers with cooperation beetwen specialists of laser physics, plasma physics and condensed matter physics. Focus is made on effects of prf, pulse length and wavelength on thin film characteristics and on plasma hydrodynamics. Interesting results on ablation and deposit processes have been obtained very recently[11].

Surface processing and cleaning of metals and various materials, including ones contaminated by radioactivity, with excimer lasers is studied in collaborations beetwen specialists of physics and metallurgists and industrial partnairs[11].

Microanalysis of various materials with excimer laser sampling associated with spectroscopic analysis is also well in progress[11].

5. The GDR 1125 CNRS "Lasermat"

In 1995, a new federative laboratory called GDR 1125 CNRS « Lasermat » and devoted to high power laser applications to material processing has been created by french CNRS. This laboratory replace two previous ones devoted to excimer laser physics (GDR 919) and high power laser applications (GDR 911). It associates 32 groups at CNRS and Universities (about 120 scientists) and 7 industrial partnairs. The idea is to associate reseach on laser beam adaptation (for various lasers), laser beam transport and material processing. Main subjects are: (a) surface alloying (experiment, theory, modelling and characterisation; (b) welding (physics of interaction, thermal and hydrodynamic process); (c) interaction of short pulse lasers with a target (plasma plume physico-chemistry and hydrodynamics, process modelling, thin film deposit)[12].

6. Conclusion

French groups at CNRS and Universities as well as industrial laboratories,often associated with these Groups, have obtained recently numerous important results on excimer lasers and applications, mainly in the frame of Eureka EU 205 European program. Association of most of the french research groups involved in high average power lasers and material processing with industrial partners in the new GDR 1125 CNRS is promising for rapid progress in the field of excimer laser material processing.

7. References

1. M. Stehle, in Excimer lasers, L.D. Laude Ed. NATO ASI Series, Series E: Applied Sciences Vol. **265** pp 15-25 (1994); also private communication (1995).
2. H. Besaucele, PhD Thesis, Louis Pasteur University, Strasbourg, France (1994); also B. Lacour private communication (1995).
3. R. Riva, M. Legentil, S. Pasquiers, V. Puech, J. Phys. D: Appl. Phys. 28, 856 (1995)
4. E. Estocq, G. Delouya, J. Bretagne, Appl. Phys. B, **56**, 209-221 (1993).
5. A. Belasry, J.P. BOoeuf L.C. Pichford, J. Appl. Phys. **74**, 1553-1567 (1993).
6. Th. Hofmann, N. Bernard, B.L. Fontaine, Ph. Delaporte, M. Sentis, B.M. Forestier, High Power Gas and Solid State Lasers, SPIE Vol. **2206**, 46-51 (1994).
7. O. Uteza, PhD Thesis , Aix-Marseille III University (1994); also O. Uteza,M; Sentis, Ph. Delaporte, B. Forestier, B. Fontaine Optics Comm., **102**, 523-531 (1994).
8. N. Bernard, PhD Thesis, Aix-Marseille II University (1995), also N. Bernard and al. 10 TH Gas Flow and Chemical Lasers, SPIE Vol. **2502**,433-439 (1995).
9. F. Chazaud, PhD Thesis, Aix-Marseille II U. (1994); also F. Chazaud, M. Sentis, L. Chinh, H. Rigneault, Ph. Delaporte, W. Marine Appl. Phys. Lett., **65,** 1626 (1994).
10. H. Tischler, Ph. Delaporte, B. Fontaine, M. Ssentis subm.it. to IEEE J. Q. E..
11. Numerous papers from french laboratories have been presented at the E-MRS 1995 spring Meeting, COLA'95, May 1995; COLA'95 proceedings (1995).
12. B.L. Fontaine, A.B. Vannes, Project of GDR CNRS LASERMAT, 1994.